FLOOD HYDROLOGY

AF788577

FLOOD HYDROLOGY

Proceedings of the International Symposium on Flood Frequency and Risk Analyses, 14–17 May 1986, Louisiana State University, Baton Rouge, U.S.A.

Edited by

VIJAY P. SINGH

Department of Civil Engineering,
Louisiana State University, Baton Rouge, U.S.A.

D. REIDEL PUBLISHING COMPANY

A MEMBER OF THE KLUWER ACADEMIC PUBLISHERS GROUP

DORDRECHT / BOSTON / LANCASTER / TOKYO

Library of Congress Cataloging in Publication Data

CIP

International Symposium on Flood Frequency and Risk Analyses (1986:
Louisiana State University, Baton Rouge)
Flood hydrology.

Includes indexes.
1. Floods—Congresses. I. Singh, V. P. (Vijay P.) II. Title.
GB1399.I58 1986a 551.48'9 87–20620
ISBN-13: 978-94-010-8255-6 e-ISBN-13: 978-94-009-3957-8
DOI: 10.1007/978-94-009-3957-8

Published by D. Reidel Publishing Company,
P.O. Box 17, 3300 AA Dordrecht, Holland.

Sold and distributed in the U.S.A. and Canada
by Kluwer Academic Publishers,
101 Philip Drive, Assinippi Park, Norwell, MA 02061, U.S.A.

In all other countries, sold and distributed
by Kluwer Academic Publishers Group,
P.O. Box 322, 3300 AH Dordrecht, Holland.

All Rights Reserved
© 1987 by D. Reidel Publishing Company, Dordrecht, Holland
Softcover reprint of the hardcover 1st edition 1987
No part of the material protected by this copyright notice may be reproduced or
utilized in any form or by any means, electronic or mechanical
including photocopying, recording or by any information storage and
retrieval system, without written permission from the copyright owner

PREFACE

Floods constitute a persistent and serious problem throughout the United States and many other parts of the world. They are responsible for losses amounting to billions of dollars and scores of deaths annually. Virtually all parts of the nation--coastal, mountainous and rural--are affected by them. Two aspects of the problem of flooding that have long been topics of scientific inquiry are flood frequency and risk analyses. Many new, even improved, techniques have recently been developed for performing these analyses. Nevertheless, actual experience points out that the frequency of say a 100-year flood, in lieu of being encountered on the average once in one hundred years, may be as little as once in 25 years. It is therefore appropriate to pause and ask where we are, where we are going and where we ought to be going with regard to the technology of flood frequency and risk analyses. One way to address these questions is to provide a forum where people from all quarters of the world can assemble, discuss and share their experience and expertise pertaining to flood frequency and risk analyses. This is what constituted the motivation for organizing the International Symposium on Flood Frequency and Risk Analyses held May 14-17, 1986, at Louisiana State University, Baton Rouge, Louisiana.

The objectives of this symposium were therefore (1) to assess the current state of the art of flood frequency and risk analyses, (2) to demonstrate applicability of flood frequency and risk models, (3) to establish complementary aspects of seemingly different models, (4) to enhance interdisciplinary interaction, (5) to discuss practice of flood frequency and risk analyses technology by federal agencies in the U.S., (6) to discuss research needs in frequency and risk analyses, and (7) to determine directions for further research.

We received an overwhelming response to our call for papers. It was indeed a difficult task to select amongst the many excellent papers that were submitted, and we regret that we could not include all of them. The sole criterion for selection of a paper was its merit in relation to Symposium objectives. The subject matter of the Symposium was divided into 20 major topics encompassing virtually all facets of flood frequency and risk analyses. Each topic comprised of an invited state-of-the-art paper and a number of contributed papers. These contributions blended naturally to evolve a synthesized body of knowledge on that topic. Extended abstracts of all the invited and contributed papers were assembled in a pre-Symposium proceedings volume. This helped stimulate discussion and exchange of ideas during the Symposium.

The papers presented at the Symposium were refereed in a manner similar to that employed for publishing a journal article. As a result, many papers did not pass the review and were therefore eliminated from inclusion in the final proceedings. The papers contained in this book, FLOOD HYDROLOGY, represent one part of the Symposium contributions. The other parts are embodied in three separate books, HYDROLOGIC FREQUENCY MODELING, REGIONAL FLOOD FREQUENCY ANALYSIS, and APPLICATION OF FREQUENCY AND RISK IN WATER RESOURCES, which are being published simultaneously. Arrangement of these books under four different titles was a natural consequence of the diversity of technical material discussed in the papers. These books can be treated almost independently, although some overlap does exist between them.

This book contains seven sections encompassing major hydrologic aspects of flood control and protection. Each section starts usually with an invited state-of-the-art paper, followed by contributed papers. Beginning with an assessment of hydrologic modeling and limitations, and a future flood research agenda for the United States, the papers go on to discuss hydrology of floods, urban and coastal flooding, streamflow simulation and forecasting, flood control and protection, and flood control programs.

The book will of of interest to researchers as well as those engaged in practice of Civil Engineering, Agricultural Engineering, Hydrology, Water Resources, Earth Resources, Forestry and Environmental Sciences. The graduate students as well as those wishing to conduct research in flood hydrology will find this book to be of particular value.

I wish to take this opportunity to express my sincere appreciation to all the members of the Organizing and Advisory Committees, and the Louisiana State University administration for their generous and timely help in the organization of the Symposium. A lack of space does not allow me to list all of them by name here. Numerous other people contributed to the Symposium in one way or another. The authors, including the invited keynote speakers, contributed to the Symposium technically and this book is a result of their efforts. The session chairmen administered the sessions in a positive and professional manner. The referees took time out from their busy schedules and reviewed the papers. Graduate students assisted in smooth conduct of the sessions. I owe my sincere gratitude to all of these individuals.

If the success of a Symposium is measured in terms of the quality of participants and presentations then most people would agree that this was a very successful Symposium. A very large number of internationally well-known people, who have long been recognized for their contributions and have long been at the forefront of hydrologic research, came to participate in the Symposium. More than 35 countries, covering the five continents and most of the countries of the world active in hydrologic research, were represented. It is hoped that long and productive personal associations will develop as a result of this Symposium.

March 1987
Baton Rouge, Louisiana

Vijay P. Singh
Symposium Director

ACKNOWLEDGEMENTS

The International Symposium on Flood Frequency and Risk Analyses was sponsored and co-sponsored by a number of organizations. The sponsors provided financial support without which the Symposium might not have come to fruition. Their financial support is gratefully acknowledged. The co-sponsors extended their help in announcing the Symposium through their journals, transactions, newsletters or magazines. This publicity helped with attendance at the Symposium, and is gratefully acknowledged. The following is a list of Symposium sponsors and co-sponsors.

SYMPOSIUM SPONSORS

Louisiana State University
- Department of Civil Engineering
- Louisiana Water Resources Research Institute

National Science Foundation
U.S. Army Research Office
U.S. Geological Survey, Louisiana District, Baton Rouge
Woodward-Clyde Consultants, Baton Rouge

SYMPOSIUM CO-SPONSORS

American Geographical Society
American Geophysical Union
American Meteorological Society
American Statistical Association
American Society of Agricultural Engineers
American Society of Civil Engineers
American Water Resources Association
Association of American Geographers
Association of State Floodplain Managers
International Association for Hydraulic Research
International Association of Hydrological Sciences
International Association of Theoretical and Applied Limnology
International Commission on Irrigation and Drainage
International Geographical Union
International Water Resources Association
National Wildlife Federation
North American Lake Management Society
Pan American Institute of Geography
Society for Risk Analysis
Soil Conservation Society of America

TABLE OF CONTENTS

CURRENT STATE OF HYDROLOGIC MODELING AND ITS LIMITATIONS

Anand Prakash
Chief Water Resources Engineer
Dames & Moore
1626 Cole Boulevard
Golden, CO 80401

ABSTRACT. This paper presents a brief description of commonly used hydrologic models. These include storm runoff models, deterministic models for generation of sequential streamflows, regression and stochastic models for extension of hydrologic data, and models for frequency analysis of floods and droughts and risk-based designs. The subjectivity and judgment involved in the estimation of input parameters for each model are indicated along with the effect of the estimated values of such parameters on the output of the model. For each model, several limitations to practical applications are described. In some cases, these limitations are attributable to inadequacy of data for calibration and verification of the model, complexity of programming algorithms, and numerousness of the input variables which have to be estimated by judgment. In others, e.g., determination of the probability of the PMF and the confidence limits of predictions for such extreme events, further research is required to develop analytical tools to yield results of desired accuracy. Areas of further research are identified to update the existing models or to develop new models to enable a practicing hydrologist to perform hydrologic analyses of water resources development projects with more certainty and better precision.

1. INTRODUCTION

Over the past 25 years or so, computer models have been developed for almost all aspects of hydrologic analyses. Some of these models have been extensively used by practicing hydrologists, whereas others have served as excellent research tools for academicians. The objective of this paper is to review state-of-the-art hydrologic models and their limitations in practical applications. During the past few years, so many models have appeared in the literature that it is almost impossible to have working knowledge of each one of them. The models proposed to be addressed in this paper may be divided into the following categories:

(i) Deterministic models for rainfall-runoff simulation

V. P. Singh (ed.), Flood Hydrology, 1–16.
© 1987 by D. Reidel Publishing Company.

(ii) Regression and stochastic models for streamflow generation

(iii) Models for frequency analysis of floods and droughts and risk-based design.

Most of the models in these categories have been used and tested extensively and perform well for the prediction of hydrologic variables in situations where sufficient site-specific hydrologic information is available for calibration and verification. The accuracy of predictions becomes almost indeterminate for ungaged watersheds or watersheds with inadequate hydrologic data. Unfortunately, a number of projects for which hydrologic analyses and predictions are required belong to this class. It is in these cases that applications of hydrologic models have serious limitations. These limitations are often camouflaged and not appreciated by the users.

2. DETERMINISTIC MODELS FOR RAINFALL-RUNOFF SIMULATION

This category includes the storm runoff (event) and sequential flow simulation models. Some examples of the storm runoff models are the Flood Hydrograph Package (HEC-1) of the U.S. Army Corps of Engineers (1981); TR-20 (Soil Conservation Service, 1969); TR-55 (Soil Conservation Service, 1975); SEDIMOT-II (University of Kentucky, 1985); and HYDROGRAPH-2 (Oklahoma Technical Press, 1985).

Some commonly known sequential flow simulation models include the Stanford Watershed Model IV (Crawford and Linsley, 1966); Hydrologic Simulation Program, HSP (Hydrocomp, 1975); the updated version of Hydrologic Simulation Program including water quality simulation, HSPF (U.S. Environmental Protection Agency, 1980); the National Weather Service River Forecast System Model (Curtis and Smith, 1976); the Sacramento Watershed Model (Burnash, 1985); and the Simulation of the Urban Runoff Process (Australian Water Resources Council, 1977).

2.1 Storm Runoff Models

All these models are based on the unit hydrograph theory and therefore on the implicit assumption that the relationship between the rainfall and runoff processes is linear. For watersheds where this relationship tends to be nonlinear, model predictions may be inaccurate and may not agree with observed data.

The unit hydrograph for a watershed can be developed by at least three different methods (U.S. Army Corps of Engineers, 1981; Prakash, 1983):

(i) Snyder's Method (Chow, 1964) which requires the estimation of the coefficients C_p and C_t. These coefficients are estimated principally by judgment and model calibration.

(ii) Clark's Method (U.S. Army Corps of Engineers, 1981) which requires the estimation of the storage coefficient, R, and the time of concentration, t_c. These parameters are estimated by judgment, empirical equations, and model calibration.

(iii) The Dimensionless Unit Hydrograph Method (Soil Conservation

Service, 1972) which requires the selection of an appropriate dimensionless hydrograph from a few generalized ones available in the literature and estimation of the lag time which is generally assumed to be 0.6 x time of concentration.

In addition to the subjectivity involved in the selection of the parameters C_p, C_t, and R and the dimensionless unit hydrograph, there are over a dozen empirical equations to estimate the time of concentration. To illustrate the wide variation in the estimates of the time of concentration, the values for a 4.8 sq. mile (12.4 sq. km.) watershed computed by different equations are shown in Table I (McCuen, et al., 1984; Soil Conservation Service, 1972).

To develop the surface runoff hydrograph, the hydrologist requires a sequence of incremental precipitation depths for convolution with the unit hydrograph ordinates. There are a number of methods used to distribute the design storm depth into smaller time increments and for sequencing the incremental precipitation depths to obtain the design flood hydrograph. Some of these methods include:

(i) A distribution and sequence for hypothetical storms incorporated in the HEC-1 computer program (U.S. Army Corps of Engineers, 1981).

(ii) Distributions based on mass curves of recorded severe storms in the region.

(iii) Distributions and sequences based on the Southwestern Division Criteria, and that given in EM 1110-2-1411 (U.S. Army Corps of Engineers, 1981 and 1952).

(iv) An empirical distribution for the probable maximum precipitation recommended by the Nuclear Regulatory Commission for the design of uranium mill tailings impoundments (Nuclear Regulatory Commission, 1983).

(v) Distributions and sequences for the local and general storms (probable maximum precipitation) recommended by the National Oceanic and Atmospheric Administration (1977).

(vi) Sequences for the probable maximum precipitation recommended by the U.S. Bureau of Reclamation and Army Corps of Engineers (Prakash, 1983).

(vii) An optimal sequence of incremental precipitation (Prakash, 1978).

There are at least four commonly accepted methods to estimate precipitation losses to convert the incremental precipitation depths to rainfall excess. These include the following:

(i) An initial abstraction followed by a uniform loss rate (U.S. Army Corps of Engineers, 1981).

(ii) The Holtan method to estimate infiltration loss (U.S. Army Corps of Engineers, 1981).

(iii) An exponentially decreasing loss rate (U.S. Army Corps of Engineers, 1981).

(iv) The loss rate defined by the Soil Conservation Service Curve Numbers (Soil Conservation Service, 1972).

The selection of a method and parameters required for the application of that method is based on calibrations and subjective judgments.

TABLE I. Times of Concentration Computed by Different Methods.*

Method	Time of concentration (min.)	References
Stream Hydraulics	53	Soil Cons. Service (1972)
Upland	60	Soil Cons. Service (1972)
Snyder	68**	Chow (1964)
Carter	38	McCuen, et al. (1984)
Federal Aviation Agency	44	McCuen, et al. (1984) with runoff coeff. = 0.77
Kirpich (Pa.)	10	McCuen, et al. (1984)
Kirpich (Tenn.)	43	McCuen, et al. (1984)
Kerby	37	McCuen, et al. (1984)
SCS Lag	147	McCuen, et al. (1984) with curve number = 76 and Manning's n = 0.045
Kinematic Wave	57	McCuen, et al. (1984) with Manning's n = 0.045 and excess rainfall rate = 6.555 in/hr (16.65 cm/hr)

*Hydraulic length = 17,000 ft (5.18 km.), watershed slope = 0.057.

**Assumes t_c = Snyder's Lag time

The cumulative effect of the aforementioned subjectivities is that two or more hydrologists may obtain significantly different design hydrographs for one and the same site. This suggests a need for extensive research work to develop calibrated hydrologic parameters and equations for typical hydrometeorological regions.

Usually the hydrologic analysis of watersheds comprised of several stream systems is performed by subdividing them into smaller subwatersheds. This subdivision is accomplished by judgment based on the scale of the available topographic maps. Runoff hydrographs are developed for each individual subwatershed, routed through interconnected stream channels, and combined to obtain the design flood hydrograph at appropriate locations. The routing may be performed by using kinematic wave approximation, Muskingum method, or dynamic wave formulation, etc. These methods require the selection of a representative cross section for a reach, appropriate Manning's n values, and the Muskingum routing coefficients x and k. In addition to the differences resulting from subjective selections of routing parameters by different hydrologists, the subdivision of watersheds into smaller subwatersheds by judgment may also result in significant changes in the design hydrographs. This is illustrated in Fig. 1. The watershed characteristics used to develop the hydrographs of Fig. 1 are summarized in Table II. A map of the subwatersheds is shown in Fig. 2. For this particular case, subdivision into small subwatersheds of 63 (0.26 sq. km.) to 1,649 (6.68 sq. km.) acres was possible because topographic maps to a scale of 1 in. = 400 ft were available. It may be argued that subdivision into such small subwatersheds is not necessary. However, similar differences in hydrographs are obtained even for larger watersheds and subwatersheds.

2.2 Sequential Flow Simulation Models

These models are useful to generate sequences of daily and monthly flows for reservoir operation studies. The major problems associated with these models are the complexities of the programming algorithms; a large number of input parameters, most of which have to be estimated by judgment; and inadequacy of available data to calibrate and verify the model for any site-specific application. The calibration of model parameters is a tedious process and requires a number of trial runs. During the verification phase, the calibration is often found to be spurious because of complex programming and a large number of subjectively estimated input parameters. As a result, the calibration process has to be repeated a number of times. The problem is further complicated by the fact that in most cases sufficient information is not available on the concurrent values of the large number of hydrologic, meteorologic, and other parameters required for model calibration and verification.

It may be desirable to conduct extensive research to develop ranges of input variables for these models which may be applicable for different climatic and hydrologic regions and to simplify the algorithms for parameter optimization.

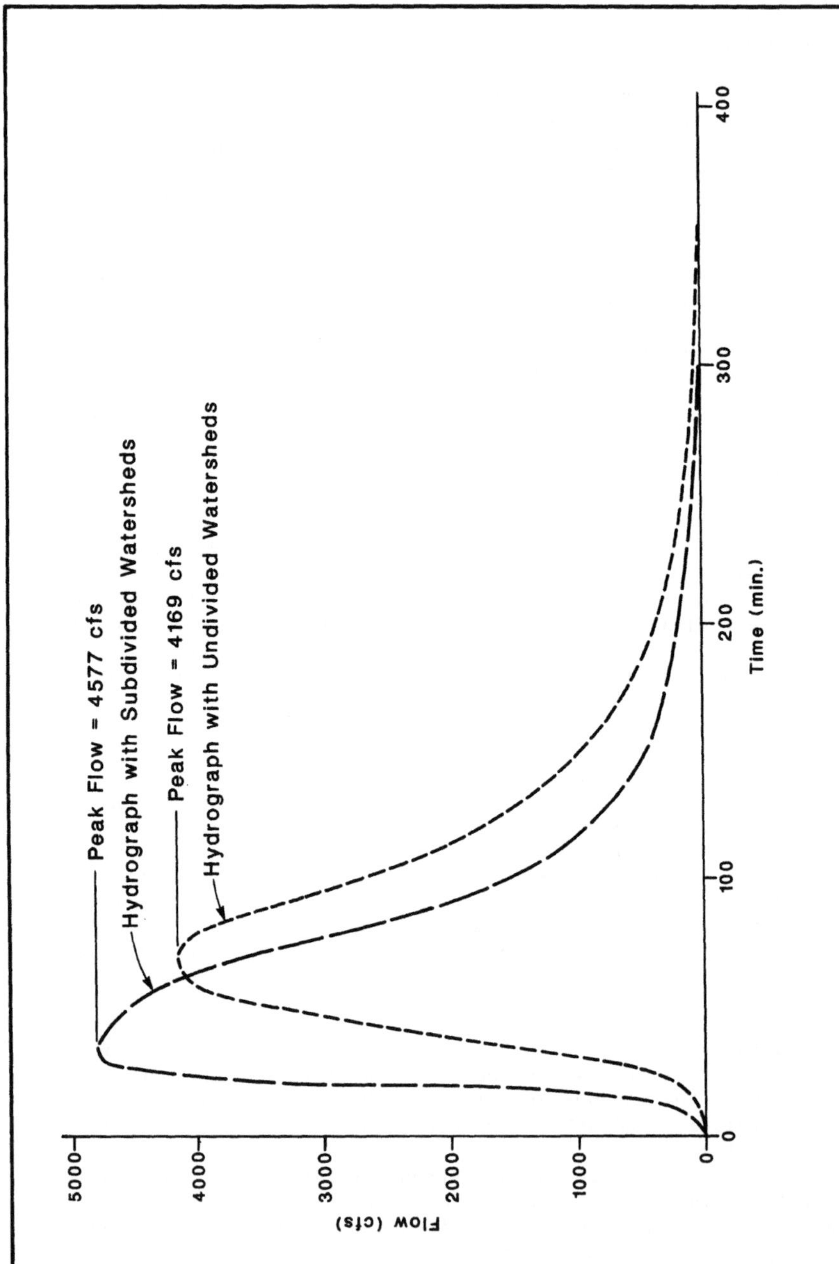

Figure 1 – Hydrographs with and without Subdivided Watersheds

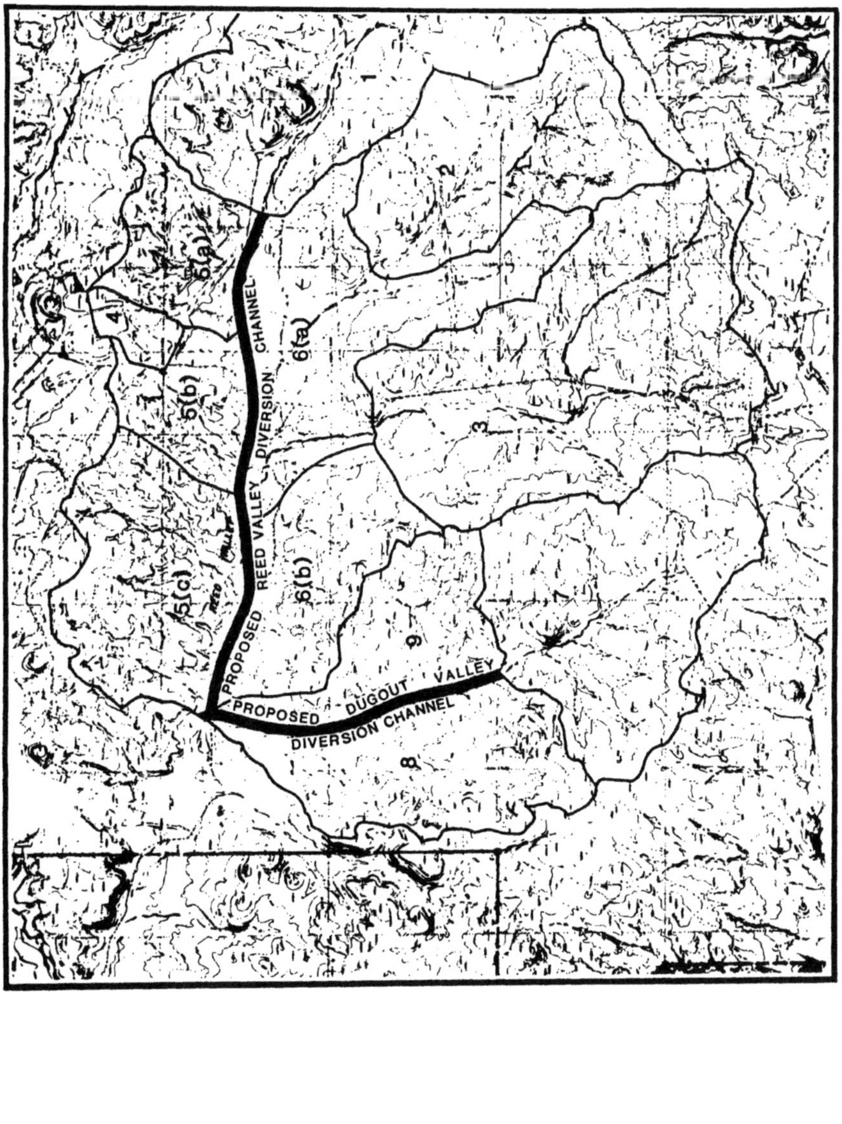

Figure 2 – Reed and Dugout Valley Sub-Watersheds

TABLE II. Watershed Characteristics

	Subdivided watersheds			Undivided watersheds		
Designation	Area (sq. mi)	Curve number	Lag time (hr)	Area (sq. mi)	Curve number	Lag time (hr)
1	2.58	79	0.46			
5(a)	0.43	79	0.16			
4	0.098	79	0.10			
5(b)	0.36	79	0.14			
2	0.94	84	0.31			
6(a)	0.94	79	0.283			
3	1.90	84	0.467			
6(b)	0.66	79	0.159			
5(c)	0.82	79	0.186			
				8.728	81	0.81
7	1.15	84	0.295			
8	0.75	79	0.21			
9	0.54	79	0.151			
				2.44	81	0.52
Total Drainage Area	11.168			11.168		

1 sq. mi = 2.59 sq. km

3. REGRESSION AND STOCHASTIC MODELS FOR STREAMFLOW GENERATION

Multiple regression, polynomial regression, and stepwise multiple regression models have been very useful in generating hydrologic data for ungaged sites and for extending records of hydrologic variables for water resource planning. Common difficulties associated with the application of these models include inadequacy of relevant data and determination of an acceptable coefficient of correlation or standard error of estimate. What should be done if, due to inadequacy of available data or otherwise, the coefficient of correlation is found to be unacceptable? This dilemma is illustrated in Table III where the monthly flows of a stream in Alaska are correlated with the total monthly precipitation and average temperature for that month. A refinement to the analysis was to include the precipitation and temperature for the previous month into the regression equation with little improvement in the results.

The multivariate regression model developed by Beard (1965) and incorporated with minor modifications into the monthly Streamflow Simulation Model, HEC-4, of the U.S. Army Corps of Engineers (1971) is very useful for stochastic generation of monthly streamflows. The model needs updating to include algorithms for stochastic generation of daily streamflows. At present, the results of this model cannot be used for operation studies of reservoirs where diurnal fluctuations in streamflows are significant.

4. MODELS FOR FREQUENCY ANALYSIS OF FLOODS AND DROUGHTS AND RISK-BASED DESIGN

The U.S. Water Resources Council (1981) has greatly simplified the task of the practicing hydrologist by recommending that the log-Pearson Type III distribution with generalized regional skew coefficients should be used for flood frequency analysis. In some cases, however, goodness-of-fit tests for the log-Pearson Type III distribution may indicate an almost unacceptable fit. To illustrate this, probabilistic analysis of the annual flood peaks of the Ohio River at Louisville, Kentucky for the period 1872-1974 is presented in Table IV (Prakash, 1977). Identification of the best-fitting probability distribution is required to estimate flood peaks of different recurrence intervals for the design of facilities within and adjacent to the floodplains of streams and also for risk evaluation, benefit-cost analysis, and ranking of structural flood control options. In many situations, the selected probability distribution or frequency curve is to be extended to the probable maximum flood which is a deterministically estimated quantity with a recurrence interval of 10^4 to 10^{12} years. This extension is inevitable if one has to use the risk-based methodology (Prakash, 1985) to determine the design-basis flood for spillways of major and medium-sized dams with moderate to significant hazard potentials (U.S. Army Corps of Engineers, 1976). None of the known probabilistic models can estimate the probability of severe events like the probable maximum flood within reasonable and well-defined confi-

TABLE III. Results of a Multiple Linear Regression Analysis (Q = a + bi + cT).

Variable	Values for				
	March	June	August	September	November
Number of data points	14	16	16	16	16
Mean of observed monthly flows (cfs-days), $\bar{Q}$	59,728.57	428,533.125	681,875.0	569,243.75	195,037.5
Mean of observed total monthly precipitation depths (inches), $\bar{i}$	1.37214	1.48625	5.42875	4.21813	1.6875
Mean of observed monthy temperatures (°F)					
$\bar{T}$	19.76429	50.7125	53.83125	47.45	24.025
a	39,167.78	-717,896.1	232,734.6	-653,968.1	93,179.58
b	4,701.537	54,488.77	25,721.53	14,831.60	33,372.02
c	713.8943	21,009.52	5,749.535	24,460.49	1,895.635
Standard error of estimate (cfs-days)	13,900	66,900	94,800	95,400	55,400
Multiple correlation coefficient	0.558	0.711	0.443	0.546	0.596

Q = total monthly flow in cfs-days (cfs-day = 0.02832 cms-day);

a = intercept of regression equation; b, c = regression coefficients;

i = total monthly precipitation in inches; and T = mean monthly temperature in degrees Fahrenheit.

TABLE IV. Probabilistic Analysis of Ohio River Flood Peaks at Louisville, Kentucky (1872-1974).

Maximum = 1,100,000 cfs, Mean = 508,214 cfs, Minimum = 237,000 cfs
Standard deviation = 135,483 cfs, Coeff. of skewness = 0.96,
kurtosis = 6.21

Probability Distribution	Estimated 100-yr peak flow (cfs)	Passes SK Test	
		Significance level %	SK Statistic
Normal	820,400	26	0.1
Log normal	910,514	33	0.09
Gumbel	981,300	25	0.1
Log-Pearson Type III	862,100	61	0.07

(1 cfs = 0.02832 cubic meters)

dence limits given a data set of about 50 years or so. To further complicate the problem, there are a number of uncertainties associated even with the deterministic estimation of the probable maximum flood (Prakash, 1983). The impact of the discrepancy in the estimation of the probability of the probable maximum flood on the selection of the design basis flood for the spillway of a dam using the risk-based methodology is illustrated in Table V (National Academy Press, 1985). Extensive effort is needed to streamline the methods for the estimation of the probable maximum flood deterministically and further research is required to develop probability distributions which could be used to assign probabilities to extreme events like the probable maximum flood. This may require development of mixed distributions, special techniques for the treatment of outliers, and sound statistical methods to estimate the confidence limits associated with events of extremely low probabilities.

Computer programs are available to perform statistical analyses of low streamflows (Prakash, 1981) to estimate the daily and the n-day low flows of streams for different return periods using different probability distributions. This information is useful in evaluating the impacts of projects on the water quality and downstream users of streamflows. It may be worthwhile to develop a model which would compute moving averages of available data, perform probabilistic analyses of low streamflows using normal, log normal, Gumbel, log-Pearson Type III, and other applicable distributions with square-root and SMEMAX transformations, etc., conduct goodness-of-fit tests, identify the best-fitting distribution, and assign confidence limits to predicted values.

To perform risk-based reservoir planning studies and to develop risk-based design criteria for water supplies, the hydrologist requires information about the magnitudes and probabilities of monthly and annual low streamflows during droughts of different severities and durations. The statistical analysis has to be coupled with algorithms for monthly reservoir operation and generation of mass curves for inflows and demands to yield reservoir sizes required to cope with droughts of different probabilities of occurrence. This information has to be transformed into a risk-based framework to facilitate the decision process.

5. CONCLUSION

Extensive research should be conducted to divide the continental United States into several hydrologic regions and subregions based on prevailing climatic and hydrologic conditions. A large number of observed runoff hydrographs should be analyzed for stream channels in each region and subregion. Through extensive hydrograph analysis, the values of all judgmental factors required for the development of storm runoff hydrographs for ungaged sites should be optimized for each region and subregion. Tables and charts should be prepared from which the required parameters for any watershed in a region or subregion may be estimated. Similar calibration studies should be conducted for a

TABLE V. Design-Basis Floods with Different Probabilities of the PMF.

Design basis flood (cfs)	Annualized construction cost (dollar/year)	Expected annual damages plus construction costs (dollars/year)	
		PMF return period = 10^4 years	PMF return period = 10^6 years
50,000	0	640,000	280,000
75,000	80,000	390,000	250,000
120,000 (PMF)	200,000	385,000	360,000
r		4.6×10^{-5}	9.2×10^{-5}
q_o		80,000	30,000
s		3×10^{-6}	3×10^{-6}

Maximum estimated dmage, M, if the dam does not fail = $200 x 10^6

Cost of rebuilding the dam and services lost, L = $20 x 10^6

E(D) = Expected annual damage

$$= M\left(-\exp[-r(q+q_o)] + \frac{r\,k}{r+s}\exp[-(r+s)\,q]\right)\Big|_{10{,}000}^{q_c} +$$

$$(M+L)\,[F(120{,}000) - F(q_c)]$$

$$k = \exp[10{,}000s - rq_o]$$

$$F(q) = 1 - \exp[-r(q+q_o)]$$

1 cfs = 0.02832 cubic meters

large number of monthly and annual streamflow hydrographs for most gaged streams in a subregion to optimize the values of the variables required for the sequential flow models. These values should be consolidated in the form of tables and charts for use by practicing hydrologists.

A multivariate regression or stochastic model should be developed to generate sequences of daily streamflows.

Extensive work is required to standardize the methods for deterministic estimation of the probable maximum flood for a watershed. Further research should focus on developing probability distributions which can be extended to estimate the probability of severe events like the probable maximum flood, without introducing indeterminate errors; on evolving techniques that treat the outliers; and on developing statistical approaches to assign confidence limits to predicted extreme events of very low probabilities.

A comprehensive probabilistic model should be developed which would compute moving averages of n-day streamflows, perform statistical analysis for low streamflows, identify the best-fitting probability distribution, and compute confidence limits of estimated low flows. A similar model should be developed for statistical analysis of droughts to perform reservoir operation studies and to develop alternative reservoir designs using risk-based methodology.

6. REFERENCES

Australian Water Resources Council 1977. ' Simulation of the Urban Runoff Process,' Technical Paper No. 26, Canberra, Australia.

Beard, L.R. 1965. 'Use of Interrelated Records to Simulate Streamflow,' Jour. of the Hydraulics Division, ASCE, Vol. 91, HY5.

Burnash, R.J. 1985. 'Sacramento Watershed Model,' California-Nevada River Forecast Center, National Weather Service, Sacramento, CA.

Chow, V.T. 1964. Handbook of Applied Hydrology, McGraw-Hill Book Co., New York, NY.

Crawford, N.H. and Linsley, R.K. 1966. 'Stanford Watershed Model IV', Digital Simulation in Hydrology, Technical Report No. 49, Department of Civil Engineering, Stanford University, Stanford, CA.

Curtis, D.C. and Smith, G.F. 1976. 'The National Weather Service River Forecast System,' Hydrol. Research Lab., National Weather Service, NOAA, Silver Spring, MD.

Hydrocomp 1975. 'Hydrocomp Simulation Programming,' Operations Manual, Fourth Ed., Palo Alto, CA.

McCuen, R.H., Wong S.L., and Rawls, W.J. 1984. 'Estimating Urban Time of Concentration,' Jour. of Hydraulic Engineering, Vol. 110, No. 7, American Society of Civil Engineers.

National Academy Press 1985. Safety of Dams, Flood and Earthquake Criteria, Washington, DC.

National Oceanic and Atmospheric Administration 1977. Probable Maximum Precipitation Estimates, Colorado River and Great Basin Drainages, Hydrometeorological Report No. 49, U.S. Dept. of Commerce.

Oklahoma Technical Press 1985. HYDROGRAPH-2, 815 Hillcrest, Stillwater, OK.

Prakash, A. 1977. 'Discussion on Flood Analysis by SMEMAX Transformation,' Jour. of the Hydraulics Division, ASCE, Vol. 103, No. HY11.

Prakash, A. 1978. 'Optimal Sequence of Incremental Precipitation,' Jour. of the Hydraulics Division, ASCE, Vol. 104, No. HY12.

Prakash, A. 1981. 'Statistical Determination of Design Low Flows,' Journal of Hydrology, Vol. 51, Elsevier Scientific Publishing Co., New York.

Prakash, A. 1983. 'Deterministic and Probabilistic Perspectives of the PMF,' Proceedings of the Conference on Frontiers in Hydraulic Engineering, ASCE/M.I.T., Cambridge, MA.

Prakash, A. 1985. 'Impacts of Risk-Based Analysis on Current Design Practices,' Engineering Foundation Conference on Risk-Based Decision Making in Water Resources, Santa Barbara, CA.

Soil Conservation Service 1969. 'Computer Program for Project Formulation Hydrology,' Technical Release 20 (TR-20), Supplement No. 1, U.S. Dept. of Agriculture.

Soil Conservation Service 1975. 'Urban Hydrology for Small Watersheds,' Technical Release 55 (TR-55), U.S. Dept. of Agriculture.

Soil Conservation Service 1972. National Engineering Handbook, Section 4, Hydrology, U.S. Dept. of Agriculture.

University of Kentucky 1985. 'Design Manual for the SEDIMOT II Hydrology and Sedimentology Model,' College of Agriculture, Dept. of Agricultural Engineering.

U.S. Army Corps of Engineers 1952. 'Standard Project Flood Determinations,' Engineering Manual 1110-2-1411, Washington, DC.

U.S. Army Corps of Engineers 1971. Monthly Streamflow Simulation, HEC-4, The Hydrologic Engineering Center, Davis, CA.

U.S. Army Corps of Engineers, 1976. 'Recommended Guidelines for Safety Inspection of Dams,' Washington, DC.

U.S. Army Corps of Engineers, 1981. Flood Hydrograph Package, HEC-1, The Hydrologic Engineering Center, Davis, CA.

U.S. Environmental Protection Agency 1980. User's Manual for Hydrologic Simulation Program - Fortran (HSPF), EPA-600/9-80-105.

U.S. Regulatory Commission 1983. 'Hydrologic Design Criteria for Tailings Retention Systems,' Staff Technical Position, WM-8201, Low Level Waste Licensing Branch.

U.S. Water Resources Council, 1981. 'Guidelines for Determining Flood Flow Frequency,' Bulletin #17B of the Hydrology Committee, Washington, DC.

A FUTURE FLOOD RESEARCH AGENDA FOR THE UNITED STATES

Stanley A. Changnon, Jr.
Chief Emeritus and Principal Scientist
Illinois State Water Survey
2204 Griffith Drive
Champaign, Illinois 61820

ABSTRACT. Flooding remains a major unresolved problem in the United States with losses mounting after 60 years of largely structural efforts to mitigate loss. Hence, a major 2-year assessment of research needs to more effectively address flood mitigation was conducted within the context of our national shift to new federalism, the National Flood Insurance Program, the shift of emphases from structural to nonstructural approaches for flood mitigation, and the evolution in national programs of emergency assistance where flooding is but one of many hazards treated. Recommendations for research centered around six general themes: 1) more attention to socio-economic-political research; 2) the program should have a goal of efficient use of flood-prone lands, not loss reduction; 3) interdisciplinary research is essential with 70% of all 115 recommended tasks being multi-disciplinary; 4) the need to develop comprehensive data banks and flood information centers; 5) future research needs to be oriented to user needs; and 6) there is a need for continuing re-assessment of flood research every 3 years.

1. INTRODUCTION

Floods are the most destructive natural hazard in the United States. Flood losses amounted to $3.8 billion in 1975, and floods cause the loss of more than 100 lives per year. Losses have been increasing at a rate of between 4 and 7% per year, in real dollars, with the losses increasing most rapidly in urban areas. It is estimated that flood losses by the year 2000 may exceed $4.3 billion (in 1975 dollars).

The United States has invested billions of dollars in flood hazard mitigation and control over the last 60 years, but the trend in flood damages continues to increase, particularly in urban and developed coastal areas. Review of the flooding problem in the United States brings forth a salient point: our approaches for controlling and mitigating flooding have not fully succeeded.

Recent major assessments of flood problems and issues have been conducted (National Research Council, 1981; National Science Foundation, 1980). The Foundation's report identified 27 broad research topics.

V. P. Singh (ed.), Flood Hydrology, 17–25.
© 1987 by D. Reidel Publishing Company.

These reports became the basis for conducting a comprehensive assessment of the research needed relating to flooding and flood mitigation. This paper describes a national blueprint of research which seeks to address two audiences: 1) the scientific and engineering communities, and 2) the organizations who fund flood research.

This assessment of the research needs, done by 15 experienced researchers on floods and reviewed by more than 50 others over a 2-year period, was performed in the context of current and future policy issues affecting flood mitigation activities. It was recognized that the research recommendations must be relevant, in terms of national policies and major issues, if the recommendations were to be properly prioritized and subsequently funded.

There are four new major national issues or trends that were accounted for. The first of these was the new federalism, as reflected by the shift in responsibilities from federal to local and state entities. The second issue was the policy of the National Flood Insurance Program. This is considered the dominant national element in both current and future flood plain activities. The third issue was the ongoing shift of emphasis from structural approaches to nonstructural approaches for flood mitigation. This includes activities such as flood plain management and zoning, coastal zone management, flood warning systems, and evacuation and relocation. The fourth major issue involved the developing national programs in emergency assistance. Flood assistance by the government now fits within a host of multi-hazard assistance activities.

2. FINDINGS AND RECOMMENDATIONS

In this research assessment (Changnon et al., 1983), five central findings emerged, along with a concern for the future of flood research. These are found as repeated expressions of needs in several of the scientific discipline assessments; in the recognition of current and future national issues and emerging policies related to floods and their mitigation; and with the recognition that the nation's flood problems are major and there is a great need to reduce the ever-moun loss to floods in the United States. This assessment identified 115 research recommentdations with 53 labeled as "critical" meaning highly urgent. These are listed and explained in detail in Changnon et al. (1983).

2.1. Inadequate Knowledge and General Priority Setting

The first major finding related to the amount of existing knowledge about floods and their mitigation. The knowledge base is very uneven. Much more is known in the physical sciences and hence about the structural approaches to flood mitigation than is known in the social sciences. Within the social sciences, such as sociology, more knowledge exists in some topical areas than in others, where major gaps of data and information exist. Hence, one major conclusion is that, in general, much more research attention, both by the scientific communities and by the funding agencies, should be given to the economic, geographic, sociological, and political scientific research than to the physically-oriented research. This greater need in the social sciences exists because they have been largely over-looked in prior years, and knowledge gains per dollar spent will be high because social research is less expensive than that in the physical/engineering areas. Hence, the payoff seems to be much greater for finding solutions with emphasis on the social-oriented interdisciplinary research. The National Science Foundation should encourage and support this type of research. This general theme of emphasis on the social scientific research also is consistent with the current national shift to nonstructural approaches to flood mitigation.

2.2. A Research Program Based on Efficient Use of Flood Lands

This assessment showed that national expertise in flood research strongly supports a philosophy that research, where possible, should have as a national goal the efficient use or enhancement of flood-prone lands, not a view of loss reduction. There is a great need for a reformulation of the goal of hazard reduction per se to be one of efficient use of flood-prone lands and the determination of socially acceptable levels of residual risk (Changnon et al., 1983). This concept is also integral to the ecological research needs where there is a need to evaluate flood mitigation in terms of natural benefits in the streams and rivers, and the flood plains, and in the wetlands that are an integral part of these systems. If this view is to be accommodated nationally, a general different order of research is needed. Our thoughts on this approach follows.

Basically the "efficiency view" addresses two broad parts of the flood hazard mitigation activities: 1) the prepardness/mitigation aspects, and 2) the recovery/restoration aspects. The research relating to warnings and emergency responses (what happens during the flood) is largely separate from the efficiency concept. Of necessity, a broad plan involving a philosophical shift to address these two concepts, or courses of research action, necessitates a long-term developmental approach. First, we must take the time necessary to develop nationally the concept of efficiency through research in certain fields (primarily in economics, ecology, sociology and political science, but also in law), while simultaneously pursuing the priority research yet needed in the warning and emergency response areas. A temporal ranking of the critical research needs would find two at the top: Evaluation of major policies and programs, and exploration of economic rationales for public and private intervention in flood hazard mitigation. These efforts should be performed first.

This view of looking at the priority and the sequencing of future research is presented in Figure 1. In a general context, it attempts to evaluate the major themes interwoven in all the recommended research. In this research scenario, three major avenues would be followed:

1) Research largely relating to the aforementioned concept of efficient use of flood lands, with its nonstructural theme

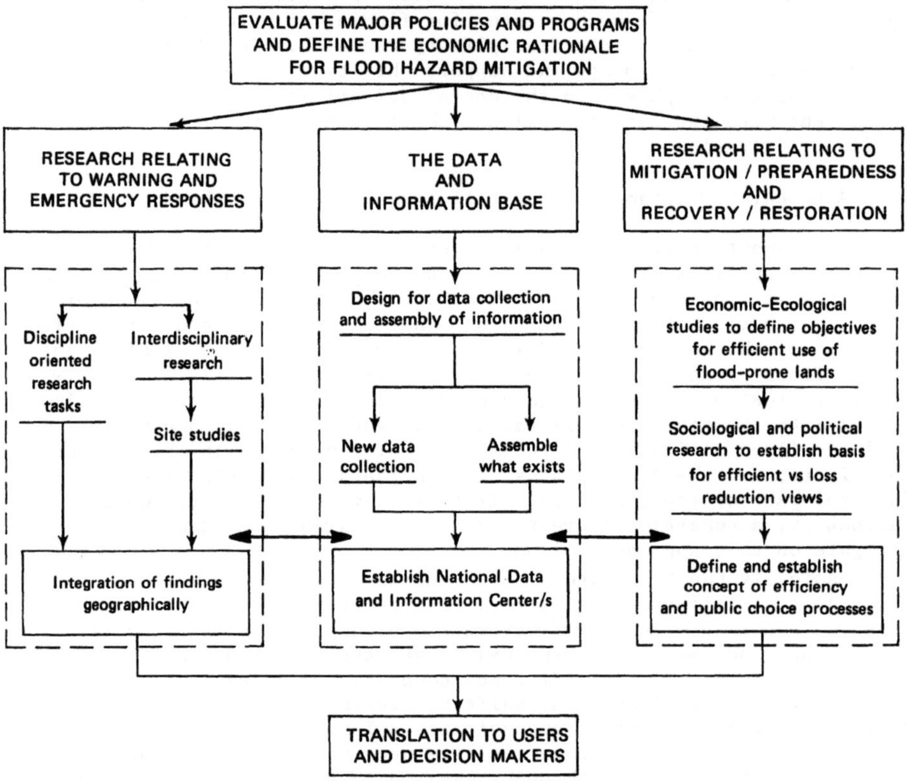

Figure 1. The major research components and one approach for sequencing of research in flood hazard mitigation

and largely focused on mitigation, prepardness, recovery and restoration;

2) Research aimed at warning and emergency responses, comprising disciplinary and interdisciplinary tasks; and

3) Development of a better flood data and information base, a major problem in all disciplines.

If this proposed program sequence were followed in setting top priority research, certain other critical research tasks should follow after 1) the evaluation of major policies and programs, 2) the development of economic rationale for public action, and 3) development of the economic case for various levels of public intervention. In the mitigation-recovery sequence (Figure 1 and Table 1), the research to be addressed first would be in the economic and ecological areas. These should include initially the redefining of the objectives of flood mitigation, and then developing better understanding of the ecological benefits of flooding.

These studies should be followed closely by high priority research in sociology and political science. It will be very important to have early study of flood mitigation management alternatives in light of scarce governmental resources, and to support this with evaluations of successful and unsuccessful flood plans. Research on the reasons for use of findings of behavioral scientists is needed along with a study of policy formulation relating to flood mitigation. Other top priority research in this chain of research relates to the long-term impacts of primary groups to families, and to individuals.

In the other basic research areas, the warning and emergency response areas (Figure 1 and Table 1), there are certain equally critical tasks to be done first and in parallel with those identified above. These include work to improve predictions of large rainfall amounts and of tropical storms. In light of growing concern and the many uncertainties about landslides, the prediction of landslides has been included in this top priority group of research needs. Research into improvements in reliable flash flood warnings is needed along with the integration of flood warning systems with other warning systems. Of particular importance is to perform case studies of the warning and emergency responses in barrier islands.

Attention to evacuation is of great importance, including studies of ways to facilitate evacuation, effects of evacuation on morbidity and mortality, and the management of emergencies. Research attention is needed relating to pre-flood planning of post-flood measures including local community responsibilities in flood mitigation.

Coupled with these top priority research areas, are three highly recommended and equally critical tasks relating to the data and information base thrust (Figure 1 and Table 1) seen as parallel to the two major research avenues. The first of these efforts is to design a basic data system for flood data, and to follow this with the procurement of economic data, and of public health data related to flooding.

This proposed research avenue for the mitigation and recovery areas places the benefit oriented research in economics and ecology first, closely followed and linked to the critical sociological research and the integral political sciences research. This general sequence will

Table 1. Recommended Program of Research in Flood Mitigation Based on a Time-Ordered Sequence

A) First Steps

1) Evaluation of Major Policies and Programs
2) Economic Rationale for Public Action
3) Economic Case for Various Levels of Public Intervention

B) Parallel Second Steps

Mitigation-Recovery Sequence

1) Redefinition of Objectives of Flood Mitigation, Effects of Flood Mitigation on Natural Stream Benefits, and Development of Proper Measures of Flood Damage
2) Flood Mitigation Management and Scarce Resources, and Successes and Failures of Flood Plans
3) Research Utilization Needs, and Process of Policy Formation
4) Long-Term Impacts to Primary Groups, Families, and Individuals

Warning-Emergency Response Sequence

1) Predictions - Large Rain Amounts
   Predictions - Tropical Storms
   Predictions - Landslides
2) Reliable Flash Flood Warnings, and Integration of Warning Systems
3) Case studies of the Barrier Islands
4) Evacuation Facilitators/Inhibitors, Effects of Evacuation Procedures, and Emergency Operation Centers
5) Pre-Flood Planning of Post-Flood Measures, and Local Community Responsibility

Data and Information Base

1) Design for a Basic Data System
2) Collection of Economic Data
3) Public Health Data for Floods

get the efficient use concept defined, as well as identification of the individual and institutional incentives to adopt the most appropriate mitigation solutions.

It is important to appreciate also, that certain research disciplines -- economics, sociology, geography, and ecology -- have research tasks with a similar over-riding theme. That is, all require considerable development of theoretical concepts in certain crucial scientific topics.

2.3. Essentiality of Interdisciplinary Research

The above theme of research oriented toward the efficient use of flood-prone lands and the scientific disciplines that it embraces, illustrates the third major conclusion: *the extreme need to integrate the research and findings across disciplines.* There is a clarion call for interdisciplinary research in the field of flood hazard mitigation with 79 of 115 recommended research tasks being interdisciplinary in nature. All of the discipline-oriented research assessments by experts called for interdisciplinary teams to work in both data collection and research.

Certain orientations were noted that offer the possibility of merging research efforts of several disciplines. One is the body of organizational theory encompassing contributions from political science and sociology. A number of the research tasks are concerned with how individuals and organizations respond to and prepare for crisis situations. Another construct is the aggregation of studies that incorporate economics, decision making and policy analysis. Much of the proposed research inherently calls for systematic examination of the geographic, economic, social, and political aspects of how public choices are made with respect to floods, and of the effects of particular public policies (see Fig. 1).

A special interdisciplinary assessment (Changnon et al., 1983) called for three studies requiring very early attention. These should be directed at 1) an evaluation of the implementation and effects of major federal policies; 2) the design of a minimal system for data collection (which is described in the next section); and 3) an analysis of the special problems raised by barrier islands along the hurricane coasts.

2.4. Data and Information Needs

The fourth major conclusion relates to flood data and information. It is evident in all disciplines that the nation lacks a comprehensive base of information about many parameters of floods, flood plain use, and the consequences of floods. The nation and the research community are faced with a key need for data within disciplines and across disciplines, and the possibility of the formation of a multidisciplinary flood data bank (Fig. 1). Public health and interdisciplinary experts recommended, for example, the formation of a multidisciplinary team to assess current data and design the components of the desired data base (Changnon et al., 1983).

As a result, the development of appropriate data banks of flood information for research is a theme reflected in Figure 1. In addition to the intense data needs for research, such information is vital to the efficient management of riverine and coastal areas subject to flooding. If broad recognition of the data/information problem were to materialize and be under serious consideration, an effort comparable to the present development of a comprehensive plan for research would likely be required. Groups of researchers, practitioners, and agency representatives could develop and design the dimensions of an adequate flood data bank.

A key final step would be to establish and institutionalize a series of flood centers, operated on a continuing basis. In some cases, as with National Weather Service and U. S. Geological Survey, expansions of present efforts would probably be most logical. In other instances, it will probably require new programs and agencies such as the Federal Emergency Management Agency or the Bureau of the Census.

2.5. Transfer of Research Results

The fifth major conclusion of this research assessment has been that research, where possible, needs to be oriented to user needs. Where possible, research should consider the transfer of information. Clearly, more is known now than is being implemented, particularly in the sociological and ecological areas. A major recommendation is to develop public understanding of the river and its flood plain as a single natural unit, that is, to redefine the public's perception of the river so that the flood plain is included.

The recommended emphasis on socioeconomic research, as opposed to physical-engineering research, reflects the needs for a cost effective approach to future research to aid in the flood mitigation issues. This, coupled with a conscious effort to focus on a view of efficient use of flood-prone lands (vs a goal of loss reduction), creates a theme of "research for applications." The research results cannot be left on the pages of scientific or legal journals; they must be aimed at application and use. Research proposals where appropriate, should identify the users and show how results will be transferred.

2.6. Ensuring Future Attention to Flood Research

A final major recommendation, based on this 2-year assessment and the changing world in which it took place, is that a review of the research needs and priorities stated in Changnon et al., 1983, should be made at least every 3 years under the direction of the National Science Foundation. From this review, a brief report assessing progress and focusing on revised priorities of the research tasks needed should be prepared and distributed -- the government agencies involved in flood research and the scientific community. Only through such continuing assessment of progress and review of priorities can realistic progress toward flood mitigation be made. Thus, a review should be conducted in 1986-1987.

3. ACKNOWLEDGEMENTS

This paper evolved from two NSF-supported grants (NSF-PAG-81-17017 and NSF ATM-8413043), but it does not necessarily reflect the views of the Foundation. The research assessment was greatly aided by the efforts of many including William C. Ackermann, J. L. Ivens, Gilbert F.

White, Helmut Landsberg, Ray K. Linsley, G. R. Marzolf, F. M. Wellings, J. W. Milliman, Thomas Drabek, Henry Caulfield, and W. A. Thomas.

4. REFERENCES

Changnon, S. A., W. C. Ackermann, J. L. Ivens. 1983. A Plan for Research for Floods and Their Mitigation in the United States. Contract Report, Illinois State Water Survey, Champaign, 226 pp.

National Research Council. 1981. Federal Water Resources Research: A Review of the Proposed Five-Year Program Plan. National Academy Sciences, Washington, DC, 83 pp.

National Science Foundation. 1980. A Report on Flood Hazard Mitigation. National Science Foundation, Washington, DC, 181 pp.

THE UK FLOOD STUDIES REPORT: CONTINUING RESPONSIBILITIES AND RESEARCH NEEDS

Max Beran
Institute of Hydrology
Wallingford
Oxon OX10 8BB
UK

ABSTRACT. The Flood Studies Report was published in 1975 following four years of research by up to 25 specialists at four research centres. Ten years on it remains unsurpassed for the scope of its recommendations, its exhaustive use of a national data set, and its translation of research methods into practical tools. Its acceptance into the UK water industry has been all but total. The main features and applications of the Report are reviewed and the importance is emphasised of a mechanism for after-sales servicing, for training, and for disseminating updates. Research into flood hydrology has continued in the UK and the continuing debates on issues raised are reported. Some special problems and responsibilities are discussed arising from applying research in an area where engineering and legal considerations may dominate.

1. INTRODUCTION AND SUMMARY

1.1 Scope of Paper

The Flood Studies Report (NERC, 1975) was published ten years ago. It contains flood estimation procedures now adopted as standard in the United Kingdom and has also provided a model that some other countries have tried to emulate. In this keynote paper I describe the institutional background to flood estimation in Britain leading up to publication of the Flood Studies Report (FSR), and also some problems of maintaining that document.

Two problem areas are highlighted. The first is mostly practical and concerns the use of the FSR procedures by non-research hydrologists. The second and more recondite set of problems relates to the conflict between the researcher's desire to innovate while retaining faith with engineering users who, because they work within cost and legal constraints, demand some stability in recommended approaches.

1.2 Geography and Historical Background

1.2.1 Geography. The United Kingdom contains no large rivers - the largest have catchments smaller than 10,000 km^2. Moreover the country occupies a location away from the zones of most intense rainfall. There is a uniform seasonal distribution of precipitation which occurs on at least 100 days per year. In the past, flooding has contributed to endemic poor drainage and disasters are rare. Communities have avoided

V. P. Singh (ed.), Flood Hydrology, 27–39.
© 1987 by D. Reidel Publishing Company.

obvious risk zones and only relatively recently have centres of population and lines of communication ventured down to the floodplain.

Despite what is a very mild experience of the flood hazard by international standards, we often represent the aim of flood hydrology as the protection of life rather than, more realistically, the relief of nuisance or the improvement of agriculture. Nowadays flood protection works require strict economic justification. However there is one area of flood estimation where the safety requirement prevails over finance; spillway design for the country's 2,000 dams.

1.2.2 History. Britain has a long history of flood protection starting in Roman times. Table I lists some of the main events relating to flood protection.

TABLE I: Main Historical Events

Year	Event	Consequence
1215	Magna Carta	"That all weres from henceforth shall be utterly put down...".
1532	Commissioners of Sewers	Reinforcement of earlier statutes and legal framework for maintenance of watercourses.
1601	Drainage of Great Level	Introduction of Dutch land drainage techniques to drain fens of East Anglia
August 1912	Rain	Heaviest areal rainfall ever recorded in UK. Extensive flooding in Norwich area.
May 1928	Louth flood	Chalk stream flood on watercourse never before flooded. More than 30 drowned.
November 1925	Dolgarrog Dam	Dam failure, 16 drowned. Led to reservoir safety legislation. Three 19th century failures, c300 drowned but cause structural not hydrological.
1930	Land Drainage Act	Current legal basis forland drainage in UK. Enables river and drainage boards to raise money for drainage improvements.
March 1947	Snowmelt + rain	Maximum flood of recent experience in much of England. Basis for many floodplain maps.
August 1952	Lynmouth flood	34 drowned; village obliterated. Basis for reappraisal of ICE 1933 report.
1968	July & September flooding	Extensive areas of south, east and London areas flooded. Indirectly responsible for FSR.
1973	Water resources Act	Set up Water Authorities with responsibilities from 'source to sink'.

While individual events often provided the design standard for schemes aimed at avoiding a repetition, it is found that all areas subsequently desire similar protection and a general upgrading of standards is adopted.

In tandem with these developments came technical advances in the sciences of hydraulics, hydrology and soil mechanics as well as in construction techniques. Advances in design flood estimation are concentrated in the more recent period. Table II traces their development in the UK.

TABLE II: Seminal Advances in Methods of Design Flood Estimation

Year	Author	Contribution
1851	Leslie	Demonstration of use of flow duration curve to quantify floods.
1862	Beardmore	Publication of book codifying flood maxima by catchment type.
1906	Lloyd-Davies	Introduction of rational formula into British practice.
1922	Williams	Time-to-peak formula from channel characteristics. Based on Indian experience.
1930	Min. of Health	Depth duration relations for rainfall.
1933	ICE	Synthesis of flood extremes into recommendations for spillway design. Simplified reservoir routing.
1936	Bilham	Statistical rainfall analysis.
1955	Richards	Publication of book containing elaborate estimation procedure based on rational formula.
1958 1960	Nash	Places Unit Hydrograph method on more scientific basis and introduces it to UK.
1965	Nash & Shaw	Use of regression on catchment characteristics to predict mean annual flood.
1975	NERC	Publication of Flood Studies Report.

2. EVENTS PRECEDING THE FLOOD STUDIES

2.1 Early Estimation Techniques

The prime mover in setting up the UK Flood Studies was the Institution of Civil Engineers (ICE). This is a professional and learned society, but additionally the ICE nominates 'panel engineers' - engineers legally empowered to design and inspect dams and spillways. Prior to the FSR an envelope curve was used to obtain the 'normal maximum flood' from catachment area (ICE, 1933). This was converted to the spillway design flood by applying a factor between one and three according to judgement.

The same envelope curve was used by some land drainage engineers for minor works but with multiplying factors below unity. To this extent both were following in the 'maximum of experience' tradition quantified a century earlier by Beardmore (1862). Rather more popular for such tasks was Lloyd-Davis's rational formula using Bilham's rainfall statistics (1936) and Bransby Williams' formula for the design storm duration (1922).

2.2 Flood Studies Committee

The ICE recognised that major advances had been made in flood hydrology since 1933; indeed fresh information (Allard et al, 1960) was incorporated into a 1960 reprint. They invited formal presentations of up-to-date techniques during the 1965 London symposium 'River Flood Hydrology' following which they set up a committee to study the adequacy of the 1933 report and to make recommendations for further data collection and research needs.

The committee report (ICE, 1967) recognised the analytical advances and was also cognisant of the considerable accumulation of river flow data. Following enabling legislation in the 1950s and 1960s the gauging station network had grown from a few dozen to over 1000 gauging stations by the late 1960s. Britain had traditionally enjoyed a very dense network of daily-read raingauges and engineers appeared wedded to rainfall based techniques even where flow data were available. The Flood Studies project was an opportunity to capitalise on this new data source and prove its value.

3. PRODUCTION OF THE FLOOD STUDIES REPORT

3.1 Constitution of Study Team

The proposal was costed by the ICE committee with recommendations for its staffing and management. They assumed that the study would be national in scope and and would present estimation methods for conventional flood protection works as well as for reservoir spillways. They envisaged a core team built up from secondment of a meteorologist, a statistical hydrologist, and rainfall-runoff and snowmelt modellers. The ICE would collaborate through membership of a Steering Committee.

Severe floods in the south-east in July and September 1968 (always the best location to prompt government action) precipitated the start of the study. The Natural Environment Research Council (NERC), the parent body of the Institute of Hydrology (IH), responded to the call and fifteen scientists and assistants were assembled at IH; a rather larger group than that envisaged by the ICE committee.

A team of five at the Meteorological Office developed depth-duration-frequency relations for rainfall. The Hydraulics Research Station was responsible for flood routing studies and a group at Newcastle University worked on snowmelt runoff.

3.2 Hydrological Research

3.2.2 Data needs. The major task of the first 18 months was the extraction of flood peaks from the water level chart records. The analysis divided into rainfall runoff and statistical studies; the former was based on 1500 events from 180 stations with catchments below 500 km^2 and the latter on about 6,000 station years of exceedance and instantaneous maxima from 530 stations. Each station was visited and graded following scrutiny of its high flow performance.

3.2.3 IH project team management. The project staff at IH were divided into three groups under Dr J.V.Sutcliffe. One group was concerned with rainfall runoff modelling, another with statistical analyses, and the third with catchment characteristic relations. A parallel study was taking place in Ireland and the data sets from the two countries were merged.

3.2.4 Report Publication. The steering committee insisted on a research report with all data included, not solely a manual of methods. Worked examples in each chapter illustrate the use of all techniques. The five volume report, Table III, was issued in 1975 priced initially at £40, subsequently at £70. The cost of the project was about £500,000 at 1975 prices.

TABLE III: Flood Studies Report Volumes

Volume number	Title	Main contents
1	Hydrological Studies	Flood statistics; data extension; ungauged site; flood volumes; hydrograph synthesis; PMF.
2	Meteorological Studies	Rainfall statistics; areal and temporal variation; PMP; snowmelt; historic events.
3	Flood Routing Studies	Theory and method choice; comparison of methods.
4	Hydrological Data	Station listings; catchment characteristics; flood event data; historic floods.
5	Maps	Rainfall statistics; soil permeability map; gauging station location

4. SOME FEATURES OF THE HYDROLOGICAL ANALYSES

4.1 Summary of Contents

Within the main division of the statistical and rainfall-runoff approaches, analysis divided further into methods applicable to gauged and to ungauged sites.

4.1.1 Ungauged site problem. A major feature of the report is the treatment of the problem of estimation at an ungauged site. In general this was achieved by relating the parameters of the various models to characteristics of the catchments. This in turn spawned a major effort to choose and derive such characteristics from topographic and thematic maps. In some cases special purpose maps had to be produced. The primary example of this is the soil map in which soils are classified according to their winter rainfall acceptance potential. Statistical and hydrograph analyses combined to assign hydrologically relevant index values to the map's five classes.

In such a way it is possible to estimate a design flood using either a statistical approach - based on a regression equation for the mean annual flood and regional 'growth' curves (Figure 1) to obtain other return periods - or a rainfall runoff approach - using a 'losses' model to convert gross to net rainfall followed by a unit hydrograph.

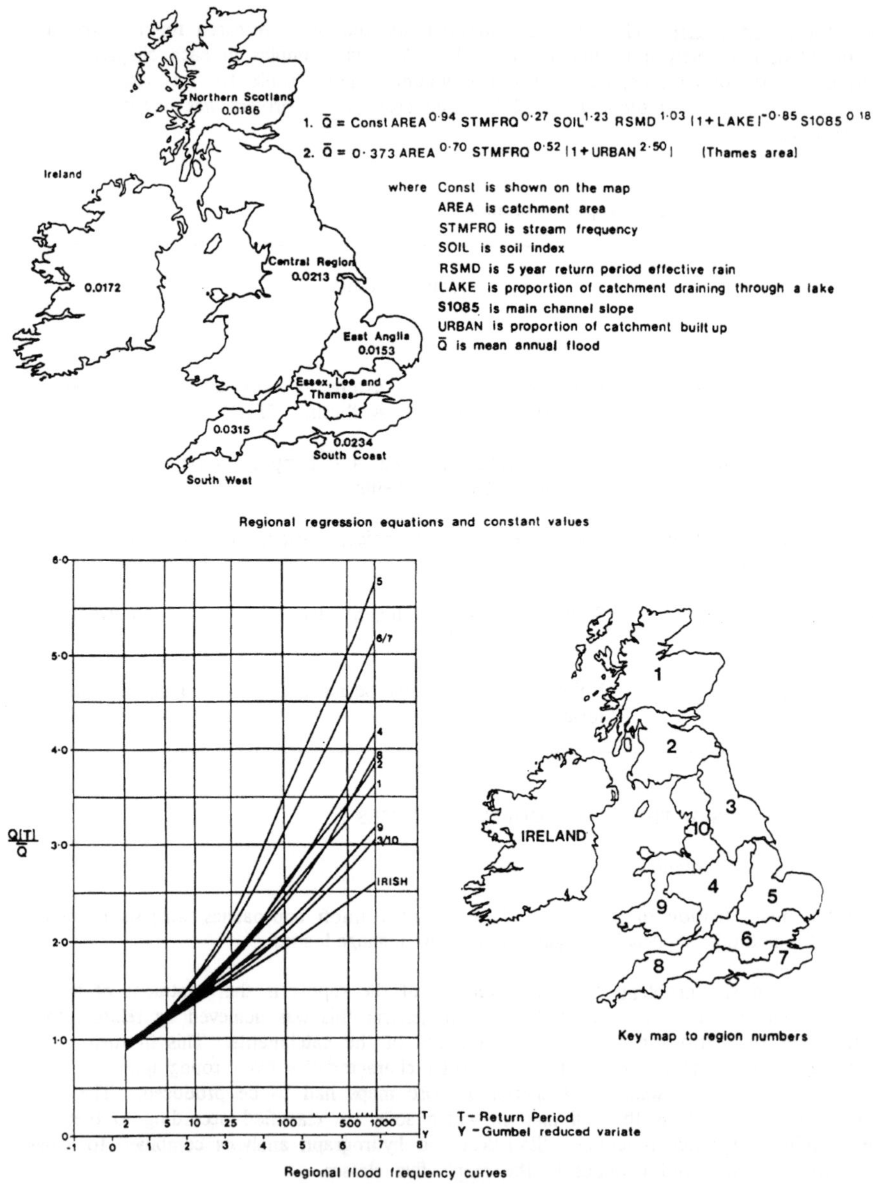

Figure 1: Statistical estimation procedure for ungauged site

4.1.2 Rainfall-runoff modelling. This comprises three sub-models: an equation for estimating losses; a unit hydrograph; and an addition for antecedent base flow. The losses depend on the soil type of the catchment and on the antecedent catchment wetness. The unit hydrograph is assumed to be of a triangular form with fixed relationship between peak discharge and duration. An important feature of the approach, and one not generally made explicit, is the determination of the correct output return period. This was achieved by a simulation exercise which investigated the sensitivities of the output hydrograph peak on each input variable.

4.1.3. Supplementary topics. The report includes procedures for extending flood data from short records, and criteria for when this is worthwhile. Consideration is given to flood volumes as well as to instantaneous peaks, the use of time series and partial duration models, and spillway design flood estimation. A non-linear storage routing model is offered as an alternative to the unit hydrograph approach.

5. REPORT MAINTENANCE

5.1 Opportunities for Discussion

Publication of the FSR was heralded by a conference in May 1975. A central issue at this conference was the relationship of the FSR to the ICE's revised guidelines on spillway design floods 'Reservoir Flood Standards'. Although the guidelines required only that inspecting engineers use the best available methods (in any case a legal requirement of any professional person), it was felt by many that there was an inescapable link with the FSR and the PMP method it recommended.

Other prominent issues raised at the conference related to the derivation of rainfall statistics and the concept of Areal Reduction Factor. Another pivotal issue concerned the relative merits of site and regional analysis - the FSR veers very much towards the latter approach.

The Flood Studies conference of May 1975 took place too soon after publication for considered technical and scientific judgement so a workshop was held in August at Birmingham University. At this the criticisms of the Meteorological volume were made more explicit and several other issues were raised in relation to the statistical approach. These revolved around local departures from the mapped and tabulated statistics.

A second major conference took place in 1980; 'The Flood Studies Report - Five Years On'. This was intended as an opportunity for practitioners to present their experiences of the use of FSR and concluded with a considerable list of topics that required further attention.

Over the intervening years there have been a number of seminars and scientific symposia at which FSR topics have been raised. A seminar held at Birmingham University in 1984 (Proceedings published in Hydrological Science Journal, Vol. 30, No. 1) focussed on criticisms of the derivation of the Figure 1 regional flood frequency curves. This has spawned new work to improve the published curves, some of which is reported on at this conference.

5.2 Flood Studies Guide

In responding to the steering committee's demand for a research report it was felt by many that the report's large bulk was not suited - indeed presented an obstacle - to the non-hydrological design engineer. To answer this point a guide to the FSR (Sutcliffe,

1978) was produced which summarises the FSR's contents and extracts the worked examples.

5.3 Training Courses

Another type of document was in preparation in the offices of a number of report users. These took the form of calculation sheets; printed pro-formas onto which the salient variables could be entered and the calculations devolved to computational assistants. In many cases these were modelled on the Institute's own course notes. Courses on the use of the FSR commenced in 1975, two per year being held for four years and an annual course since. The courses were held initially under the auspices of the Water Industry's own training scheme although with IH staff instructors.

Additional seminars have been held at the offices of individual authorities, often tailored to their particular needs. One example of the latter was a course for Road Construction Units which concentrated on the problem of small and urbanising catchments and treated peak discharge estimation only. Another example was a day-course for panel engineers which concentrated on the philosophy and application of the PMP and PMF approaches.

A more recent venture has been a three-week International Course intended for hydrologists from developing countries and with the object of teaching the specific skills needed to produce national or regional flood and low flow estimation manuals. A counterpart study on low flows was completed and reported on in 1980 (IH, 1980).

5.4 Supplementary Report Series

As soon as the original research was complete members of the study team started working on refinements to the techniques. Topics under review included a check on the Areal Reduction Factor estimates and a stricter focus on special types of catchment. These included small (<30 km^2) catchments, urbanised basins, and very permeable catchments.

In order to communicate the results of these further developments it was decided to issue a series of short loose-leaf reports. Every FSR purchaser can subscribe £12-50 and receive future Flood Study Supplementary Reports (FSSR). Since the second printing of the FSR all purchasers are automatically issued with the FSSR as a fifth volume. The editorial style of the FSSRs is that research details are not presented; only a statement of the problem, the final results, and a worked example where appropriate. To date 15 FSSRs have been issued which treat the topics listed above plus others including very low return period floods, a review of the regression relations for unit hydrograph and loss calculation, a comparison of the FSR procedure and the Rational Formula, suggestions for the use of local data, and modifications to the PMF procedure for reservoirs in 'cascade'.

5.5 Calculation Aids

Some important computer aids to the use of the report have been developed. An example is the digitisation by the Meteorological Office of the various maps that accompanied Volume II of the report. Starting from a list of map coordinates representing the boundary of the catchment a computer listing is provided which tabulates the rainfall corresponding to a standard set of durations and return periods, including the Probable Maximum Flood (Keers and Westcott, 1977). Developments of the Flood Routing procedures from Volume 3 are now available commercially from Hydraulics Research Ltd (Price, 1980). The 'Wallingford' storm sewer design procedure is also available in package form. This uses FSR and subsequently developed urban hydrological procedures as input to a pipe routing algorithm.

6. RESEARCH NEEDS

6.1 Recent Investigations

6.1.1 General studies. Research aimed at extending and improving the Flood Studies techniques has continued since 1975. Many of the topics have been reported in earlier sections. One important area not mentioned is the use that has been made of the techniques of digital cartography to speed up extraction and enlarge the range of catchment characteristics. Terrain models have been used as a base from which to derive new morphometric measures (Heerdegen and Beran, 1982), to generate an objectively defined channel network, and to determine the catchment boundary.

Another trend that has been evident since publication of the FSR has been the particularisation of techniques to suit specific types of catchment or classes of problem. Urban areas, highly permeable catchments, very small catchments - these have all been referred to previously. Also in this category are substantial areas of low-lying flat land that exists in Britain. Situated below sea level or below leveed rivers these areas require artificial drainage, in particular by pumping. FSR methods do not apply to such catchments so a programme of data collection and instrumentation has been instituted in order to provide rational bases for their design and operation (Beran, 1982).

6.1.2 Rainfall-runoff studies. The direction taken with rainfall-runoff modelling has been primarily one of acquiring more and better data and improving data handling facilities. This has led to some positive advances in unit hydrograph derivation in which groups of events are analysed simultaneously (Boorman and Reed, 1981). The importance of the spatial variability of rainfall over a catchment has been recognised and has led to improved averaging techniques (Jones, 1983). Closer scrutiny of flood event data has led to improvements in the soil mapping - in other words departures from the hydrological predictions have fed back into soil science (Farquharson et al., 1978). The basic published prediction equations for estimating unit hydrograph and loss model parameters at ungauged sites have been recomputed with the improved data set and the revisions are now being issued to report users (Boorman, 1985).

Although primarily a flood prediction technique the FSR rainfall-runoff model has been used in some real time applications for flood forecasting. This is considered in Reed (1984) as part of a broader review of practical flow forecasting procedures in UK use.

6.1.3 Statistical studies. The regional flood and rainfall analyses of the FSR use the station year approach to combine data from different gauges. Suspicions have been voiced that this approach will lead to underestimation of quantiles because of correlation in the data. A theoretical study demonstrated that, under the assumption of a homoscedastic form of interstation dependence, the levels of correlation found in UK flood data would not lead to any bias (IH, 1982). A more recent study has confirmed this conclusion under less restrictive assumptions (Hosking, 1985).

6.2 Future Research

The main area of current concern is the accuracy of the regional flood frequency curves of Figure 1. The published curves were obtained graphically from regional flood maxima and tests performed on a computer version of the procedure revealed it to be relatively inefficient and biassed (Hosking et al., 1985). Recent tests (Arnell and Gabriele, 1985) give a less unfavourable picture but it is considered important to re-evaluate the curves with more objective procedures.

Alternative regional flood estimators that have been considered are (a) General extreme value estimated by probability weighted moments (PWM); (b) Wakeby distribution estimated by PWM; and (c) two-component extreme value distribution estimated by maximum likelihood. Simulation is used to evaluate these alternatives using tests described at this conference by Arnell and Beran (1986).

Equally important is the problem of designating a homogeneous region or data subset prior to regionalising. This work is also described in a paper submitted to this conference by Wiltshire and Beran (1986).

7. PROBABLE MAXIMUM FLOOD

Extreme flood estimation is an issue of great sensitivity due to the legal and professional overtones associated with spillway design and dam safety in general. With some reluctance panel engineers have turned away from the ICE (1933) envelope curve towards the FSR recommendations. These are based on the unit hydrograph losses model with suitable adjustments to reflect the extreme conditions. The starting point for the calculation is an estimate of the probable maximum rainfall obtained from maps and tables in the Meteorological Studies volume.

Some critics have claimed that the new estimates have led to considerable expenditure in upgrading spillways to cope with the new standard. We have undertaken a large number of comparisons and conclude that the new and old standards are comparable overall with some tendency towards lower design floods on small catchments and higher ones on large catchments. The expense referred to is due to the lack of symmetry in the situation - increased standards lead to expense but reduced standards do not lead to any corresponding saving. Research into the PMP flood is progressing on two fronts.

PMP 'Probability'. Although the concepts of return period and PMP are incompatible, engineers have asked for some idea of the frequency of a PMP as an event within a region. This has been translated into a question about regional maxima: for example what is the return period of a 250 mm storm at any point over northern England? Figure 2 shows the type of diagram that is currently being drawn to answer this point. The upper lines relate to group maxima from a number of raingauges in a region. Empirical and theoretical models are under development to relate the point and regional maxima for different rain regimes, region sizes, and durations.

PMP Storm Definition. The current PMP procedure produces a very artificial event; symmetrical in time, with a very high mid-point intensity, and uniform in space. Users have asked for design storms that resemble actual large events more closely. Use is to be made of radar rainfall data to devise more reasonable storm temporal and spatial patterns for PMF studies.

8. CONCLUSIONS

A report which is intended for external use cannot be issued and then forgotten by its authors. The UK was fortunate in that research management changed at a crucial

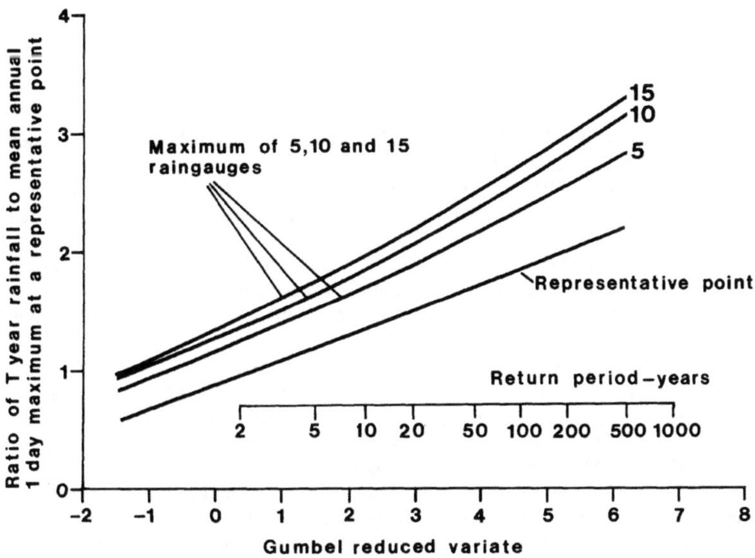

Figure 2 Frequency curves for point and regional daily rainfall maxima from 20 raingauges in 20,000 km^2.

moment to permit a group to remain dedicated to maintenance tasks. Those involved have had to be prepared to renounce some of their intellectual freedom of exploring avenues of their own choosing. This is not a major imposition as the problems thrown up by the users contain adequate meat to satisfy any engineering hydrologist. It is true that there is an empirical flavour to much of the research and development work that has to be done but increasingly process studies are finding footholds in design problems.

To be more specific about some of the lessons learned; the steering committee decided wisely that the report should embrace more than just recommended procedures. The high price does not appear to have acted as a barrier to use - the report has reached all offices where one would expect to find it. Despite its 10 year age it still has a very high profile among younger working hydrologists, due in part to its use as a source document for theory and data, and in part to the continued updates and training courses.

Subsequent research and development continues to reach users through updates in the form of Supplementary Reports. However the time is fast approaching where the scale of changes, prompted by discoveries of better methods and the accumulation of a lot of extra flood data, must lead us to a more fundamental revision.

Simple lessons are difficult to draw from such a disparate topic as the review of experience with a flood report. A simple 'one-liner' conclusion is that it is decidedly a 'good thing' for a country to have a single compendium of estimation techniques. The fear that such a document tends to stultify research is no justified.

9. ACKNOWLEDGEMENTS

This paper has been prepared as part of the contract to the Ministry of Agriculture, Fisheries and Food.

10. REFERENCES

Allard, W., Glasspoole, J. and Wolf, P.O. 1960. 'Floods in the British Isles'. Proceedings of the Institution of Civil Engineers, **15**, 119-144.
Arnell, N.W. and Beran, M.A. 1986. 'Testing the suitability of the Two-component extreme value distribution for regional flood estimation'. Symposium on Flood Frequency and Risk Analysis. Baton Rouge. Reidel.
Arnell, N.W. and Gabriele, S. 1985. 'Regional flood frequency analysis with the TCEV distribution: an assessment using computer simulation experiments'. Proceedings of the Seminar on Combined Efficiency of Direct and Indirect Estimates for Point and Regional Flood Prediction. Perugia, Italy.
Beardmore, N. 1862. Manual of Hydrology. Waterlow and Sons, London.
Beran, M.A. 1982. 'Aspects of flood hydrology of the pumped fenland catchments of Britain'. In Polders of the World Symposium, **1**, 643-652. Lelystad, Holland.
Bilham, E.G. 1936. 'Classification of heavy falls of rain in short periods'. British Rainfall, 1935. pp 262-280. Meteorological Office, London.
Boorman, D.B. 1985. A 'review of the Flood Studies Report rainfall-runoff model parameter estimation equations'. Institute of Hydrology Report **94**. Wallingford, 58p.
Boorman, D.B. and Reed, D.W. 1981. 'Derivation of a catchment average unit hydrograph'. Institute of Hydrology Report **71**. Wallingford.
Farquharson, F.A.K.. Mackney, D., Newson, M.D. and Thomason, A.J. 1978. 'Estimation of runoff potential of river catchments from soil survey'. Special Survey **11**. Soil Survey, Harpenden.
Heerdegen, R. and Beran, M.A. 1982. 'Quantifying source areas through land surface curvature and shape'. Journal of Hydrology, **57**, 359-373.
Hosking, J.R.M. 1985. 'The effect of inter-site dependence on regional flood frequency analysis'. Presented at the Fall A.G.U. meeting, San Francisco.
Hosking, J.R.M., Wallis, J.R. and Wood, E.F. 1985. 'An appraisal of the regional flood frequency procedure in the UK Flood Studies Report'. Hydrological Sciences Journal, **30**, 85-89.
Institute of Hydrology. 1980. Low Flow Studies Report. Wallingford.
Institute of Hydrology. 1982. Plotting positions for dependent samples. Research Report 1978-81. pp 58-59.
Institution of Civil Engineers. 1933 Report of the Committee on floods in relation to reservoir practive. London. 1933. (Revised and reprinted in 1960 under the name Floods in relation to reservoir practice.)
Institution of Civil Engineers. 1967. 'Flood studies for the United Kingdom'. Report of the Committee on Floods in the United Kingdom. London. 38p.
Institution of Civil Engineers. 1975. Flood Studies Conference. Thomas Telford Publishing, London.

Jones, S.B. 1983 'The estimation of catchment average point rainfall profiles'. Institute of Hydrology Report **87**. Wallingford, 34p.
Keers, J.F. and Westcott, P. 1977. 'A computer based model for design rainfall in the UK'. Meteorological Office Scientific Paper **36**. Meteorological Office, Bracknell.
Leslie, J. 1851. 'On a mode of computation whereby flood water may be excluded from a set of gaugings taken at regular intervals, and therefore, including floods'. Minutes of Proceedings of the Institution of Civil Engineers, **10**, p.327.
Lloyd-David, D.E. 1906. 'The elimination of storm water from sewerage systems'. Minutes of the Proceedings of the Institution of Civil Engineers, **164**, 41-67.
Ministry of Health. 1930 Report of Departmental Committee on Rainfall and Runoff. Journal of the Institution of Municipal and Country Engineers, **56**, pp. 1172-1176. (Reprinted as Appendix B in Escritt, L.B. 1950. Surface Water Sewerage. London.)
Nash, J.E. 1958. 'Determining runoff from rainfall'. Proceedings of the Institution of Civil Engineers, **10**, 163-184.
Nash, J.E. 1960. 'A unit hydrograph study, with particular reference to British catchments'. Proceedings of the Institution of Civil Engineers, **17**, 249-282.
Nash, J.E. and Shaw, B.L. 1966. 'Flood frequency as a function of catchment characteristics'. Symposium on River Flood Hydrology, London 1965. Institution of Civil Engineers, 115-135.
Natural Environment Research Council, 1975. Flood Studies Report. 5 vols. London.
Price, R.K. 1980. FLOUT - A river catchment flood model. Hydraulics Research Station Report **IT 168**. Wallingford.
Reed, D.W. 1985. 'A review of British flood forecasting practice'. Institute of Hydrology Report **90**. Wallingford, 113p.
Richards, B.D. 1955. Flood estimation and control. Chapman and Hall, London.
Sutcliffe, J.V. 1978. 'Methods of flood estimation: a guide to the Flood Studies Report'. Institute of Hydrology Report **49**. Wallingford.
Williams, G. Bransby, 1922. 'Flood discharge and the dimension of spillways in India'. The Engineer, **134**, 321-322.
Wiltshire, S.E. and Beran, M.A. 1986. 'Multivariate techniques for the identification of homogeneous flood frequency regions'. Symposium on Flood Frequency and Risk Analysis, Baton Rouge. Reidel.

MORE FREQUENT FLOODING IN LOUISIANA: CLIMATIC VARIABILITY?

Robert A. Muller
John D. McLaughlin
Department of Geography & Anthropology
Louisiana State University, Baton Rouge, LA, 70803, USA.

ABSTRACT. Climatic events associated with the Louisiana floods of 1982-1983 are analyzed in the context of the long-term climatic regime of the region. Regional flooding between 1973 and 1983 has been the greatest and highest of record, but can these same objectives be applied to the climatic events responsible for these floods?

In southeastern Louisiana there are rarely more than 40 years of river stage and discharge data available for analyses of longterm trends of flooding frequencies and intensities, and these periods of record are not long enough to evaluate the possible effects of climatic variability. Furthermore, individual cooperative station precipitation data are neither very representative of rainfall received over drainage basins, nor do they represent homogeneous data sets over extended time periods.

To overcome these deficiencies, divisional climatic data are used in this study. Temperature and precipitation data from 1931 to date from selected cooperative stations have been combined by the National Climatic Data Center (NCDC) to represent the Eastcentral Climatic Division of Louisiana. We have compiled divisional data from 1899 to 1930, and merged it with the NCDC data set to produce a climatic data series of almost 100 years.

A modified Thornthwaite water-budget model is used to estimate monthly divisional moisture surpluses, or precipitation available for groundwater recharge and runoff, for the division. Interpretations of time-series graphs of the moisture surplus indicate that the 11-year period between 1973 and 1983 was the wettest in these terms of the entire data series. Another unusually wet period occurred during the late 1940s, and a much earlier one during portions of the first two decades of the 20th century. The recent flooding is clearly associated with the wettest period of the last 100 years.

Correlation analyses of flood events on regional rivers and the monthly moisture surpluses for the climatic division show only a modest relationship. Studies of the precipitation patterns of flood events, however, show that excessive rainfalls are normally limited to areas much smaller than the climatic division. Hence, flood series on small to intermediate sized drainage basins undoubtedly reflect the

V. P. Singh (ed.), Flood Hydrology, 41–56.
© 1987 by D. Reidel Publishing Company.

spotty or localized patterns of excessive rainfalls, and may or may not reflect regional climatic variability or change. The climatic divisional data are more representative of regions and probably are a better index of regional climatic variability associated with changing atmospheric circulation patterns.

INTRODUCTION

Floods are a way of life in Louisiana, especially on the alluvial lowlands and deltaic plain of the Mississippi River, and on the coastal wetlands that border the Gulf of Mexico. These two regions include about 52,000 square kilometers (20,000 square miles) of lowlands and wetlands, with a population of about 2,100,000 people. Not everyone living in these regions is directly subject to flooding, of course, but extensive floods occur almost every year, and the damages associated with the flooding have continued to increase during recent decades.

The floods of southern Louisiana can be grouped conveniently into four basic types. There are the massive floods associated with the Mississippi River and its distributaries, notably the Atchafalaya. In recent decades the engineering works of the U.S. Army Corps of Engineers, including systems of artificial levees, diversion channels, and floodways, have restricted Mississippi River floods largely to backwater flooding of rural agricultural lands and swamps.

Secondly, there are the floods along the regional river systems that flow across Louisiana. Floods have been particularly frequent and costly along the Amite and Pearl rivers in southeastern Louisiana, for example, during the last 12 years. Thirdly, there are the floods across the coastal wetlands due to storm surges associated with tropical storms and hurricanes. The storm surges are normally up to 1.5 to 3 meters (five to ten feet) above sea level, but the storm surge associated with Hurricane Camille in 1969 reached 6.7 meters (22) feet along the adjacent Mississippi Gulf Coast. These storm surges have claimed numerous fatalities in the past, and quick and orderly evacuations are necessary in preventing great losses of life in the threatened areas. Finally, there is the local flash flooding increasingly common to drainage systems in urban areas; flash flooding is especially threatening to some areas of New Orleans where all of the runoff must be pumped over levees enclosing the metropolitan area that is almost entirely below sea level.

This paper focuses on the regional flooding problem, and does not treat floods associated with the Mississippi River, coastal storm surges, and urban flash flooding. The Amite River system is used as an example of regional flooding. River stage data are available for only about 40 years, and climatic data for a larger region for nearly 100 years are utilized to help evaluate the possible impacts of climate variation on the recent increase of flooding frequency and intensity along the Amite.

FLOODING ON THE AMITE RIVER SYSTEM

Figure 1 shows that the Amite River drainage basin has a north to south orientation just to the east of the Mississippi River. The river rises in southwestern Mississippi and flows generally southward across the coastal plain physiographic province to the east of Baton Rouge to drain into Lake Maurepas which connects through Lake Pontchartrain to the Gulf of Mexico. The maximum elevation in the drainage basin is about 150 meters (490 feet), and the length of the main channel is roughly 200 kilometers (124 miles). The total area above the gauging station at Denham Springs is 3,315 square kilometers (1,280 square miles).

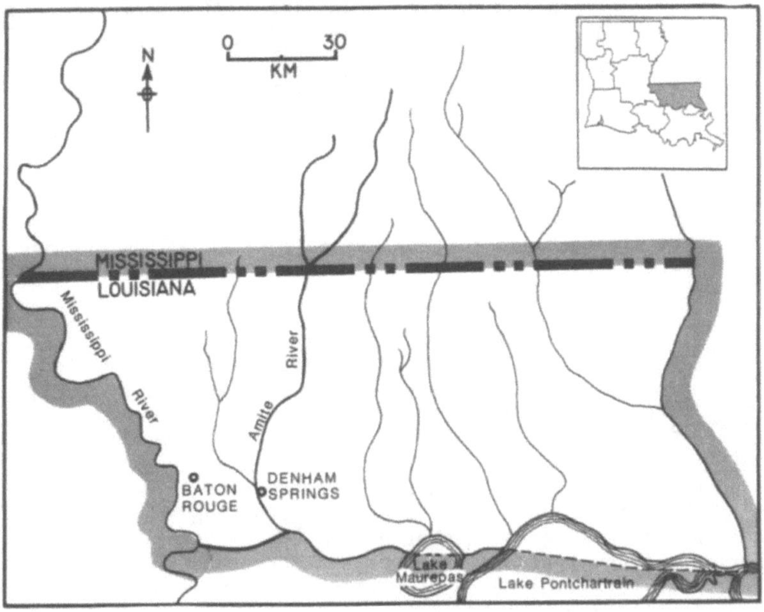

Figure 1. Location of the Amite River drainage basin and the Eastcentral Climatic Division of Louisiana.

The annual peak discharge series of the Amite River at Denham Springs, return periods as estimated by the United States Geological Survey (USGS), and peak stages for events with return periods greater than 10 years, are shown in Figure 2 (Muller & Faiers, 1984). The figure shows the clustering of unusually high stages beginning with 1973, and the two events in the Aprils of 1977 and 1983 when discharges exceeded 2,832 cms (100,000 cfs), when peak stages exceeded 12.5 meters (41 feet) - flood stage at Denham Springs is 8.8 meters (29 feet) - and when the estimated return periods approached 100 years. During these

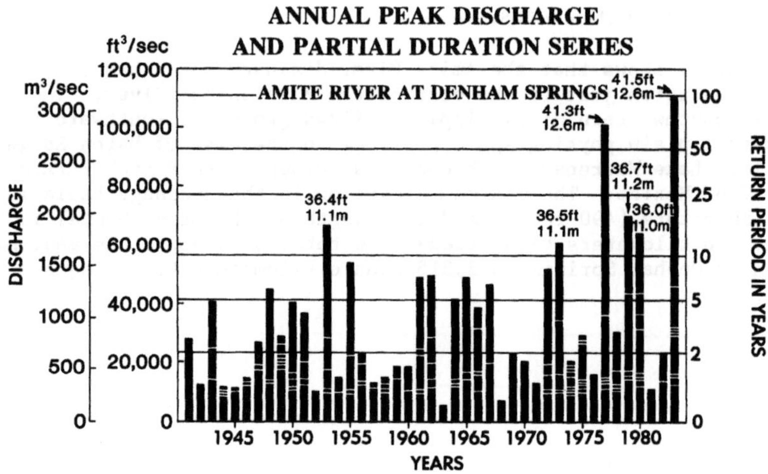

Figure 2. Annual peak discharge and partial duration series, and selected peak river stages for the Amite River at Denham Springs (Muller & Faiers, 1984).

two events there was widespread flooding at Denham Springs, Baton Rouge, and elsewhere across the lower Amite basin. These flood events ultimately forced state agencies to draw up plans for a costly flood protection reservoir on the Amite River upstream in the vicinity of the gauging station at Darlington (Baton Rouge State Times, Dec. 6, 1985).

Unfortunately, the April, 1977 and 1983, events were not the only floods on the Amite River during the 11 years from 1973 through 1983. USGS data shows that there were 19 events totalling 44 days when the Amite was at flood stage or higher during the 11 years. During the 12-year period between 1961 and 1972, on the other hand, there were only 9 flood events totalling 26 days, with the maximum stage during this period of just 10.8 meters (35.4 feet) on April 17, 1967. Figure 2 shows that flooding on the Amite River at Denham Springs were also less common during the 1940s and 1950s as well.

During the 31 years between 1941 through 1972,the peak stage according to Figure 2 was 11.1 meters (36.4 feet) in 1953, but between 1973 and 1983 this stage was exceeded during four events. During the 12 years between 1961 and 1972 there was an average of 0.75 flood events and 2.2 flood days per year. But during the 11 years from 1973 through 1983, there were 1.7 flood events and 4 flood days per year on the average.

Is the clustering of frequent and intense flood events in the 1970s and early 1980s due to climatic variability or change, or possibly due to land use changes? At the present time a small proportion of the drainage basin in the vicinity of Baton Rouge and Denham Springs has been subject to suburban development with storm sewers and some concrete-lined channels. It is unlikely that the land use changes cover enough area to affect peak discharges for the entire drainage basin so greatly, and this investigation is restricted to the questions of climatic variability.

CLIMATIC DATA

Homogeneous longterm records of temperature and precipitation suitable for rigorous climatic analyses are uncommon in the United States. National Weather Service (NWS) personnel staff most of the first-order weather stations, of which there are less than 200 in the 48 states. Hourly observations are made at the first-order stations which are located mostly at airports to provide the necessary data for the safe operation of commercial and private aircraft. These first-order stations are relocated as the airports are remodeled or rebuilt at new sites, so temperature and precipitation data from the first-order stations are rarely suitible for analyses of longterm climatic series. The NWS first-order station at Metropolitan Airport at Baton Rouge is at the western edge of the Amite River drainage basin.

Beginning in the late 19th century, the Weather Bureau, now the NWS, supported an extensive network of cooperative climatic stations where precipitation and sometimes maximum and minimum temperatures are observed once each day. There have been more than 12,000 of these stations in the United States staffed by unpaid cooperative observers. Most of these stations rarely remain located at the same site for more than 10 to 20 years, and with some notable exceptions, the precipitation data are not especially suitable for the identification of the climatology of precipitation variability through time.

At the present time there are only 9 official rainfall stations within the Amite River drainage above Denham Springs; in 1961 there were only 6 stations. In addition, analyses of storm rainfall producing flood events shows that the individual cooperative station data are not necessarily representative of basin-wide rainfall. Hence, it is difficult to find longterm precipitation records that are continuous, homogeneous, and representative of the flood events.

To partially overcome these difficulties, divisional climatic data were employed in this analysis. The National Climatic Data Center (NCDC) has divided each of the states into normally about 8 to 10 climatic divisions, and they have published divisional temperature and precipitation data from 1931 to date. The Louisiana Office of State Climatology (LOSC) has compiled divisional data from 1889 through 1930, and the two data sets have been merged in order to study the longterm climatic variability of this region.

Figure 1 includes all areas of the Eastcentral Climatic Division of Louisiana. The division is bounded by the Mississippi River to the west and the Pearl River to the east, the Mississippi border to the north, and Lakes Maurepas and Pontchartrain to the south. Temperature and precipitation data from as few as 9 and as many as 13 stations have been combined to obtain the monthly divisional precipitation average. These stations are located throughout this relatively flat division, and the divisional data should be more representative of climatic variabililty through the decades than the records of any single station.

WATER-BUDGET ANALYSES

The magnitude of flood events is, of course, not entirely dependent upon precipitation alone, and it is well recognized that antecedent moisture conditions greatly affect the magnitudes of storm peaks. For example, a heavy rainfall in summer is not as likely to generate as much runoff and flooding as an equal rainfall in winter because much of the summer rainfall will go into soil moisture and groundwater storage or go directly back to the atmosphere as evapotranspiration. In winter, the storage capacities tend to be full and evapotranspiration rates are low, so much of the heavy rainfall in the Amite River basin becomes runoff directly.

Figure 3 is a graphical representation of the some of the components of the continuous monthly water budget model used in this analysis of climatic variability (Muller, 1981, and Muller & Thompson, 1986). In the figure, PE represents model estimates of monthly potential evapotranspiration. PE is an estimate of the amount of water that would evaporate and transpire from a vegetation-covered landscape with no shortages of water for the evaporation and transpiration processes; PE, therefore, is a climatic index of energy availability and moisture demand. Evapotranspiration, or ET, represents model estimates of the return of soil moisture and precipitation back to the atmosphere together. The moisture deficit, D, represents water that would have evaporated and transpired if it had been available.

The final component in Figure 3, the moisture surplus, S, represents precipitation that does not go into evapotranspiration or soil moisture storage, but is instead available for groundwater recharge and runoff. The moisture surplus is much more closely correlated with runoff and streamflow than raw precipitation, and it is the primary water-budget component for this analysis. We recognize that use of the Thornthwaite model for estimation of daily PE is not appropriate, but that the model reasonably estimates the seasonal regimes of PE that affect the generation of runoff in southern Louisiana. Similar water-budget models are employed as the basic framework of the Palmer Drought Index (Palmer, 1965), and for analyses of wetter and drier moisture variability across the United States by climatic divisions and states (Karl & Knight, 1985).

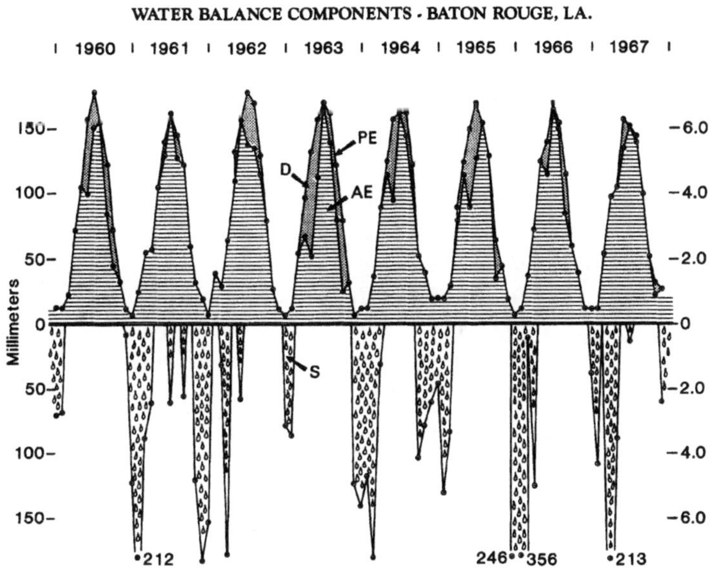

Figure 3. Continuous monthly water-budget components for Baton Rouge, Louisiana. PE = potential evapotranspiration, AE = evapotranspiration, D = moisture deficit, and S = moisture surplus (Muller, 1981).

CLIMATIC VARIABILITY OF THE MOISTURE SURPLUS IN THE EASTCENTRAL CLIMATIC DIVISION

Figure 4 is a histogram showing the annual moisture surplus series from 1889 through 1984 for the Eastcentral Climatic Division of Louisiana. The variability of the annual surplus, even in this humid if not wet subtropical climate is considerable, with the wettest year, 1961, having an estimated divisional surplus of 1,219 mm (48 inches), and the driest year, 1904, a surplus of only 190 mm (7.5 inches). There are 96 years in the data series, but three of the ten largest annual surpluses have occurred since 1972, and seven of the 20 largest annual events have occurred during the 11 years between 1973 and 1983. This variability is large enough and persistent enough to have significant environmental and economic impacts over the area of the climatic division.

A 5-year weighted moving average is also shown in Figure 4 in order to highlight climatic variability and the clustering of wetter and drier years in the Eastcentral Climatic Division. The annual moving average has ranged between 844 mm (33.2 inches) in 1947, and 387 mm (15.2 inches) in 1895, and it shows extended wetter periods in

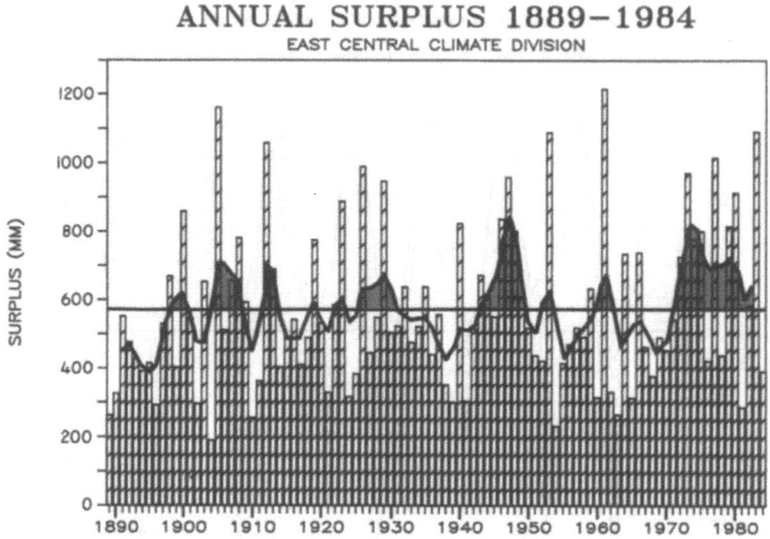

Figure 4. Annual moisture surplus series and weighted 5-year moving average for Eastcentral Climatic Division of Louisiana.

the 1970s and early 1980s, in the late 1940s, and between 1904 and 1914.

On the average most of the annual surplus in the Eastcentral Climate Division is generated during the six-month winter-spring wet season from December through May. Figure 5 shows the histogram of this six-month wet season series, and the 5-season weighted moving average in addition; summer and autumn months when surpluses are relatively uncommon are eliminated from this data set. The largest two winter-spring surplus totals occurred in 1983 and 1980; the 1983 season includes December of 1982 and the first five months of 1983. Indeed, six of the 12 largest seasonal events occurred during the 11-year period between 1973 and 1983, and the 5-season weighted moving average peaked in 1973.

On the Amite River at Denham Springs there have been more flood days during April than during any other month. Figure 6 shows the histogram and the 5-month weighted moving averages for the surpluses generated during Aprils for the Eastcentral Climatic Division. Again, the largest two monthly surpluses occurred in 1983 and 1980, and five of the eight largest surpluses occurred during the 11 years between 1973 and 1983. The 5-month weighted moving average peaked during 1980.

These analyses of April, winter-spring, and annual moisture surpluses in the Eastcentral Climatic Division show that the 11-year

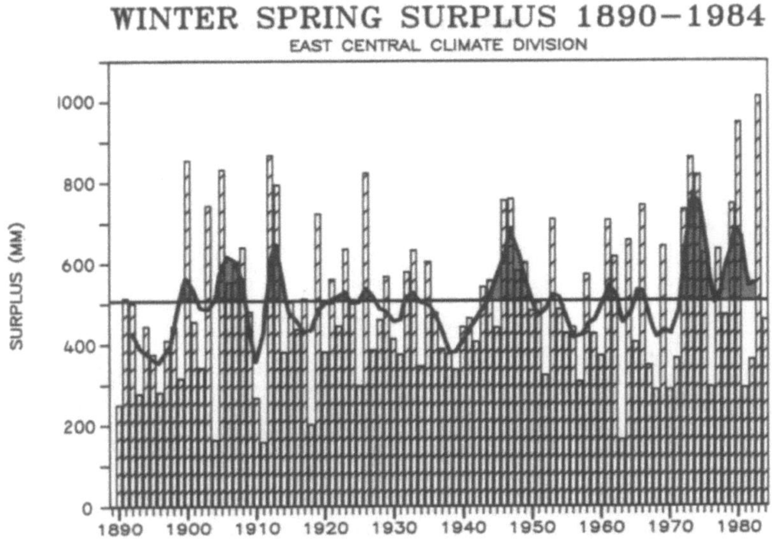

Figure 5. Winter-spring moisture surplus series and weighted 5-season moving average for Eastcentral Climatic Division of Louisiana.

period between 1973 and 1983 was essentially the wettest of the climatic record. More frequent and intense flooding of the regional drainage systems, such as the Amite, during this wetter period should not be surprising or unexpected. These analyses suggest that the more frequent and intense flooding along the Amite River may well be due to climatic variability, although land use changes and geomorphological factors may also be present.

Figures 4 through 6 also suggest that runs of wetter and drier years and associated clustering of flood events are characteristic of the climate of southeastern Louisiana. The figures suggest, for example, that clusters of wetter periods in the Eastcentral Climatic Division occurred during the first 10 to 15 years of this century, during the late 1940s, and again between 1973 and 1983, with the latest period clearly the wettest of the three in these terms. These wetter periods are associated with global and regional atmospheric circulation pattern fluctuations that are barely recognized and not very well understood. For example, El Nino years and Southern Oscillation reversals over the eastern Southern Pacific Ocean appear to be associated with increased storminess and rainfall along the northern Gulf Coast (Douglas & Englehart, 1981), and more frequent winter precipitation along the western and northern Gulf Coast in the 1970s has been associated with more frequent polar frontal activity than during the 1950s (Faiers, 1986). Figure 7 also shows that from an

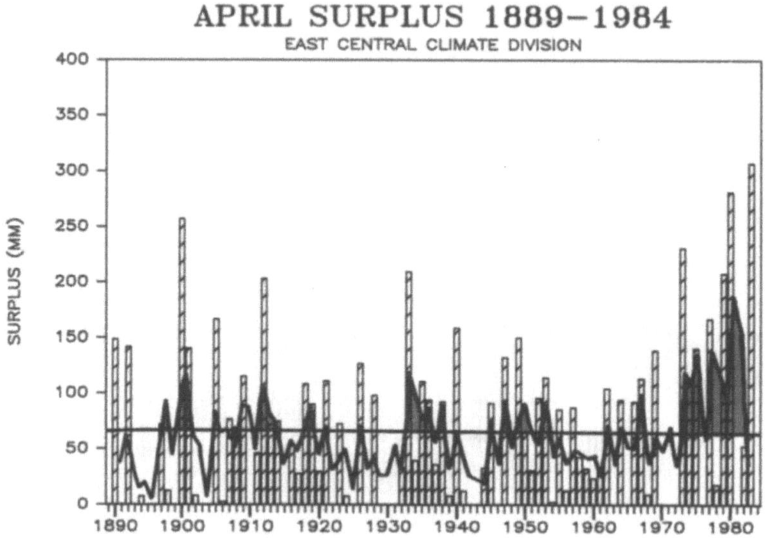

Figure 6. April moisture surplus series and weighted 5-month moving average for Eastcentral Climatic Division of Louisiana.

atmospheric circulation perspective, there is considerable variability of stormy weather types during recent decades at New Orleans, and that stormy weather was indeed more frequent during the later years of the 1970s (Muller & Willis, 1983). Hence, there is also atmospheric circulation evidence for more frequent stormy weather during much of the 1973 through 1983 period of excessive flooding along the Amite River.

FLOODING ON THE AMITE RIVER AND THE DIVISIONAL MOISTURE SURPLUS

The monthly divisional moisture surplus is undoubtedly a useful regional measure of climatic variability relative to water resources and flooding potential, but how closely is it related to flood events within the climatic division? Figure 8 is a scatter diagram of the monthly divisional surplus on the horizontal scale and peak stage of the Amite River at Denham Springs during the 28 flood events between 1961 and 1983. There is a large scatter in the diagram, and the correlation coefficient is only 0.27. The scatter diagram also shows that there were three flood crests on the Amite River during months when the divisional surplus was 0. Interestingly, these three events occurred during late summer and early fall, and two were produced by tropical storms from the Gulf of Mexico.

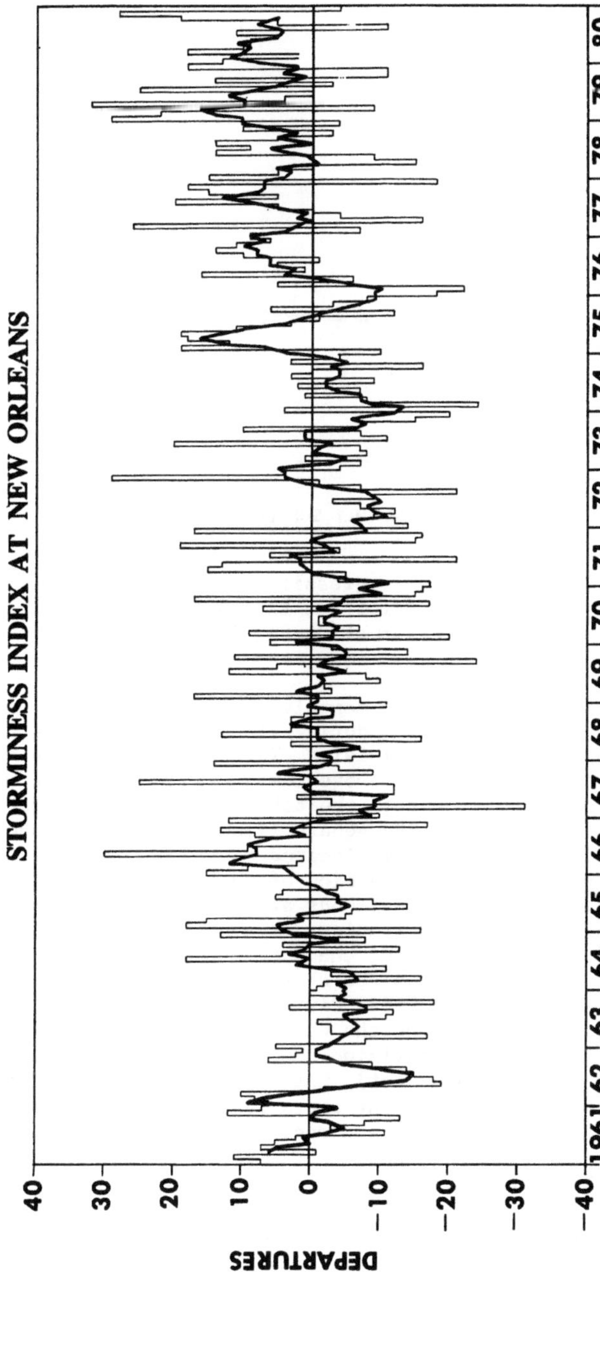

Figure 7. Storminess Index and 5-month moving average for New Orleans. The index represents the monthly departures of stormy weather types from their respective monthly averages (Muller & Willis, 1983).

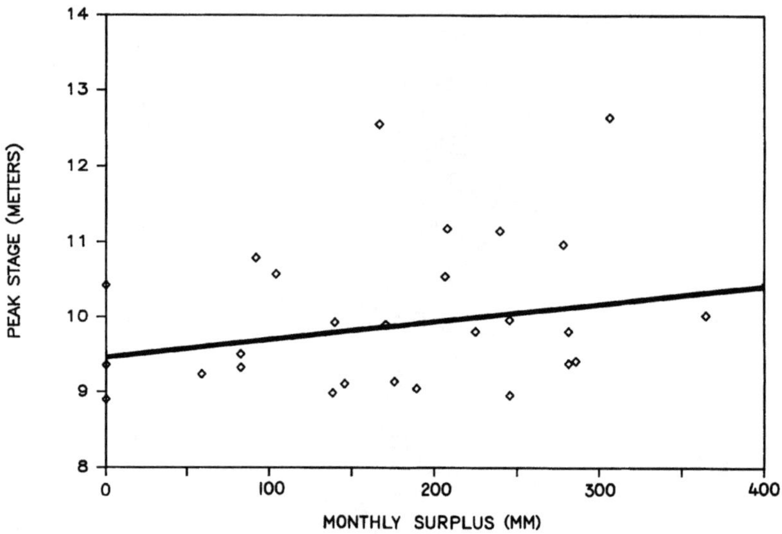

Figure 8. Scatter diagram and regression line for relationship between monthly moisture surplus for the Eastcentral Climatic Division and flood peaks on the Amite River at Denham Springs.

It is not surprising that the monthly divisional surplus is not highly correlated with peak stages during flood events along a particular drainage basin within the division. In this example, the Amite River drainage basin in Louisiana above the gauging station at Denham Springs includes only 10 percent of the total area of the climatic division, with an additional 1,290 square kilometers (499 square miles) of the Amite River drainage located to the north in Mississippi in the Southwest Climatic Division of that state. Excessive precipitation totals that produce floods are often relatively localized, and basin storm precipitation may be more or less than the divisional precipitation during the same event.

Figure 9, for example, shows the storm precipitation for Hurricane Hilda during October 6-8, 1964. Divisional rainfall for October was only 147 mm (5.8 inches), and the water-budget model predicted a moisture surplus of 0 for the division. But the flood stage on October 7 climbed to 10.4 meters (34.2 feet), the ninth highest event between 1961 and 1985. The figure shows that storm rainfall in the Amite River basin was 305 mm (12 inches) at one cooperative station, and the calculated basinwide storm rainfall above Denham Springs averaged 229 mm (9.0 inches). The excessive flood-producing rainfall was restricted to areas much smaller than the entire climatic division. Storm precipitation maps of each of the flood events shows that the excessive rainfalls producing floods are rarely so widespread that the entire

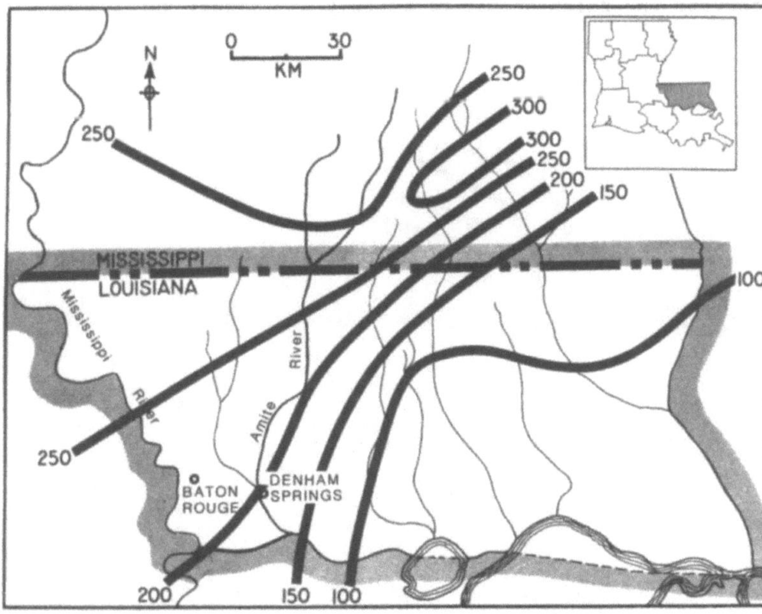

Figure 9. Storm precipitation in millimeters for Eastcentral Climatic Division during Hurricane Hilda, October 6-8, 1964.

climatic division is affected by the excessive rains. An analysis of flood events beginning with 1951 along five of the rivers flowing southward across the Eastcentral Climatic Division shows that there were only 11 days when all five rivers were above their respective flood stages at the same time (Goldberg, 1985).

One additional example is shown in Figure 10 where the storm precipitation pattern for the minor flood event on December 28, 1982, is shown. The Amite River peak at Denham Springs rose to only 9.0 meters (29.4 feet), barely 0.12 meter (0.4 foot) above flood stage. Nevertheless, the divisional surplus for December, 1982, amounted to 246 mm (9.7 inches). Figure 10 shows that storm precipitation over the drainage basin averaged only 99 mm (3.9 inches), and that some areas in the southeastern sections of the division received less than 12 mm (0.5 inch) of rainfall. This event, however, was the second flood event of the month along the Amite River. The first flood occurred on December 5-7, when the flood peak at Denham Springs climbed to 10.0 meters (32.7 feet) in association with a basin-wide rainfall of 180 mm (7.1 inches). Hence, the very large divisional surplus was produced by two flood events, and 5 more rain events that did not generate flood stages on the Amite River directly. The frequency and timing of events also disturbs the relationship of monthly divisional surplus to flood peaks.

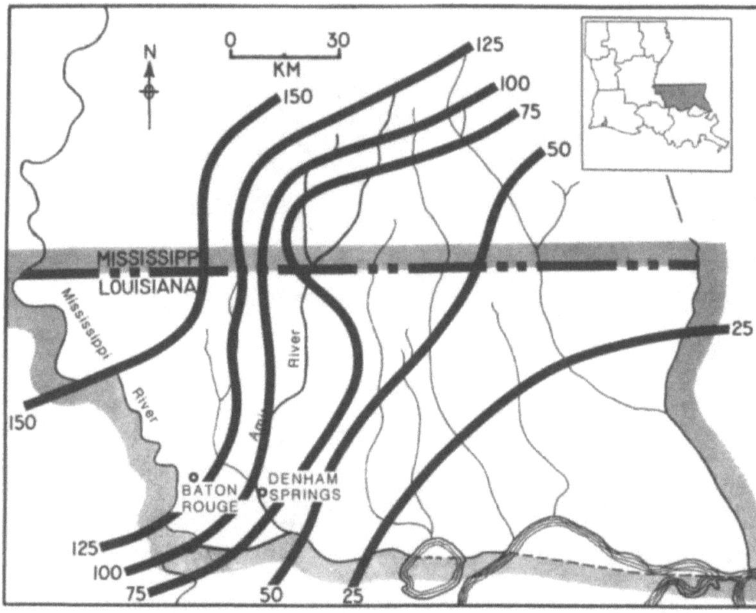

Figure 10. Storm precipitation in millimeters for Eastcentral Climatic Division during December 23-28, 1982.

SUMMARY AND CONCLUSIONS

The flood record of the Amite River at Denham Springs goes back a little less than 50 years, but the record shows more frequent and much higher flooding in much of the 1970s and the early 1980s. The temperature and precipitation averages for the Eastcentral Climatic Division of Louisiana, which includes most of the Amite River drainage, goes back to 1888, almost 100 years of record. The divisional data were entered into a water-budget model producing monthly estimates of the moisture surplus averaged for the division. This record shows that the 11-year period between 1973 and 1983 was extremely wet most of the time, generating many of the largest monthly and seasonal moisture surpluses of the nearly 100-year record. The divisional data series also suggests that there tends to be clusters of much wetter than average periods, separated by distinctly drier periods. It is likely that the clusters of wetter and drier periods are related to global and regional atmospheric circulation anomalies, such as the Southern Oscillation and El Nino phenomena in ways that are not well understood at this time.

Analyses of the storm precipitation patterns for the flood events along the Amite River suggest that the excessive rainfalls generating flood events are often rather localized, and rarely occur in flood-producing quantities over the entire climatic region. Hence, the annual or partial flood series of drainage basins of intermediate size are significantly affected by local patterns of excessive rainfalls, especially the larger events. The divisional averages of precipitation and moisture surplus departures are indices of regional wetness and dryness, and are undoubtedly more closely related to global and continental atmospheric circulation and climatic variability investigations that have received much attention in the last few years. Water resource planners and engineers charged with flood protection need to be aware of this climatic variability and its affects on flood frequencies and intensities.

REFERENCES

Baton Rouge State Times, Dec. 6, 1985.

Douglas, A.V. & Englehart, P.J. 1981. 'On a Statistical Relationship between Autumn Rainfall in the Central Equatorial Pacific and Subsequent Winter Precipitation in Florida.' Monthly Weather Review, Vol. 109, pp. 2378-2382.

Faiers, G.E. 1986. 'The Formation and Geographic Relocation of January Diurnal Precipitation Patterns in Louisiana and Southeastern Texas.' Unpublished PhD dissertation, Louisiana State University, Baton Rouge, LA.

Goldberg, A. 1985. 'A Flood Inventory and Comparison of Five Rivers in Southeast Louisiana 1950-1983.' Unpublished manuscript, Dept. of Geography & Anthropology, Louisiana State University, Baton Rouge, LA.

Karl, T.R. & Knight, R.W. 1985. 'Atlas of Monthly Palmer Moisture Anomaly Indices (1895-1930) and (1931-1983) for the Contiguous United States.' Historical Climatology Series 3-8 and 3-9, National Climatic Data Center, NOAA, Asheville, NC.

Muller, R.A. 1981. 'The Water Budget as a Tool for Inventory and Analysis of Factors Affecting Variability and Change of River Regimen', in: J.C. Ma & A.G. Noble (Editors), The Environment: Chinese and American Views, Methuen & Co., London, pp. 171-186.

Muller, R.A. and Faiers, G.E. (Editors), 1984. A Climatic Perspective of Louisiana Floods 1982-1983, Dept. of Geography & Anthropology, Louisiana State University, Baton Rouge, LA, pp. 48.

Muller, R.A. & Thompson, R., 1986. The Climatic Water Budget, in: J. E. Oliver and R. W. Fairbridge (Editors), The Encyclopedia of Climatology, Van Nostrand Reinhold Co., New York, pp. 914-921.

Muller, R.A. & Willis, J.E. 1983. 'New Orleans Weather 1961-1980: A Climatology by Means of Synoptic Weather Types.' Misc. Publ. 83-1, School of Geoscience, Louisiana State University, Baton Rouge, LA, 70 pp.

Palmer, W.C. 1965. 'Meteorological Drought.' Research Paper No. 45, U.S. Weather Bureau, 58pp.

FLOODS OF MARCH 1982, FORT WAYNE, INDIANA

Dale R. Glatfelter
U.S. Dept. of the Interior
Geological Survey
Water Resources Division
Indianapolis, Indiana 46254

Edwin H. Chin
U.S. Dept. of Commerce, National
Oceanic and Atmospheric Admin.
National Weather Service
Silver Spring, Maryland 20910

ABSTRACT. Rapid melting of a snowpack containing up to 6 inches (15 centimeters) of water equivalent, which occurred with a moderate rainfall of up to 3 inches (8 centimeters), caused major flooding in March 1982 in northern Indiana, northwestern Ohio, northeastern Illinois, and southern Michigan. Particular attention is given in the paper to the Maumee River basin, where flooding on most large streams was the worst since the devastating flood of 1913. In Fort Wayne, Indiana, flooding of the Maumee River and its tributaries, the St. Marys and the St. Joseph Rivers, damaged 1,500 homes and 100 businesses, forced the evacuation of 9,000 people, and caused over $50 million in damage. A major flood-fighting effort prevented additional damage.

INTRODUCTION

In March 1982, moderate rainfall of up to 3 in. (8 cm), which occurred with melting of a snowpack containing up to 6 in. (15 cm) of water equivalent, caused major flooding in northern Indiana, northwestern Ohio, northeastern Illinois, and southern Michigan. The floods caused at least six deaths and forced the evacuation of about 15,000 people. Peak discharges recorded at several gaging stations in the Wabash River, St. Joseph River, River Raisin, Maumee River, and Kankakee River basins have recurrence intervals of from 50 to greater than 100 years (Glatfelter, Butch, and Stewart, 1984).

Flooding in March 1982 on most large streams in the Maumee River basin in northeastern Indiana and northwestern Ohio was the worst since the devastating flood of March 1913. The Maumee River basin was the scene of major flood-fighting efforts, particularly in the vicinity of Fort Wayne, Indiana. The peak stage, 25.93 ft (7.903 m), recorded March 17 at the gaging station on the Maumee River at Fort Wayne, Indiana (04182900), was only 0.2 ft (0.06 m) lower than the March 1913 peak, 26.1 ft (7.96 m), which was the worst flood known in the area. Flooding in Fort Wayne was compounded because the river remained above flood stage of 15.0 ft (4.57 m) from March 12 through 26. The prolonged high stage saturated and strained the dikes protecting the city.

V. P. Singh (ed.), Flood Hydrology, 57–67.
© 1987 by D. Reidel Publishing Company.

Data collected by the National Weather Service document the severity and the sequence of meteorological conditions that provided the potential for and triggered the floods. Streamflow data were collected by the U.S. Geological Survey at 19 gaging stations and partial-record sites in the Maumee River basin (fig. 1, table I).

METEOROLOGICAL CONDITIONS

Meteorological Setting

Above-average autumnal precipitation resulted in moist soil conditions at the onset of the first significant snowfall on December 17 over northern Indiana, northwestern Ohio, northeastern Illinois, and southern Michigan. Additional snowfall the remainder of the month produced a snow cover of 6 to 15 in. (15 to 38 cm) throughout the region. Snow cover decreased the first week of January 1982 as temperatures rose to above 40° F (4° C). Soil moisture increased as the unfrozen ground absorbed some of the melted snow. Record snowfall and low temperatures prevailed during the remainder of January. High winds removed snow cover from unprotected locations and caused heavy drifting in shielded areas. Exposed ground froze quickly as temperatures plunged to record lows of -10° to -20° F (-23° to -29° C) at many locations January 10 and remained near or below 0° F (-18° C) for almost 2 days. Another mass of cold air accompanied by high winds and temperatures near -20° F (-29° C) moved into the region January 17 and froze exposed soils to depths of up to 3 ft (0.9 m). Moderate to heavy rainfall January 23 and 30 saturated and compacted the snowpack. Subzero temperatures after each rain formed an ice layer at least 1 in. (3 cm) thick between the snow and the ground surface. Heavy snowfall January 31 to February 10 and temperatures below 32° F (0° C) maintained an extensive snowpack. Temperatures below -10° F (-23° C) were recorded at most locations February 10. Snow depth decreased by monthend as moderating temperatures and rainfall compacted the snowpack. Additional snow and rain March 1-9 added to a snowpack that already contained a high water equivalent.

Antecedent precipitation for the period December 1981 through February 1982 for the Northeast Indiana and Northwest Ohio NWS (National Weather Service) climatic divisions, which contain the Maumee River basin, was about 34 percent and 30 percent above normal. Daily temperature, snow depth, and water equivalent data for the NWS station at Fort Wayne, Indiana, represent the conditions that existed in the Maumee River basin from December 1981 through March 1982 (fig. 2). At Fort Wayne an all-time record of 71.2 in. (180.8 cm) of snow fell during the winter of 1981-82 through March 15. Extremely low temperatures and high water equivalent of an excessive winter snowfall combined to provide the potential for the March 1982 floods.

Even though the March 1982 floods were significant, the flooding could have been much worse had the snowmelt and coincident rainfall occurred in February when water equivalent of the snowpack was much higher. For example, a record snow depth of 20 in. (51 cm) with a water equivalent of 4.2 in (10.7 cm) was observed on February 9 at Fort Wayne.

Figure 1.– Maumee River basin and locations of stations within the basin.

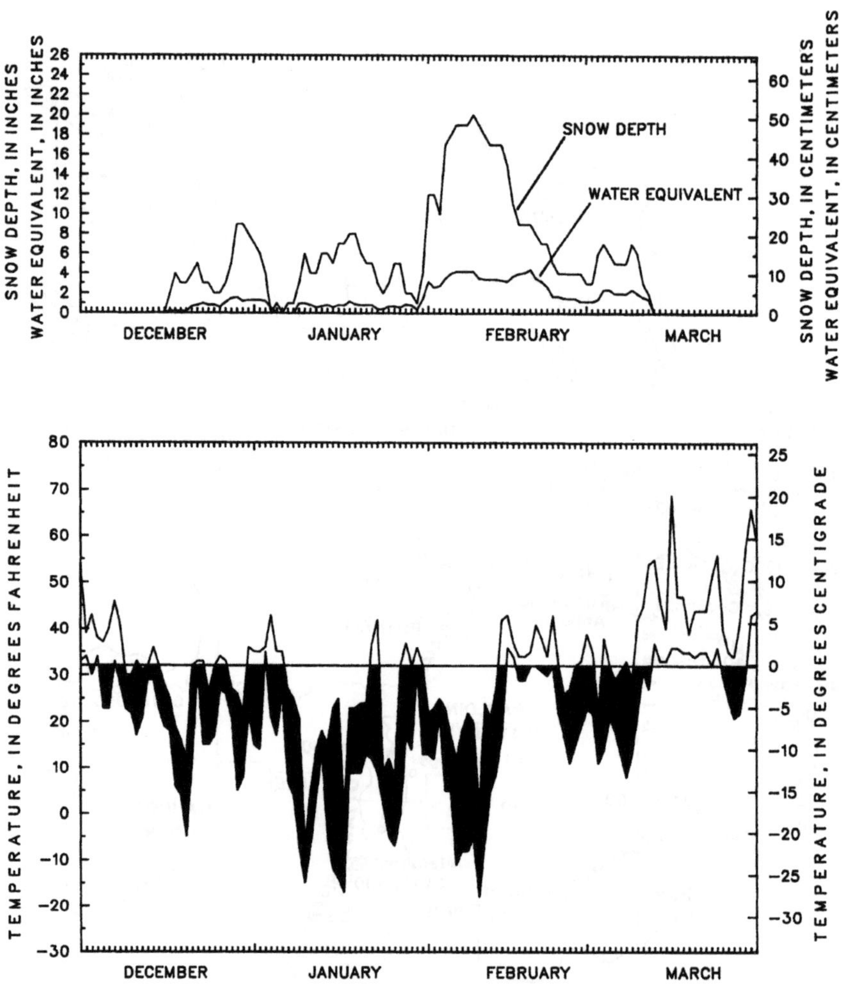

Figure 2. —— Daily temperature, snow depth, and water equivalent, December 1981 through March 1982, Fort Wayne, Indiana. Temperatures at or below freezing are shaded.

General thawing reduced the snow depth to 15 in. (38 cm) and the water equivalent of the snowpack to 3.2 in. (8.1 cm) by February 15. Light rain and snow February 16-19 compacted the snow depth to 9 in. (23 cm) but increased the water equivalent to 4.4 in. (11.2 cm) by February 19. Gradual melting of the snowpack reduced the snow depth to 4 in. (10 cm) and the water equivalent to 1.1 in. (2.8 cm) by February 28. Snow, or snow mixed with rain, on March 2, 4, and 8 increased the snow depth to 7 in. (18 cm) and the water equivalent to 2.4 in. (6.1 cm) by the morning of March 9.

The water content of the snow cover over the Maumee River basin from Fort Wayne to Lake Erie on March 9 varied from less than 1 in. (3 cm) to greater than 5 in. (13 cm). Water equivalent of the snowpack in the northwestern part of the basin drained by the St. Joseph and Tiffin Rivers was about 3.5 in. (9 cm); water equivalent of the snowpack in the southwestern part of the basin drained by the St. Marys and Auglaize Rivers was about 2.5 in. (6 cm).

Weather Patterns, March 5-20

During the period March 5-8 the upper air flow over the North American continent was characterized by the presence of high amplitude waves and a deep trough progressing through just east of the Rocky Mountains to the Central Plains and then to the Appalachian Mountains. Large amounts of Arctic air were being driven into the Midwest, causing temperatures to drop much below normal over the region. For example, during the period March 5-9 the average daily temperature was 10° F (6° C) below normal at Fort Wayne, Indiana. Thawing of the snow cover was retarded by this cold spell.

The amplitude of the upper air wave pattern began to decrease on March 9, and circulation became much more zonal by the morning of March 10 as cold advection ceased over the region. At the surface, an area of high atmospheric pressure (high) over the northeastern seaboard of the United States strengthened on the morning of March 10 causing southeasterly winds to prevail over the Midwest. This combination of circulation features initiated a warming trend, and the average daily temperature over most of the region increased more than 10° F (6° C) from March 9 to March 10. At 1900 hours March 10 a weak warm front was moving through the region as southerly flow brought in a modified maritime air mass that had originated over the Gulf of Mexico. Surface temperature differences across the front were from 5° to 10° F (3° to 6° C). By 0700 hours March 11 Ohio and southeastern Indiana were in the warm sector as the warm front had passed and a cold front approached from the northwest. By 1900 hours March 11 this weak cold front had moved out and the region was under a high. Surface winds were northerly. The high progressed to the northeast by 0700 hours March 12.

At 0700 hours March 12 there was a well-defined wind flow pattern associated with a storm in southern Canada. Surface temperatures in the central and eastern part of the country increased substantially as the storm in Canada moved eastward, while a warm front associated with warm, moist air from the Gulf of Mexico advanced northward. The warm front across central Illinois, Indiana, and Ohio produced a striking contrast of temperatures at 1900 hours March 12 as temperatures ranged from 68° F (20° C) south of the front to 36° F (2° C) north of the front. After passage of the warm front through the region, temperatures exceeded 50° F (10° C) at most locations. At Fort Wayne, Indiana, temperatures increased from 27° F (-3° C) on the morning of March 12 to 55° F (13° C) following passage of the warm front. Lower level winds over the region continued to be southerly until the morning of March 13 when passage of a cold front caused a shift in wind direction to westerly.

The increased temperatures and light to moderate rainfall associated with the weather patterns of March 9-13 caused the snowpack over most of the region to rapidly melt. Surface runoff from the melting snowpack and rainfall was nearly 100 percent because the saturated ground was frozen. Much of the snow in the Maumee River basin had melted by 0700 hours March 13. The decrease in snow depth at the Fort Wayne NWS station associated with the increase of temperature from March 9 through March 13 is shown in figure 2. On March 9 at Fort Wayne, Indiana, there was 7 in. (18 cm) of snow on the ground with a water equivalent of 2.4 in. (6.1 cm). By the morning of March 13, this snow cover had completely melted.

Total rainfall for the period March 10-20 over the region was moderate. Most of the Maumee River basin had a total rainfall of 1.5 to 2.0 in. (3.8 to 5.1 cm), but several locations received amounts of up to 2.5 in. (6.4 cm). Although the rainfall in the 11-day period March 10-20 was only a secondary factor, it did contribute to the flooding. Streams that peaked early due to rapid snowmelt had slower recessions or additional rises because of the rainfall. Flood peaks on those streams that rose later because of delayed snowmelt or that characteristically have a broad flood peak were also increased by the rainfall. At Fort Wayne, Indiana, the Maumee River was kept above flood stage of 15.0 ft (4.57 m) from March 12 through March 26 by both runoff from the melting snow and by rainfall.

DESCRIPTION OF THE FLOODS

Magnitude and Frequency

Flood frequency is generally expressed in terms of the probability of occurrence of floods of a given magnitude (discharge). The probability of occurrence is the percent chance of a given flood magnitude being exceeded in any 1 year. The recurrence interval (the reciprocal of the probability of occurrence multiplied by 100) is the average number of years between exceedances of a given flood magnitude. The recurrence interval is an average interval, and the occurrence of floods is random in time; no schedule of regularity is implied. The occurrence of a flood having a 50-year recurrence interval (2-percent change of occurrence) is no guarantee, therefore, that a flood of equal or greater magnitude will not occur the following year, or even the following week.

Discharge-frequency determinations in Indiana are coordinated by a memorandum of understanding among the U.S. Geological Survey, the U.S. Soil Conservation Service, the U.S. Army Corps of Engineers, and the

Indiana Department of Natural Resources to assure consistency of estimates among the agencies. Flood-frequency values have not been coordinated for all streams in Indiana. For those streams having coordinated values, the recurrence intervals given in table I for the floods of March 1982 are estimates from Indiana Department of Natural Resources (1981). For streams where flood frequencies have not been coordinated and at least 10 years of peak data have been collected, a log-Pearson type III statistical analysis was done by techniques described in U.S. Water Resources Council (1981) to determine recurrence intervals of the peaks.

Maumee River Basin

Although a large portion of the Midwest was affected by the March 1982 floods, no area was hit harder than the Maumee River basin in Indiana and Ohio. Flooding in the basin was the worst since the historic flood of March 1913. Rapid melting of a snowpack containing as much as 5 in. (13 cm) of water equivalent combined with moderate rainfall to cause flooding on major tributaries as well as the mainstem Maumee River. At Fort Wayne, Indiana, 1,500 homes and 100 businesses were damaged, 9,000 people were evacuated, and over $50 million in damages was reported.

Data on drainage area, period of record, maximum flood previously recorded, and the March 1982 flood are presented in table I for 19 continuous-record stations and crest-stage partial-record sites in the Maumee River basin. Locations of the gaged sites in the Maumee River basin are shown in figure 1. Peak discharges recorded in March 1982 at gaged sites on the St. Joseph River, Cedar Creek, the Maumee River, and the Tiffin River have recurrence intervals of from 50 to greater than 100 years.

A narrow belt along the western boundary of the basin is drained by the St. Marys and the St. Joseph Rivers (fig. 1). The St. Marys River, flowing from the south, and the St. Joseph River, flowing from the north, combine to form the Maumee River at Fort Wayne. At Defiance, Ohio, two more major tributaries enter the Maumee River, the Auglaize River flowing from the south and the Tiffin River flowing from the north. From Defiance, the Maumee River continues its northeasterly flow into Lake Erie.

The St. Marys River, a southern tributary, is gaged at Decatur, Indiana (04181500), and near Fort Wayne, Indiana (04182000). The discharge hydrograph for the St. Marys River at Decatur (fig. 3) shows the rapid rise in streamflow from March 10 through March 13. Peak discharge of 10,900 ft^3/s (309 m^3/s) at the Decatur gage was recorded at 0100 hours March 14. At the Fort Wayne gage, the peak was 12,600 ft^3/s (357 m^3/s) and also occurred on March 14, 3 days before the peak on the St. Joseph River. The recurrence intervals of the peaks on the St. Marys River range from 20 to 25 years.

Discharge from the St. Joseph River and its major tributary, Cedar Creek, also contributed to the flooding in Fort Wayne. Headwaters of the St. Joseph River and Cedar Creek both are in the area where the water equivalent was over 4 in. (10 cm). Melting of the dense snowpack

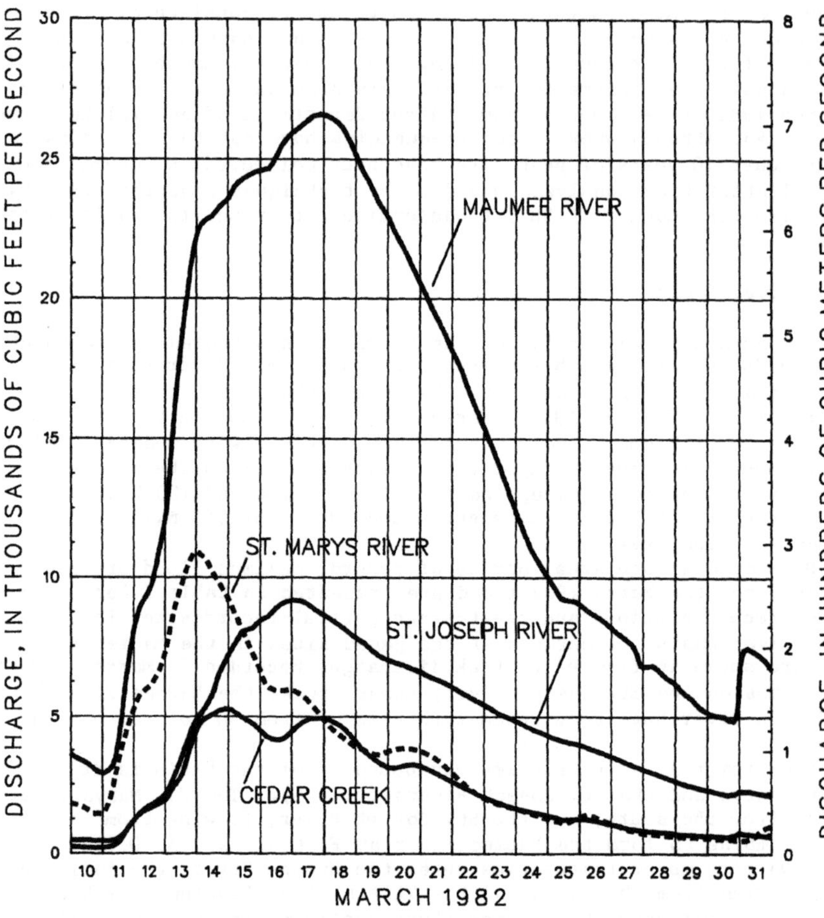

Figure 3. -- Discharge hydrographs, St. Joseph River near Newville, Indiana; Cedar Creek near Cedarville, Indiana; St. Marys River at Decatur, Indiana; and Maumee River at New Haven, Indiana.

in combination with moderate rainfall March 13 and 16 produced high peak discharges of the streams. The St. Joseph River is gaged near Newville, Indiana (04178000), and at Cedarville, Indiana (04179000). The discharge hydrograph for the gaging station near Newville (fig. 3) shows a broad crest caused by slower melting of the snowpack. Peak discharge of 9,190 ft^3/s (260 m^3/s) at the Newville gage was recorded at 0100 hours March 17. The gaging station at Cedarville was inundated during the flood, but high-water marks recovered at the site were used to determine the peak discharge of 14,500 ft^3/s (411 m^3/s). Recurrence intervals are 50 years for the peak discharge at the Newville gage and greater than 100 years for the peak discharge at the Cedarville gage.

Cedar Creek enters the St. Joseph River downstream from the Cedarville gage (fig. 1). The crest-stage partial-record site on Cedar Creek at Auburn, Indiana (04179500), recorded a peak discharge of 2,100 ft^3/s (59.5 m^3/s), which has a recurrence interval of greater than 100 years. The gaging station on Cedar Creek near Cedarville, Indiana (04180000), recorded a peak discharge at 2100 hours March 14 of 5,340 ft^3/s (151 m^3/s), which has a recurrence interval of 45 years. March 14 is the same date the peak occurred on the St. Marys River. The discharge hydrograph of the Cedarville gage (fig. 3) shows the March 14 peak, but also shows another peak of about 5,000 ft^3/s (142 m^3/s) occurring at 2200 hours March 17, the same date the St. Joseph River crested. The prolonged flooding in the Fort Wayne area was caused not only by the magnitude of the peak discharges on the St. Marys River, Cedar Creek, and the St. Joseph River, but also by the timing of the peaks.

The peak stage of 25.93 ft (7.903 m) on March 17, 1982, at the gaging station on the Maumee River at Fort Wayne, Indiana (04182900), was only 0.2 ft (0.06 m) lower than the devastating flood of March 1913, the worst flood known in the area. The flooding in Fort Wayne was compounded because the river remained above flood stage of 15.0 ft (4.57 m) from March 12 through March 26. The prolonged high stage saturated and strained the dikes protecting the city.

The gaging station on the Maumee River at New Haven, Indiana (04183000), was inundated during the flood. Highwater marks recovered at the site were used to determine the peak discharge of 26,600 ft^3/s (753 m^3/s) which has a recurrence interval of 80 years. The discharge hydrograph of the station (fig. 3) was constructed using recorded and observed gage-height data. Based on the observations of gage height, the peak discharge occurred late March 17 or early March 18. The broad crest is the result of the magnitude and timing of the peaks on the St. Marys River, Cedar Creek, and the St. Joseph River.

Flooding along the Maumee River downstream from the confluence of the St. Marys and the St. Joseph Rivers at Fort Wayne, Indiana, received national attention and prompted a Presidential visit. The efforts of thousands of volunteers, many of them school children, who filled, transported, and placed sandbags to build or reinforce dikes, prevented much damage to the city. Damage still exceeded $50 million in Allen County, Indiana, mostly in Fort Wayne.

Table I.--Summary of peak stage and discharge data, Maumee River basin

[To convert to metric: multiply ft^3/s by 0.02832 to obtain m^3/s, multiply ft by 0.3048 to obtain m, multiply mi^2 by 2.59 to obtain km^2]

Station number	Station name	Contributing drainage area (mi^2)	Period of record (water years)	Maximum flood previously recorded: Date	Maximum flood previously recorded: Gage height (ft)	Maximum flood previously recorded: Discharge (ft^3/s)	Maximum flood previously recorded: Discharge [(ft^3/s)/mi^2]	Maximum during 1982 flood: Date	Maximum during 1982 flood: Gage height (ft)	Maximum during 1982 flood: Discharge (ft^3/s)	Maximum during 1982 flood: Discharge [(ft^3/s)/mi^2]	Recurrence interval (years)
04177720	Fish Cr at Hamilton, Ind.	37.5	1970-82	03-23-78	10.79	497	13	03-17-82	11.52	603	16	35
04178000	St. Joseph R nr Newville, Ind.	610	1947-82	04-06-50	17.05	9,710	16	03-17-82	17.96	9,190	15	[a]50
04179000	St. Joseph R at Cedarville, Ind.	763	1956-82	03-24-78 05-01-56	18.62 -----	----- 10,100	13	03-17-82	21.94	14,500	19	[a]>100
04179500	Cedar Cr at Auburn, Ind.	87.3	1943-82	04-05-50	9.90	1,520	17	03-14-82	10.63	2,100	24	[a]>100
04179510	Cecil Metcalf ditch nr Auburn, Ind.	.78	1973-82	06-13-81	-----	90	115	03-14-82	10.50	140	179	20
04180000	Cedar Cr nr Cedarville, Ind.	270	1947-82	04-05-50	11.67	4,870	18	03-14-82	12.98	5,340	20	[a]45
04181500	St. Marys R at Decatur, Ind.	621	1947-82	02-10-59	24.22	11,300	18	03-14-82	24.40	10,900	18	[a]20
04182000	St. Marys R nr Fort Wayne, Ind.	762	1931-82	02-11-59	19.42	13,600	18	03-14-82	19.66	12,600	17	[a]25
04182590	Harber ditch at Fort Wayne, Ind.	21.9	1965-82	06-13-81	11.67	916	42	03-14-82	12.25	900	41	10
04182900	Maumee R at Fort Wayne, Ind.	1,926	1907-82	03-26-13	26.10	-----	---	03-17-82	25.93	------	---	----
04183000	Maumee R at New Haven, Ind.	1,967	1947-82	03-24-78	23.58	22,400	11	03-17-82	25.49	26,600	14	[a]80
04183500	Maumee R at Antwerp, Ohio	2,129	1922-82	05-20-43	20.29	26,200	12	03-17-82	21.70	26,100	12	[a]40
04185000	Tiffin R at Stryker, Ohio	410	1922-28 1941-82	04-25-50	15.45	6,640	16	03-15-82	18.36	7,760	19	50
04186500	Auglaize R nr Fort Jennings, Ohio	332	1922-36 1941-82	01-23-59	20.30	12,000	36	03-13-82	15.05	5,850	18	<5
04187500	Ottawa R at Allentown, Ohio	160	1924-35 1943-82	01-22-59	10.88	7,740	48	03-12-82	8.70	3,640	23	<5
04189000	Blanchard R nr Findlay, Ohio	346	1924-36 1941-82	06-14-81	17.43	13,000	38	03-13-82	12.35	6,320	18	<5
04191500	Auglaize R nr Defiance, Ohio	2,318	1916-82	02-13-59 02-16-50 02-12-59	27.65 ----- -----	Ice jam 52,500 52,500	 23 23	03-14-82	27.39	52,300	23	20
04192500	Maumee R nr Defiance, Ohio	5,545	1925-36 1939-75 1979-82	02-16-50	13.70	87,100	16	03-15-82	15.87	104,000	19	>100
04193500	Maumee R at Waterville, Ohio	6,330	1900-01 1922-36 1939-82	02-16-50	14.52	94,000	15	03-15-82	17.18	120,000	19	90

[a]From Indiana Department of Natural Resources (1981).

SUMMARY

Flooding in March 1982 in northern Indiana, northwestern Ohio, northeastern Illinois, and southern Michigan was preceded by above-average autumnal precipitation; frozen, saturated ground; and rapid melting of a record winter snowfall. By early March, a snowpack containing water equivalent of up to 6 in. (15 cm) covered most of the area. Rainfall March 10-12 and the passage of a warm front through the area March 12-13 triggered rapid melting of the snowpack and generated the floods. Subsequent rainfall March 15-20 maintained high streamflow and produced additional peaks on some streams.

Flooding in the Maumee River basin in Indiana and Ohio was the worst since 1913. Particular attention is given in the paper to the Fort Wayne, Indiana, area where 1,500 homes and 100 businesses were damaged, 9,000 people were evacuated, and over $50 million in damage was reported. Major flood-fighting efforts in the City prevented millions of dollars of additional damages.

ACKNOWLEDGEMENTS

Estimates of damage and additional information were supplied by the U.S. Army Corps of Engineers, Detroit District, and by the Federal Emergency Management Agency, Region V.

REFERENCES

Glatfelter, Dale R., Butch, Gerard K., and Stewart, James A., 1984, Floods of March 1982, Indiana, Michigan, and Ohio: U.S. Geological Survey Water-Resources Investigations Report 83-4201, 40 p.

Indiana Department of Natural Resources, 1981, Coordinated discharges of selected streams in Indiana: Indiana Department of Natural Resources, Division of Water.

U.S. Water Resources Council, 1981, Guidelines for determining flood flow frequency: Hydrology Committee, Bulletin 17B, 28 p., 14 appendixes.

FLOOD ESTIMATION FOR AN UNGAGED FLOODPLAIN

Dr. Jing S. Wei, Research Assistant Professor, Center for Wetland Resources, Louisiana State University, Baton Rouge, LA 70803, U.S.A.

Dr. Flora Chu Wang, Professor, Department of Marine Sciences, Louisiana State University, Baton Rouge, LA 70803, U.S.A. and

Julie A. Amft, Research Associate, Coastal Ecology Institute, Louisiana State University Baton Rouge, LA 70803, U.S.A

ABSTRACT. Many rivers in south Louisiana flow into coastal embayments and terminate in broad, low-lying marsh systems with various branching distributaries and extensive floodplains. The western Terrebonne Parish marsh study area, located in south-central Louisiana, is a riverine and tidally dominated floodplain. Flooding in the region is driven mainly by astronomical tides, river discharge, and meteorological winds. There is no long-term discharge gaging station existing in the area.

The objective of this paper is to utilize the stage and flow records at nearby stations for the estimation of tidal and discharge boundary conditions to be used for computer flooding simulations. The analysis attempts to correlate the in-situ short-term field measurements with the available long-term records elsewhere. A backwater flooding event was simulated. The dynamic response of floodplains to such an event is discussed briefly.

INTRODUCTION

The Louisiana coastal floodplain is a part of the Gulf of Mexico coastal plain physiographic province, comprised of the Chenier Plain and the Mississippi River Deltaic Plain. This coastal region is primarily a river-lake-estuary system connected by a network of river tributaries, intracoastal waterways, numerous lakes, and relatively shallow estuarine embayments.

The swamp and marsh floodplains are considered to be an essential component of the natural renewable resources affording habitats in their shallow vegetated areas providing nursery grounds for many organisms, and playing an important role in the the supply of nutrients to the bay systems. The area is also rich in non-renewable resources (oil and gas), and serves an important regional center for the petrochemical industry.

The hydrologic, hydrodynamic, and associated sedimentation processes of the coastal floodplain are driven mainly by three mechanisms: (1) astronomical tides, (2) riverine discharge, and (3) meteorological

V. P. Singh (ed.), Flood Hydrology, 69–79.
© 1987 by D. Reidel Publishing Company.

winds. The daily flooding and ebbing of the astronomical tides and the resulting discharge patterns in the area are fairly predictable. For less common events, riverine inflow and sediment content are increased substantially and cause the system response to change considerably; particularly, lowlands adjacent to the river become inundated and sediments are deposited over the swamp and marshland. During extreme events, when the passage of a storm occurs conjunctively with high river discharge, the behavior of the area may be dramatically altered. Such an event may result in extensive flooding of lowlands, redistributing of the deltaic deposits, and reshaping of the marsh floodplains.

STUDY AREA AND ITS VICINITY

The drainage basins in the Mississippi Deltaic Plain region are interdistributary basins formed by shifts of major river courses. The Atchafalaya River Basin, which is relatively young and dynamic, is a floodplain for Mississippi and Red River floodwaters (Fig. 1). The Atchafalaya River receives one-third of the combined flow of the Mississippi River and Red River, and is actively building a new delta in shallow Atchafalaya Bay (Wang, 1984). The diversion of fresh water and sediments from the Mississippi River to the Atchafalaya River is undoubtedly a dominant factor in shaping the Louisiana coastal floodplain and its associated environment (Coleman, 1976).

The Atchafalaya River Basin is confined within the protection levee system (Fig. 1). The protection levees were constructed 15 miles (24 km) apart and parallel to the Atchafalaya River course. The Avoca Island levee, approximately 13 miles (20 km) in length, was also built south of Morgan City to protect the area east of the city from the backwater flooding (Fig. 1).

The building of a new delta in Atchafalaya Bay has resulted in the lengthening of the Atchafalaya River course and has raised the stage about 2 feet (0.6 m) at the end of the existing levee. To compensate for this rise and to continue backwater flooding protection, a 14,000 ft (4.25 km) levee extension, from the end of the existing levee to the location about halfway between Big Horn Bayou and Deer Island Bayou (Fig. 2), has been proposed (USACOE, 1981).

The 160 square mile (410 km^2) western Terrebonne Parish drainage basin is an extensive marsh floodplain. It is immediately adjacent to the Lower Atchafalaya River and Atchafalaya Bay (Fig. 2). The area is subjected to both backwater flooding from the Atchafalaya River and tidal flooding from the Gulf of Mexico (Wang and Scarlatos, 1982).

Concern has been expressed that the extension of the existing levee would cut off the normal amount of freshwater and nourishing sediments presently discharged by the river into the backwater area, which in turn could result in accelerated marsh deterioration. In assessing the wetland management issue, it is necessary to understand the manner in which the basin operates as a natural system and to identify trends and their relationship to man-induced processes.

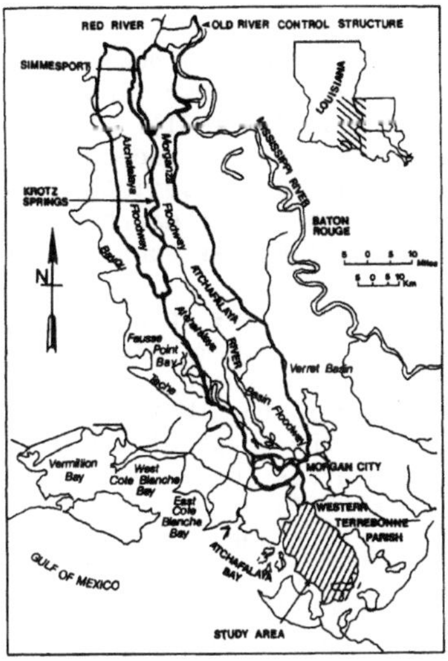

Fig. 1 Map of Atchafalaya River Basin.

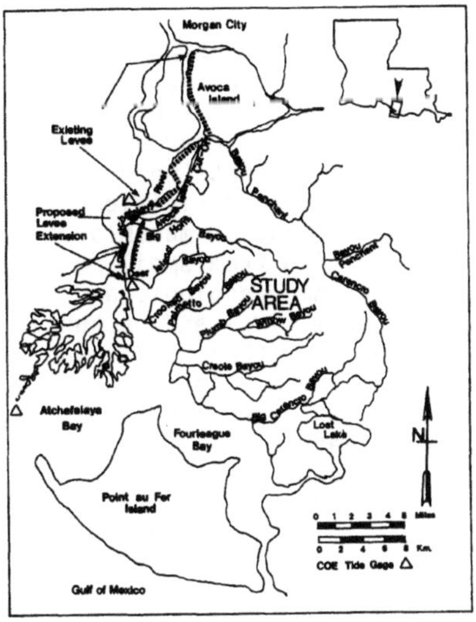

Fig. 2 Map of the Study Area - WTPM (Wang, 1987).

STUDY OBJECTIVES

The overall objective of the study is to evaluate whether the proposed levee extension will have any significant effect, beneficial or detrimental, on the marshland. This requires: (1) Investigation of the local hydrologic regime and the area's hydrodynamics; (2) Quantification of flow distribution and marsh inundation by application of a computer model of the wetland system; (3) Initiation of a low-cost meso-scale field sampling program to collect the input data for model calibration; (4) Analysis of the stage and flow records of our sampling sites (short-term) and determination of their relationship with the available records (long-term) for model verification; (5) Simulation of the dynamic behavior of marshlands at the present and future conditions under various estimated backwater flooding events and (6) Predication of the changes of marsh productivity due to changes in water circulation and sediment distribution resulting from the proposed levee extension.

This paper summarizes only the analysis of stage and flow records for use in the flooding simulation. The work is focused on two kinds of boundary conditions, astronomical tides and river discharge, which are used to quantify the effects of levee extension on marsh hydrology and the area's hydrodynamics (Wang, 1987).

MODELING OF MARSH FLOODPLAIN

The Western Terrebonne Parish Marsh (WTPM) is a depressional watershed. The area is primarily a low-lying swamp and marsh floodplain that is at most a few feet above mean sea level (USACOE, 1981). These wetlands are drained by numerous meandering bayous and have undergone some changes over the past two decades. The changes are associated with natural and man-induced activities, such as the break-up of marshes by storms, the replacement of marshlands by oil and gas access canals, and impoundment resulting from spoil bank construction.

To quantify the hydrologic and hydrodynamic processes in such an area resulting from tidal and backwater flooding, the Storm Water Management Model/SWMM (EPA, 1975) was used to simulate the dynamic behavior of the system. The solution scheme, used in the SWMM model, requires that a two-dimensional spatial domain of the physical study area be discretized into a number of subdrainage areas (catchments) and represented by a one-dimensional junction-channel (storage-flow) hydraulic network. Figure 3 shows the physical abstraction and the schematic network for the study area (Wang, 1987).

The SWMM model employs a finite-difference numerical solution to the channel-mean equations of continuity and momentum. The swamp and marsh floodplains function as storages of water volume rather than sources of momentum (Ward, 1985). The theory, assumption, and computational method used in the SWMM model are described elsewhere (EPA, 1975).

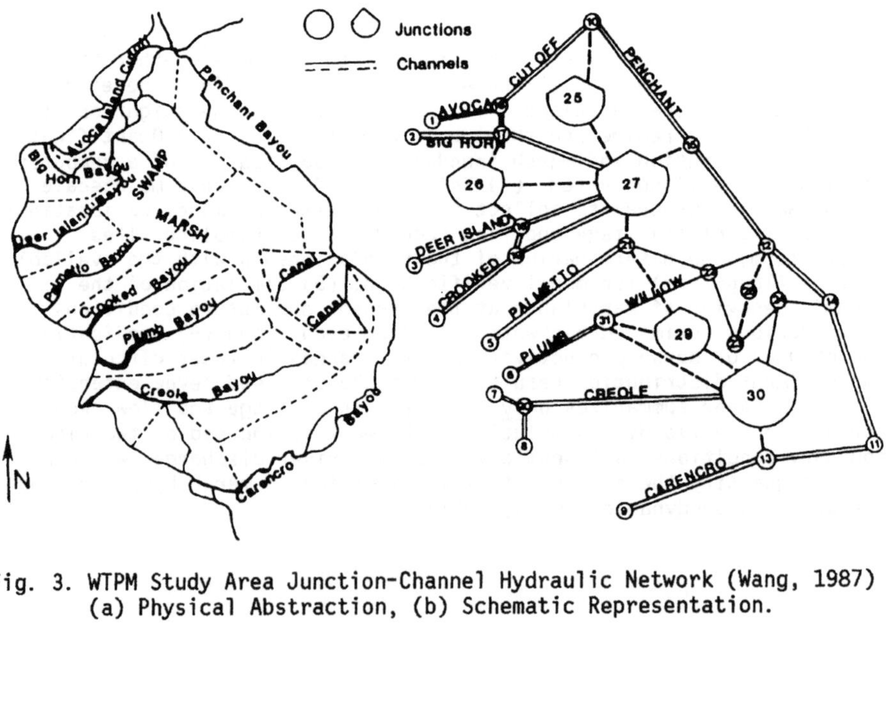

Fig. 3. WTPM Study Area Junction-Channel Hydraulic Network (Wang, 1987) (a) Physical Abstraction, (b) Schematic Representation.

SHORT-TERM DATA COLLECTION

The number of junctions and channels conceptualized in the network design (i.e. resolution) is determined largely by the degree of accuracy desired for the study objective (Zaghloul, 1981), monetary resource constraints, and the input availability of information for the area. There is no long-term continuous stream flow gaging station within the entire study area. There are a few nearby tide gage stations at which recent data are available upon request from the U.S. Army Engineer New Orleans District.

A low-cost meso-scale field program was implemented in October 1982 (Amft and Wang, 1983; Wang et al., 1985). Most of the sampling stations were established at or near the entrances of the bayous and at study area boundaries. A quarterly sampling schedule was chosen to measure the spatial and temporal variability of input parameters.

Each 2 to 3 day field trip involves intensive measurement of (1) relative tidal water levels, (2) channel velocities and cross-sectional areas, (3) sediment concentrations, and (4) wind speed and direction at specified sampling stations. A total of nine field trips were conducted from October 1982 to April 1985. Table 1 lists the dates and locations of the sampling trips conducted in the WTPM study area.

Table 1. Field stations and sampling schedule for the Western Terrebonne Parish Marsh study area (Wang et al., 1985).

Junction Number (Fig.3b)	Station Name (Fig.3a)	Sampling Schedule								
		1982	1983			1984				1985
		Oct 29-30	Mar 14-15	Jul 7-8	Oct 9-11	Jan 28-30	Mar 16-18	Jun 4-6	Nov 14-16	Apr 19-20
1	Avoca Island	S						S		P
2	Big Horn							S		
3	Deer Island	P	P				S	P		
4	Crooked	S								
5	Palmetto	S			P		S			
6	Plumb	P			S	P				
7	Creole	S			S		S			
9	Carencro			P			P			
10	Penchant North		S							
12	Penchant Split								P	
13	Carencro Lake			S			S			
15	Kent		P				P			
17	Little Horn									S
21	Palmetto Central				S				S	
22	Palmetto Canal								S	
31	Willow						P			

P = Primary station; and S = Secondary station.

LONG-TERM DATA CORRELATION

In the western Terrebonne Parish marsh, the backwater flooding problem is largely controlled by the peak discharges of the Lower Atchafalaya River, the tidal fluctuations in Atchafalaya Bay, and the regional frontal weather activity. These controlling factors vary both spatially and temporally. Their relative strength governs a particular flooding event and produces flooding patterns.

The "snapshot" data collected in the study area provide baseline information to test the computer model. Time and support limitations prohibit a continued monitoring program to include all flooding events. In order to predict future flooding events, an attempt to correlate the field data on water stage, river discharge, and sediment concentration, with the data collected (published and unpublished) by federal and state agencies (U. S. Geological Survey, U. S. Army Engineer New Orleans District, Corps of Engineer Waterways Experiment Station, and Louisiana Department of Transportation and Development) is made.

Extrapolation of Stage Records for Tidal Boundary Conditions

The water flow in many of the study area bayous is often tidally influenced. The stage record is complicated by the fact that backwater from Atchafalaya Bay affects water levels in the Lower Atchafalaya River along the entire western boundary. The occurrence of high tides downstream in the bay may also create high water stages upstream in the river, because the freshwater cannot drain away as rapidly as during low tides (Godin, 1985).

Due to the lack of benchmark stations in the large and relatively inaccessible study area, only relative water levels were measured at the sampling sites. The USACOE installed several tide gages from the mouth of Lower Atchafalaya River to the Gulf of Mexico where absolute water levels, referred to National Geodetic Vertical Datum (NGVD) are recorded (USACOE, 1973-1982). Selected stage records at Sweet Bay Lake, Deer Island Bayou, and Eugene Island (Fig. 2) which correspond to the field sampling dates are shown in Figure 4a. In comparison, Figure 4b displays the relative water levels (short-term field measurements) and the absolute water stage (USACOE long-term records) at Deer Island Bayou for two representative trips.

By examining the water level records (Fig. 4a), it is apparent that during periods of low river discharge (fall and winter), tidal influence becomes more important upstream; while in high river flow seasons (spring and summer) water stages are raised and tidal amplitude is reduced. The agreement of the two independent water level data sets (Fig. 4b) ensures that the in-situ relative water levels can be tied in with the NGVD datum plane to extrapolate the tidal boundary conditions at any point along the western boundary for simulation of flooding events.

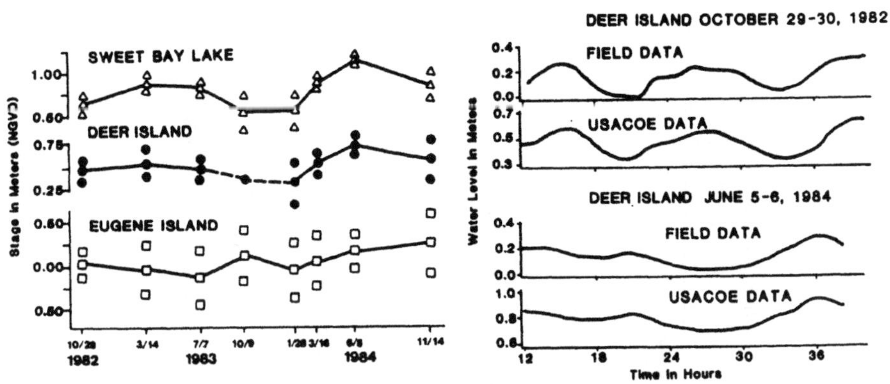

Fig. 4 Long Term and Short-Term Water Stage Records
(a) Selected Water Stages at USACOE Stations,
(b) Comparison of Water Levels between Field Data and USACOE Data at Deer Island Bayou.

Prediction of Flooding Events

Due to the environmental setting and resource constraints, it was not practical to field-measure the Atchafalaya River discharge distributed to the study area. Historical records for stations upstream (Simmesport, Krotz Springs, and Morgan City; see Fig. 1) were used to construct discharge-frequency curves for estimation of the magnitude of flood discharge that may occur in future events.

Daily records (since 1928) of Atchafalaya River discharge at Simmesport and Krotz Springs, about 125 and 90 river miles (200 and 144 km) upstream of the study area, were available. Intermittent (approximately biweekly) discharge data at Morgan City; 15 river miles (24 km) upstream, were recorded (USGS, 1973-1984). The discharge-frequency curves (Fig. 5) were derived from these recorded data.

A comparison of 1964-75 and 1973-82 frequency curves at Simmesport indicates that the magnitude of low flow has increased and high flow remains the same during these periods. Two major floods occurred during the 1973-1982 period (in the spring of 1973 and 1979). These events may be responsible for the same magnitude of high flow at both Morgan City and Simmesport stations (Figs. 5a and 5b).

In this study, it is assumed that the 1973-1984 period of record is most representative of the present condition in the study area. The discharge-frequency curve at Morgan City is used to predict the future flooding events. The magnitude of a high flow event, corresponding to a 90-percent frequency of occurence (Fig. 5a), is about 450,000 cfs (12,500 cms). By examining the historical daily discharge records, it was found that this magnitude of flood occurred on May 2, 1979. By noting the stage records at Sweet Bay Lake, Deer Island, and Eugene Island for the same date, the tidal boundary conditions for the simulation of the high flow scenario can be stipulated.

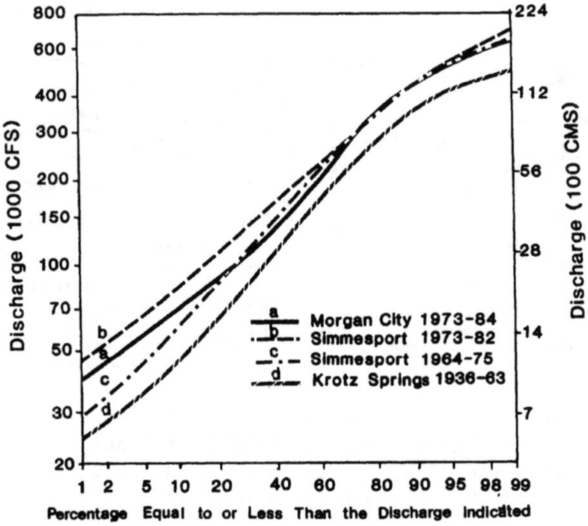

Fig. 5 Discharge-Frequency Curves for the Atchafalaya River at
(a) Morgan City 1973-84 (USGS data),
(b) Simmesport 1973-82 (USACOE data),
(c) Simmesport 1964-75 (Wells and Demas, 1977)
(d) Krotz Springs 1936-63 (Wells and Demas, 1977)

Relation of Flow and Sediment Discharges

Sediment transported by the Atchafalaya River moves in suspension and along the bottom of the river as bedload. This study refers only to suspended sediment ranging in size from fine sand to clay. Suspended sediment concentrations measured in the study area during the field trips were pooled together and used in combination with corresponding discharges to determine their relationship.

In-situ measurements of sediment concentration for all stations in the study area were relatively low, only in the range of 15 mg/l to 150 mg/l. In contrast, the recorded suspended sediment concentration of the Atchafalaya River at Morgan City (USGS, 1973-1984) varied from 100 to 450 mg/l. It appears that an increasing amount of suspended sediment has passed through the Lower Atchafalaya River outlet into the Atchafalaya Bay.

Suspended sediment loads generally increase as the river discharge increases and decrease as the flow decreases. An exponential relationship between flow and sediment discharge for the study area was derived (Fig. 6c). Similar relations of flow and sediment discharge of the Atchafalaya River at Simmesport (Letter, 1982) and Morgan City (USGS, 1973-84) are all shown in Figure 6 for comparison. The scatter of the WTPM data points illustrate the spatial and temporal variability of several environmental parameters within the study area.

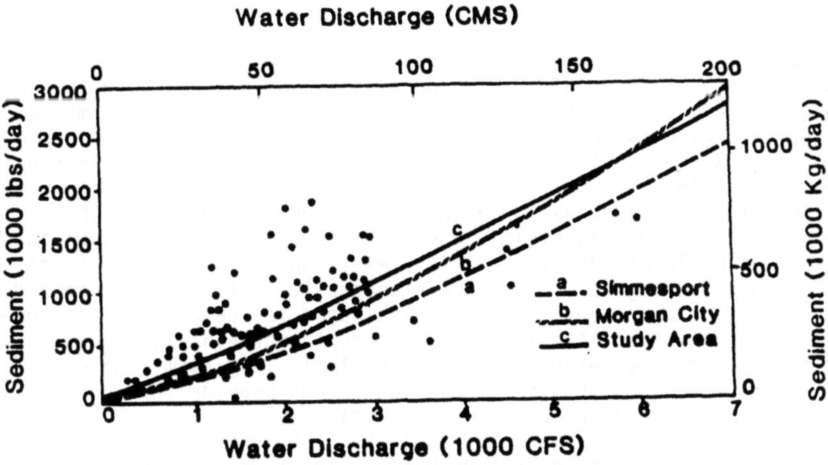

Fig. 6 Flow and Sediment Discharge Relationship
(a) Atchafalaya River at Simmesport (Letter, 1982),
(b) Atchafalaya River at Morgan City (USGS, 1973-84),
(c) Western Terrebonne Parish Marsh (Wang et al., 1985).

SIMULATION OF FLOODING DYNAMICS

Computer simulation allows "observation" of the response of events taking place in a dynamic system. This includes the technique of varying boundary conditions and system parameters to obtain a realistic result. The SWMM model has been extensively calibrated and verified for the WTPM study area (Wang, 1987). Boundary conditions for the points of entry to the study area can be either river discharges or tidal stages. The local inflow hydrograph generated from rainfall-runoff processes can also be entered into the interior of the study area (EPA, 1975).

A high flow simulation, representing a typical backwater flooding event and corresponding to a time of high river discharge and moderate tidal influence (corresponding to the May 1979 flood), was conducted. Boundary conditions for river discharge at Junctions 1 and 2 and tidal boundary conditions for Junction 3 through 10 were used. A quasi-steady state was reached by the sixth day simulation period.

The simulated results of flood and ebb tide water surface elevations and discharge patterns were obtained. The water elevation contour reveals that the drainage basin exhibits a surface gradient from the northwest to the southeast. The marsh and swamp catchments have been filled during flood tide. As the tide ebbs, the area is drained primarily by the central and southern bayous. This demonstrates the flushing characteristics of a tidally influenced watershed. The circulation patterns indicate that most of the river inflows are conveyed by Avoca Island Cutoff, the largest conduit at the northwest corner of the basin, and the flow is then distributed to the swamps and marshes via bayous, canals and overland sheet flow.

CONCLUSIONS

The western Terrebonne Parish marsh study area, a part of the coastal Louisiana floodplain, is a depressional watershed characterized by large areas of low-lying swamps and marshes, reversing water movement through meandering bayous and sluggish sheet flow over the marsh surface. Seasonal rainfall and flooding events govern the dynamic response of the floodplain.

There is no continuous long-term stream gaging station in this large and relatively inaccessible area. Flood estimations in such an ungaged floodplain are needed as this area is undergoing both natural and man-induced changes. This paper attempts to utilize the stage and flow records at the nearby stations established for several streamflow and climatic networks. The analysis intends to correlate the in-situ short-term field measurements with the available long-term records. The estimated river discharges and tidal stages thus provide the boundary conditions for simulations of various flooding events.

A mathematical model was used to investigate and predict the dynamic response of such an ungaged basin. The model has been calibrated and verified for use in flooding simulations (Wang, 1987). A flooding scenario, representing a typical backwater flooding event in the area, was simulated and discussed in this paper.

The flooding simulation and its dynamic response are then used to guide the design of other flooding scenarios to reflect the changes of the physical system imposed by man. The overall objective is to evaluate the changes of hydrologic regime and the area's hydrodynamics before and after the proposed levee extension (Wang, 1987). The predicted results of wetland dynamics in coastal Louisiana will then be used for environmental impact assessments in the area.

ACKNOWLEDGEMENTS

This paper is based on the results of research supported, jointly, by the Board of Regents' Research and Development Program for the Louisiana State University System and by the Louisiana Sea Grant College Program, the National Oceanic and Atmospheric Administration, U. S. Department of Commerce,

The authors wish to thank their research team members: Leila Gosselink, Cynthia A. Moncreiff, and P. D. Scarlatos for their contributions throughout this research project. Our thanks are also extended to Clara Coleman, Joseph V. Letter, and William H. McAnally of the Waterways Experiment Station, USACOE; Stuart Brehm, Bill Garrett, and Venetia Blazer of the New Orleans District, USACOE; and George Arcement and Fred Lee of the Water Resources Division, USGS, for the tide and discharge data.

Louisiana State University Coastal Ecology Institute Publication No. LSU-CEI-86-8.

REFERENCES

Amft, J. A., and Wang, F. C. "A Meso-Scale Field Monitoring Technique," in: R. J. Varnell (Ed.), Water Quality and Wetland Management Conference, New Orleans, LA., Aug., 1983, pp. 223-241.

Coleman, J. M. Deltas: Processes of Deposition and Models for Exploration. Continuing Education Publishing Co., Inc., Champaign, IL., 1976.

Environmental Protection Agency. "Storm Water Management Model User's Manual, Version II," Environmental Technology Series, EPA 67012-75-017, USEPA, Cincinnati, OH, 1975.

Godin, G. "Modification of River Tides by the Discharge," Journal of Waterway, Port, Coastal and Ocean Engineering, American Society of Civil Engineers, Vol. 111, No. 2, 1986, pp. 257-274.

Letter, J. V. "The Atchafalaya River Delta: Extrapolation of Delta Growth," Report 3, HL-82-15, U. S. Army Engineer District, New Orleans, LA. and Waterways Experiment Station, Vicksburg, MS., 1982.

U. S. Army Corps of Engineers. "Stages and Discharges of the Mississippi River and Tributaries and other Watersheds in the New Orleans District," Published annually. USACOE, New Orleans, LA., 1973-1982.

U. S. Army Corps of Engineers. "A Feasibility Report: Environmental Impact Statement on the Atchafalaya Basin Flooding System, Louisiana," Vol 1, U. S. Army Engineer District, New Orleans, LA., 1981.

U. S. Geological Survey. "Water Resources Data, Louisiana," Published annually. USGS and Louisiana Department of Transportation and Development, Baton Rouge, LA., 1973-1984.

Wang, F. C. "The Dynamics of River-Bay-Delta System," Journal of Geophysical Research, OCEANS, American Geophysical Union, Vol. 84, No. 15, 1984, pp. 8054-8060.

Wang, F. C. "Effects of Levee Extension on Marsh Flooding," Journal of Water Resources Planning and Management, American Society of Civil Engineers, Vol. 113, No. 2, 1987, pp. 1536-1545.

Wang, F. C., and Scarlatos, P. D. "Influence of Weather Patterns on Wetland Floodings," in: A. J. Johnson and R. A Clark (Editors), International Symposium on Hydrometeorology, American Water Resources Association, Denver, CO, 1982, pp. 373-379.

Wang, F. C., Amft, J. A., Wei, J. S., and Moncreiff, C. A. "A Study of the Effects of Man's Activity on Wetland Floodings," Final Report to the Board of Regents' Research and Development Program, State of Louisiana, Center for Wetland Resources, Louisiana State University, Baton Rouge, LA., 1985.

Ward, G. H. "Marsh Enhancement by Freshwater Diversion," Journal of Water Resources Planning and Management, American Society of Civil Engineers, Vol 111, No. 1, 1986, pp. 1-23.

Wells, F. C., and Demas, C. R. "Hydrology and Water Quality of the Atchafalaya River Basin," Water Resources Technical Report, No.14, Department of Transportation and Development, State of Louisiana, and the U. S. Geological Survey, Department of the Interior, 1977.

Zaghloul, N. A. "SWMM Model and Level of Discretization," Journal of Hydraulic Engineering, American Society of Civil Engineers, Vol. 107, No. 11, 1981, pp. 1536-1545.

DIFFUSIVE FLOOD WAVES IN LARGE RIVERS

Willi H. Hager
Dr., Sen. Res. Engr.
Chaire de Constructions Hydrauliques,
Département de Génie Civil, DGC

Jean-Jacques Droux
Mathematics Engr.
Département de Mathématiques,

Ecole Polytechnique Fédérale de Lausanne, EPFL,
CH-1015 Lausanne, Switzerland

ABSTRACT

The diffusive flow pattern of flood waves in large rivers is investigated using the zero-inertia wave approximation. The river geometry is approximated by a simple substitute river corresponding to a wide rectangular, prismatic channel of uniformly distributed wall roughness. Although the local river geometry is overlooked, the main geometrical characteristics are retained as average values. The results obtained by such a simple approach thus may not be applied to local river zones, but only to the overall discharge characteristics. The upstream river hydrograph corresponds to a single-peaked flood wave, while the lateral inflow to the river is uniformly distributed. Both of the hydrographs are normalized to the same time to peak.

The governing convection-diffusion equation for river discharge as a function of space and time is solved with an implicit numerical scheme. Typical results are graphically presented; in particular, the maximum discharge as a function of the longitudinal coordinate as well as the corresponding time to peak at arbitrary locations are investigated.

Owing to the generalized approach, the diffusive pattern of flood waves in large rivers can be analysed straightforward, once the basic parameters such as the characteristics of the substitute river and the upstream and lateral inflows quantified. The model presented thus enables a rapid prediction of all discharge quantities and applies particularly to preliminary design informations. Therefore, it is also of interest with respect to the frequency and risk analyses of floods in rivers.

The results are applied to a large flood wave in the upper Rhine valley. The comparison of the predictions with the observations indicates that the present approach is able to model the main discharge characteristics. Furthermore, it is found that the reaches considered must be long in order that the diffusive flow pattern appears. Conditions in terms of basic variables are given, for which the wave diffusion can be overlooked and the zero-inertia wave approximation may be substituted with the kinematic wave approximation.

V. P. Singh (ed.), Flood Hydrology, 81–95.
© 1987 by D. Reidel Publishing Company.

INTRODUCTION

Flood waves in rivers may cause serious losses of lifes and amounts of money. The main reasons for floods in large rivers are heavy rains, snowmelting, dam breaching and land subsidence, among others. The physics of flood routing has been founded in the 19th century particularly by the French hydraulicians de Saint-Venant and Boussinesq (Rouse and Ince, 1957). Further contributions have been forwarded by Forchheimer and Massau, Stocker and Henderson, the latter providing a review of literature, (Henderson, 1966).

Unsteady flows in rivers are usually treated by the **one-dimensional, gradually varied flow equations** according to de Saint-Venant. For local river zones as inundation reaches, in which the transverse variation of velocity and flow depth becomes important, two-dimensional approaches as presented by (Cunge, 1975) should be applied. However, for long reaches, a river is well characterized by a streamtube, for which the hydraulic model equations according to de Saint-Venant apply. Based upon the mass and the longitudinal momentum balances, a system of first order, **hyperbolic** partial differential equations (PDE) for the unknows velocity and flow depth as functions of space and time must be solved, thereby imposing appropriate initial and boundary conditions.

With the exception of river reaches in which either the bottom slope, the river width, the lateral inflow or the bottom roughness are rapidly changing, a large river is characterized by the **subcritical** flow condition. Based upon a scaling approach, it can be demonstrated that the total change of velocity with respect to time, $dV/dt=\partial V/\partial t+V\cdot\partial V/\partial X$, is much smaller than the bottom slope, see e.g. (Henderson, 1966; Hager et al., 1986). The longitudinal momentum equation thus simplifies considerably, expressing that the spatial change of the free surface slope equals to the difference of bottom slope and frictional slope, see later Eq. (3). The new system of equations then corresponds to a parabolic PDE, and is refered to as **zero-inertia wave approximation.** In contrast to the de Saint-Venant approach, disturbances are only propagated in the downstream direction. Furthermore, shocks as may appear for hyperbolic PDE are not developing (Whitham, 1974).

Flood routing in rivers over long reaches is time-consuming and expensive. However, regarding flood forecasting, aspects of flood frequency analysis and risk problems, flood routing is indispensable. In particular, the hydraulic engineer is often confronted with questions of how a flood waves propagates along a reach, or what the maximum velocity, discharge and flow depths will be at a cross-section of interest. Aspects such as overtopping and flooding then may be analysed. Owing to the hydraulic and hydrologic complexities by which flood waves are governed, these questions can satisfactorily be answered only by solving the unsteady flow equations. However, since the river characteristics (such as the river bed, the local roughness distribution and the lateral inflows) are basic input properties troublesome to quantify, and since the governing equations offer difficulties regarding the solution procedure, no **preliminary approach** seems to exist by which the most important flow characteristics can be predicted. More precisely, it would be of certain engineering interest to provide an elementary approach, by which the pattern of flood wave routing in large rivers is outlined, once the input parameters are specified.

The purpose of this investigation is to present a simple flood routing approach, based on the zero-inertia wave aproximation, by which the most important flow properties are retained. Using plausible simplifications regarding the substitute river, the upstream and lateral inflows, and by accounting for the non-linearity of the governing equations, a model is presented which can be applied to arbitrary flow situations with a fair degree of accuracy. The flood characteristics may then be estimated at an arbitrary location of interest by an elementary, straightforward procedure. The method presented is particularly useful for preliminary design purposes, and may serve as a decisive basis with respect to more complex numerical methods. Furthermore, hydraulic questions related to the flood frequency and risk analysis can be answered without refering to sophisticated flood routing investigations.

GOVERNING EQUATIONS

If the flow characteristics in a long river are considered, the transverse flow components may approximately be overlooked. To the lowest order of approximation, the river then becomes a single streamline in which the mean flow depth, h, and the average velocity V=Q/A represent the actual, complex flow pattern. Q is discharge and A is cross-sectional area; both parameters may vary with the longitudinal coordinate x and with time t.

The geometry of a river is quite complex; besides river forelands and sand banks, the river topography may seriously change due to significant turbulence along the bottom and due to sediment transport. However, again considering long distances, the **substitute river** to be considered shall be of rectangular cross-section having average width B, average bottom slope S_0 and average roughness coefficient K=1/n according to Manning-Strickler's formula

$$S_f = \frac{V^2}{K^2 h^{4/3}} \tag{1}$$

in which the hydraulic radius has been replaced by the flow depth, corresponding to **shallow water** flow conditions. Since the river width is spatially increasing while, at the same time, the river bottom slope decreases except for local discontinuities, the above assumptions regarding a prismatic substitute river are justified (Hager, 1984).

Flood waves in large rivers are characterized by a maximum velocity and maximum flow depth such that $F_{max}=(V/(gh)^{1/2})_{max}<1$ in which F is the Froude number. Consequently, the dynamical effects play an insignificant role in the prediction of the flow behaviour. The governing flow equations according to de Saint-Venant (Liggett, 1975) thus may be simplified to read for the rectangular substitute river (Henderson, 1966; Jansen, 1979)

$$\frac{\partial h}{\partial t} + \frac{\partial q}{\partial x} = p \quad , \tag{2}$$

$$\frac{\partial h}{\partial x} = S_o - S_f \quad . \tag{3}$$

Eq. (2) corresponds to the full mass balance, q being discharge per unit width, and p lateral in- or outflow, p=p(x,t). Eq. (3) is the **zero-inertia wave approximation** with S_o as bottom slope (assumed constant herein). Compared to the full de Saint-Venant equations, Eqs. (2,3) constitute a system of parabolic partial differential equations (PDE). This is easily realized upon eliminating $(\partial^2 h/\partial x \partial t)$ from the two of the differentiated relations, the result being

$$q_{xx} = \frac{2qq_t}{K^2h^{10/3}} - \frac{10}{3}\frac{q^2h_t}{K^2h^{13/3}} + p_x \tag{4}$$

or, when eliminating h_t with Eq. (2) and $h^{13/3}$ with $S_o - S_f = 0$

$$q_{xx} = \frac{2S_o q_t}{q} - \frac{10}{3}S_o\left(\frac{K^2S_o}{q^2}\right)^{3/10}(p-q_x) + p_x \quad . \tag{5}$$

Indeed, $S_o = S_f$ corresponds to the **kinematic wave approximation** (Woolhiser and Liggett, 1967) which can only be applied provided **F** <1 and $(h/(S_oL))_{max} \ll 1$, L being a typical river reach. However, since the approximation introduced does not change significantly the governing flow equation, and since the results will be much easier to handle, this is an accepted procedure compared to the other approximations regarding the substitute river.

The **discharge load** to be considered shall be realistic and simple at the same time. At the upstream end of the river reach (x=0) the inflow hydrograph is approximated as **single-peaked.** For heavy flood waves, peak discharge q* is much larger than the base flow q_b; asymptotically $q_b/q^* \to 0$. Let t* be time to peak, q(t=t*)=q*, and r a hydrograph shape parameter; then the Maxwellian time distribution

$$Q = (Te^{1-T})^r \tag{6}$$

reflects a typical inflow hydrograph. Q=q/q* and T=t/t* are normalized discharge and time, respectively. Figure 1 shows Q(T) for r=2 and r=10; it is seen that $\partial Q/\partial T(T=0)=0$, providing a continuous transition of the hydrograph for T<0 and T>0.

The **lateral inflow** to a river is particularly concentrated at river junctions. Other sources are the infiltration and smaller streams running into the main river. The most simple approach considers a spatially uniform inflow which smears over local discontinuities, whence p=p(t) only. Assuming again a Maxwellian time distribution for the lateral inflow, it is reasonable to consider the analogon to Eq. (6), i.e.

$$P = (Te^{1-T})^r \quad , \tag{7}$$

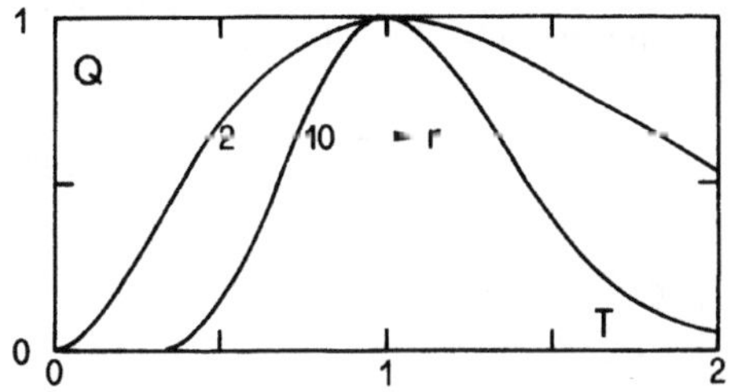

Fig. 1 : Normalized inflow hydrograph Q(0,T) according to Eq. (6) for typical hydrograph shape parameters r.

$P=p/p^*$ being relative lateral inflow discharge. Eqs. (6,7) are normalized to the same typical time to peak, t^*; approximately a heavy storm yields comparable runoff features regarding the river and the lateral. Although not precisely found in actual catchment areas, the present physical model accounts for the significant effects and thus allows an approximate prediction of the **overall flow characteristics.**

NON-DIMENSIONAL FORMULATION

For uniform flow conditions $(S_o=S_f,\ \partial/\partial t\equiv 0,\ p\equiv 0)$ Eq. (1) yields $q^*=KS_o^{1/2}h^{*5/3}$ in which h^* is maximum flow depth with respect to q^*. Let

$$X=\frac{h^*x}{t^*q^*}\quad,\quad T=\frac{t}{t^*}\quad,\quad Q=\frac{q}{q^*}\quad,\quad \sigma=\frac{h^{*1/3}}{t^*KS_o^{3/2}}\quad,\quad \omega=\frac{t^*p^*}{h^*}\tag{8}$$

be non-dimensional longitudinal coordinate, normalized time, relative discharge per unit width, diffusion number and relative lateral inflow, respectively. Then, with $p_x\equiv 0$ and upon accounting for Eqs. (6,7), the governing flow equation (5) transforms into

$$\sigma QQ_{XX}=2Q_T+\frac{10}{3}Q^{2/5}[Q_X-\omega P]\tag{9}$$

This is a non-linear **diffusion-convection equation** for non-dimensional discharge $Q=Q(X,T)$. It will be solved imposing the initial condition $Q(X,T=0)=0$ (base flow much smaller than peak discharge) and the upstream boundary condition $Q(X=0,T)=(T\exp(1-T))^r$ according to Eq. (6). As downstream boundary condition, one might impose a lake, a weir as a part of a hydroelectrical plant or an empty dam. In order not to fix a particular flow configuration, but to leave open possible flow development, an infinitely long river is considered, for which $Q(X\to\infty,T)\to 0$. Consequently, the diffusive effects of a flood wave can be studied without external disturbance. Moreover, the results will be applicable to quite arbitrary flow configurations as are found in reality. Fig. 2 shows the substitute river and the discharge loads located at the upstream boundary and at either side of the river.

It is interesting to note that, for $\sigma \to 0$, the parabolic PDE (9) becomes hyperbolic and then corresponds to the kinematic wave approximation (Hager et al., 1986).

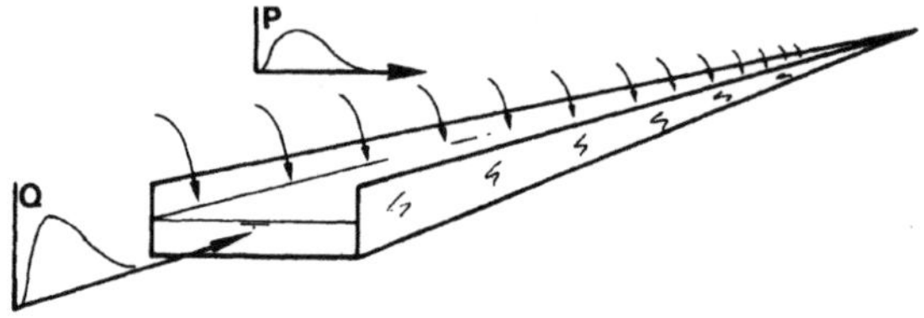

Fig. 2 : Schematized flow configuration along substitute river.

NUMERICAL INTEGRATION

When solving numerically Eq. (9), the downstream boundary condition $Q(X \to \infty, T) \to 0$ must be replaced by $Q(X \to X_d, T) \to 0$ thereby choosing X_d large enough with respect to the maximum river length X. In particular, the domains $0<X<10$ and $0<T<10$ have been chosen, and $X_d=15$.

The **computational procedure** consists in using a splitting method (Richtmyer and Morton, 1967) applied to the non-linear convection-diffusion Eq. (9). By writing Eq. (9) as

$$2Q_T = -\frac{10}{3}\frac{5}{7}(Q^{7/5})_X + \left[\sigma Q Q_{XX} + \frac{10}{3}Q^{2/5}\omega P\right] , \tag{10}$$

the **hyperbolic** portion ($\sigma=0$) without the source term ($\omega=0$) corresponds to

$$2Q_T = [f(Q)]_X \tag{11}$$

in which $f(Q)=-(50/21)Q^{7/5}$. The **parabolic** portion of Eq. (9) is

$$2Q_T = \sigma Q Q_{XX} + \frac{10}{3}Q^{2/5}\omega P \quad , \tag{12}$$

thereby including the source term.

Knowing the solution Q according to Eq. (9) at time T, the solution at time $(T+\Delta T)$ is computed by successively solving the hyperbolic Eq. (11) and the parabolic Eq. (12), thereby using the solution of Eq. (11) as starting condition at time T for the parabolic Eq. (12) between T and $(T+\Delta T)$.

Hyperbolic Equation

Eq. (11) is solved by the conventional finite difference scheme

$$2\,\frac{Q_i^{j+1}-Q_i^j}{\Delta T} = \frac{f(Q_i^j)-f(Q_{i-1}^j)}{\Delta X} \tag{13}$$

with $Q_i=Q_0(X_i)$, $X_i=i\Delta X$ $(i=0,\pm1,\pm2,\ldots)$ and $T_j=j\Delta T$ $(j=0,1,2,\ldots)$. This scheme is stable provided $(3/5)\Delta X \geq \max_{(X,T)}(Q^{2/5}(X,T))\Delta T$ (Courant criterion).

The local derivative $\partial f/\partial X$ is approximated by a backward step, because the flow propagates in the flow direction. The error of the computational scheme is $(\Delta X+\Delta T)$, provided the stability condition is fulfilled. Furthermore, the unknown Q_i^{j+1} is **explicitely** given in Eq. (13).

Parabolic Equation

Eq. (12) is solved by the standard Θ-method (Lapidus and Pinder, 1982)

$$\Theta\sigma Q_i^{j+1}\left(\frac{Q_{i+1}^{j+1}-2Q_i^{j+1}+Q_{i-1}^{j+1}}{(\Delta X)^2}\right) + (1-\Theta)\sigma Q_i^j\left(\frac{Q_{i+1}^{j}-2Q_i^{j}+Q_{i-1}^{j}}{(\Delta X)^2}\right)$$

$$= 2\left(\frac{Q_i^{j+1}-Q_i^j}{\Delta T}\right) - \Theta\frac{10\omega}{3}(Q^{2/5}P)_i^{j+1} - (1-\Theta)\frac{10\omega}{3}(Q^{2/5}P)_i^j \qquad (14)$$

in which $0\leq\Theta\leq1$.

For $\Theta=0$ (forward Euler step), the scheme is explicit and the approximation is of order $((\Delta X)^2+(\Delta T))$. For $0<\Theta\leq1$, the scheme is implicit, however. Particular values of Θ are $\Theta=1/2$ (Crank-Nicolson) in which the approximation is of order $((\Delta X)^2+(\Delta T)^2)$, and $\Theta=1$ (backward Euler step) in which the approximation is of order $((\Delta X)^2+(\Delta T))$. For $1/2\leq\Theta\leq1$, the scheme is **unconditionally** stable (Raviart and Thomas, 1983).

It should be noted that, for $0<\Theta\leq1$, the scheme as given in Eq. (14) is not only implicit but also **non-linear.** However, upon considering small time steps ΔT, the finally chosen computational scheme

$$\Theta\sigma Q_i^{j}\left(\frac{Q_{i+1}^{j+1}-2Q_i^{j+1}+Q_{i-1}^{j+1}}{(\Delta X)^2}\right) + (1-\Theta)\sigma Q_i^j\left(\frac{Q_{i+1}^{j}-2Q_i^{j}+Q_{i-1}^{j}}{(\Delta X)^2}\right)$$

$$= 2\left(\frac{Q_i^{j+1}-Q_i^j}{\Delta T}\right) - \frac{10\omega}{3}(Q^{2/5}P)_i^j \qquad (15)$$

becomes linear in Q^{j+1}.

Splitting Method

Suppose that Q_i^j for all i are known at a certain time $T=j\Delta T$. In order to solve for Q_i^{j+1}, the first computational step is to consider the hyperbolic Eq. (13). The result is denoted by $\tilde{Q}_i$ in which the diffusive flow character is overlooked. In the second computational step, Eq. (15) is solved by considering $Q_1^j = \tilde{Q}_i$, whence by accounting for the solution of the hyperbolic equation as a starting condition.

The error introduced by the splitting method is equal to the sum of the errors originating from the hyperbolic and parabolic schemes. To the first order of approximation, the error is $(\Delta X+\Delta T)$ for all Θ.

NUMERICAL RESULTS

The program itself is able to accept T_d, X_d, ΔT, ΔX, σ, ω, r and Θ as basic variables. Four different tests have been carried out in order to check several aspects of the solutions. These include the solution of the hyperbolic equation (comparison with analytical solution), the linearized version of parabolic PDE (see also Jansen, 1979; Hager et al., 1986), the total volume flux, $V=\int[QX,T)dT]_0^{X_{max}}$, and the comparison to the solutions with (Hager et al., 1986), whose computational approach is quite different from the present. Of all these tests it is concluded that the numerically computed solutions differ but slightly from the exact solutions. Furthermore, the finally accepted values are $\Delta T=(3/10)\Delta X$, $\Delta T=1/20$, $\Theta=1$ (backward Euler step). For any other $\Theta>1/2$, the stability of the solutions has been (slightly) poorer and the errors introduced by the splitting method have grown with time T.

Fig. 3 presents the hydrographs Q(T) for the locii X=0,2,4,6,8 and 10, $0<T<10$. The parameters have been chosen as r=2 and r=10 (inflow hydrograph shape factor), σ=0.01, 0.05, 0.10, and for ω=0,0.1,0.5,1.0. Once the inflow hydrograph Q(0,T), r, σ and the lateral inflow parameter ω known from observational data, or from hydrological calculations of the upstream reach, the temporal development of the flood wave can easily be predicted.

DISCUSSION OF SOLUTIONS

Consider first the solutions for $\omega=0$, figure 3a). It is noted that the diffusive flow characteristics become significant already for small distances X. The larger X, the smaller the maximum discharge $Q_{max}=Q_{max}(X)$. Evidently, Q_{max} remains independent of X, $Q_{max}=Q^*$, for $\sigma=0$, corresponding to the kinematic wave approximation (Lighthill and Whitham, 1955). However, **shocks** then develop at the leading side of the flood wave. These shocks are absent in the diffusion wave approximation. Since the inclusion of shock wave modelling may complicate the numerical approach (Borah et al., 1980) the diffusion wave approximation is easier to control from this point of view. Wall friction thus is the significant parameter that leads to continuous solutions. The larger σ, the more the hydrographs Q(T) are damped, and the slower the wave propagation becomes at a fixed location X. Therefore, the wave steepness reduces with growing σ and X.

From Eq. $(8)_4$ it is seen that σ depends primarily on the average river bottom slope, S_0; for large S_0 as found in mountainous regions, σ becomes very small; simultaneously $K=1/n$ and t^* decrease for heavy

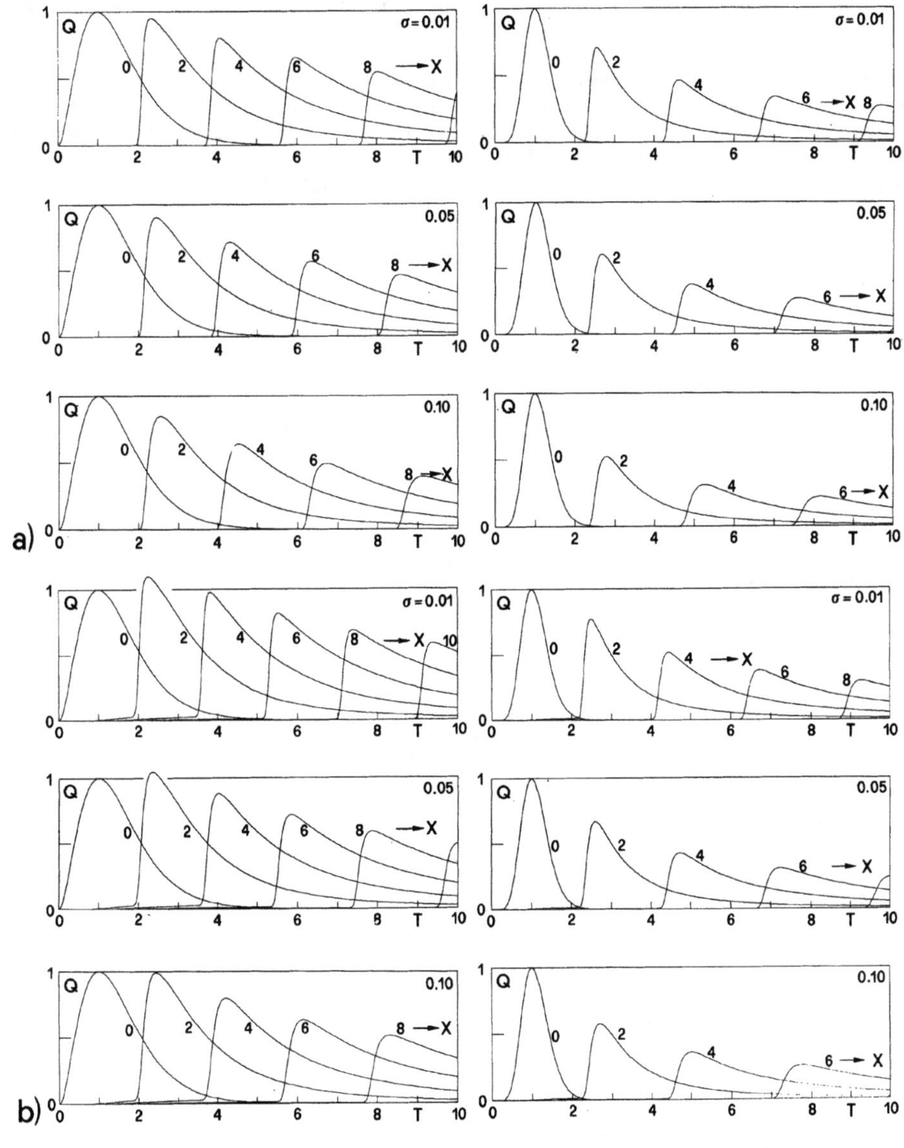

Fig. 3 : Hydrographs Q(T) at various locations X and diffusion characteristics σ, a) $\omega=0$; b $\omega=0.1$. $r=2$ (left), $r=10$ (right).

thunderstorms in small catchment areas. Consider, as an example, $h^* \simeq 1m$, $t^*=1h$, $K=10m^{1/3}s^{-1}$ (n=0.1) $S_o=0.15$ for which $\sigma_c=5 \cdot 10^{-4}$. In contrast, typical munbers in a larger river are $h^* \simeq 5m$, $t^*=15h$, $K=30m^{1/3}s^{-1}$ (n=0.033), $S_o=0.1\%$, for which $\sigma_r=0.03$. Consequently, σ_r has an order about 10^2 larger than σ_c in a small catchment area. The diffusive effects in the latter thus may be overlooked (Raudkivi, 1979).

On the other hand, σ becomes large for small t^* and large h^*; the **dynamic** wave effects then become important, and the full de Saint-Venant formulation must be accounted for. At present, no general upper limits have been recommended for the application of the diffusive wave approximation. Herein, $\sigma=0.10$ has been chosen as the maximum wave diffusion parameter.

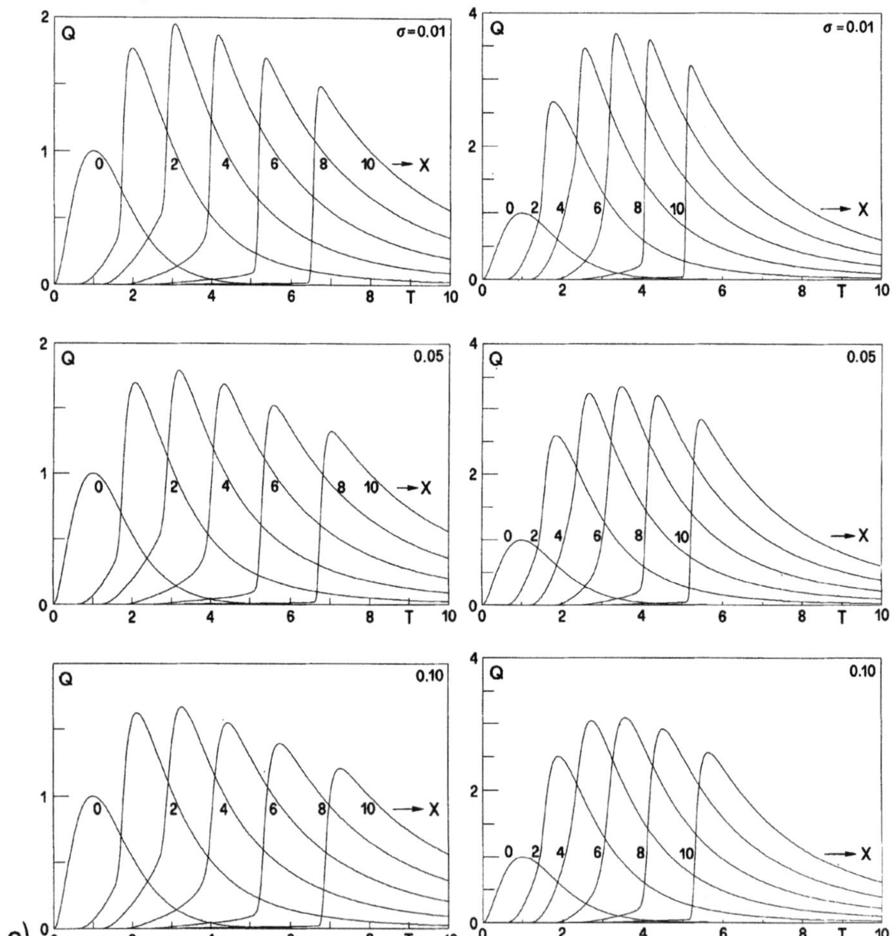

Fig. 3 : idem, c) r=2, ω=0.5 (left) and ω=1 (right).

The sufficient condition for wave diffusion to appear is the longitudinal reach X considered. According to Eq. $(8)_1$, X depends significantly on $V^*=q^*/h^*$ and t^*, V^* being the peak inflow velocity. It is seen from the plots that, for $X<1$ and $\sigma<0.1$, Q_{max} remains almost unchanged, $Q_{max} \simeq Q^*=1$. Therefore, diffusive flood effects are insignificant if

i) $\sigma<10^{-4}$, application of kinematic wave approximation,
ii) $X<1$, inflow hydrograph nearly unchanged.

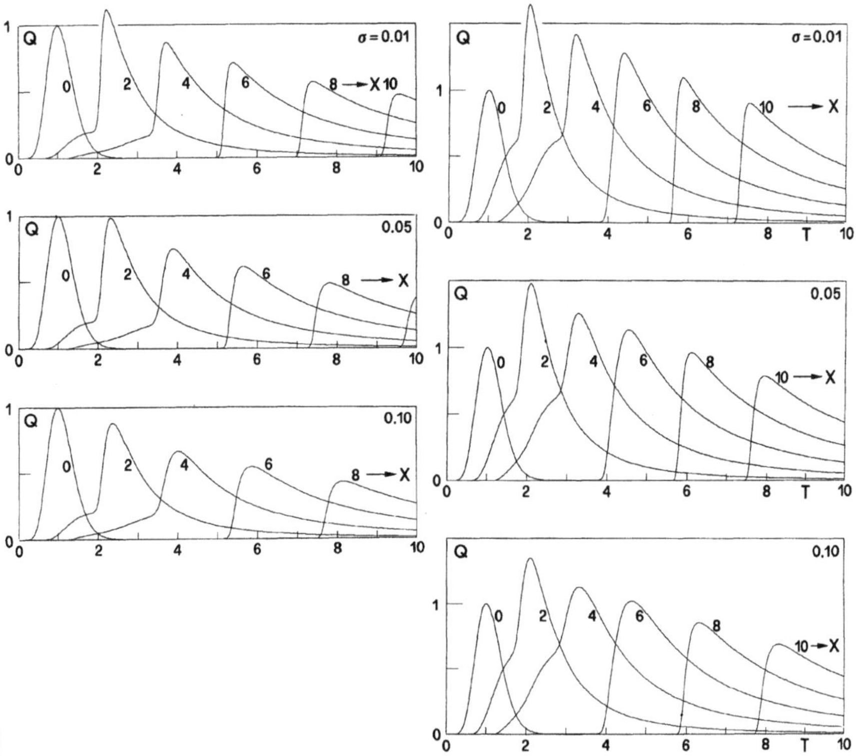

d)

Fig. 3 : idem; r=10, ω=0.5 (left), ω=1.0 (right).

The numbers given must be considered as orders of magnitude. However, using the basic data (K,S_0,q^*,t^*,h^*) regarding the substitute river, the above criteria allow a simple insight into the overall phenomena of the diffusive flow pattern.

The effect of $\omega>0$ does not change qualitatively the results regarding $Q(X,T)$. However, since the propagation speed C of diffusive waves depends on the flow velocity $V=q/h$, and since $V=KS_0^{1/2}h^{2/3}$ at the wave crests $(\partial h/\partial X=0)$ $C \sim h^{2/3} \sim q^{2/5}$. Increasing discharge thus makes waves moving faster. As is seen from the plots for $\omega=0.5$ and $\omega=1$, the lateral inflow influences the rising hydrograph only for $X<8$ $(r=2)$ and $X<6$ $(r=10)$, respectively.

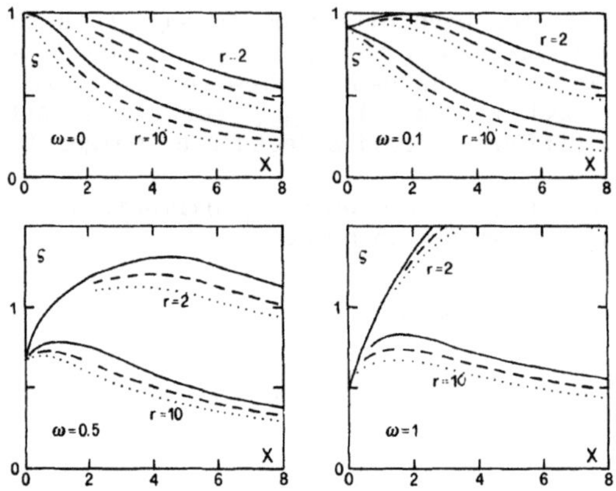

Fig. 4 : $\zeta=Q_{max}/(1+\omega)$ as a function of X and ω for (——) σ=0.01, (---) σ=0.05, (•••) σ=0.10.

Of particular engineering interest is the **maximum discharge** $Q_{max}=q_{max}/q^*$ as a function of distance X, and the corresponding time-locii $X_{max}(T)$. Figure 4 shows $\zeta=Q_{max}/(1+\omega)$ as a function of X for various σ and ω. It is seen that, for X>2, curves for various σ are nearly parallel. For equal σ, ω and X, curves for r=10 are lower than curves for the hydrograph shape parameter r=2.

Fig. 5 shows an evaluation of time to peak $T_{max}=t_{max}/t^*$ at which $Q=Q_{max}$ as a function of the longitudinal coordinate X. By letting $\tau=T_{max}/X$, the plot is correlated with X for various hydrograph shape parameters r, diffusion parameters σ and lateral inflow coefficients ω.

It is seen that the effect of σ decreases with increasing ω. Also, it is noted that τ is nearly independant of X for small ω, while $\tau \sim X^{-1}$ for large ω, provided X>2. For smaller X, 0<X<2, τ increases rapidly with decreasing X. For X→0, for which $T_{max}=1$, the relative time to peak becomes $\tau(X\to 0)\to\infty$. However, these results are not of particular interest since the effects of σ may be overlooked, and hydrographs are easier computed using the kinematic wave approximation (σ=0).

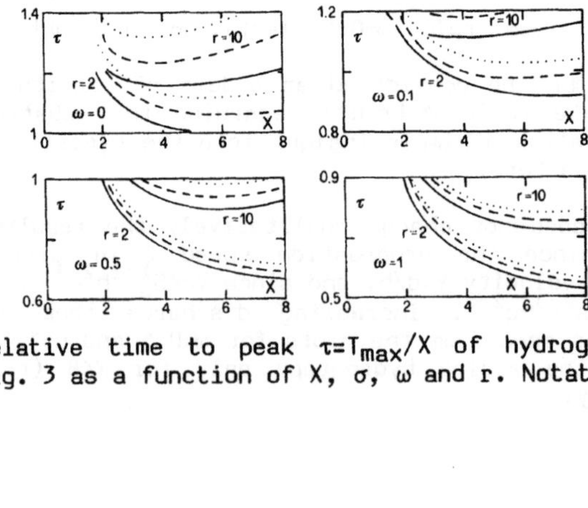

Fig. 5 : Relative time to peak $\tau=T_{max}/X$ of hydrographs according to Fig. 3 as a function of X, σ, ω and r. Notation as in Fig. 4.

APPLICATION

The present method of computation is applied to a typical flood as appeared in 1954 on the upper Rhine river in Switzerland. The **objective** persued herein is twofold, namely illustrating the simplicity of the approach as well as comparison of the prediction with the observation. The example chosen is particular in so far as the same river reach has already been examined by (Hager et al., 1986); consequently, the basic river data can be taken over.

The reach to be considered is situated at the North Eastern part of Switzerland, bounded some 5 km upstream of Chur at Felsberg (x=0) and at Schmitter, some 10 km upstream of the inflow to the Lake of Constance. The complete reach has a length of Δx=85 km; the bottom slope is gradually varying from $S_0 \approx 0.3$ % (x=0) to $S_0 \approx 0.1$ % (x=85 km), the average being S_0=0.2 %. The average roughness coefficient of the river has been estimated as $K=34\ m^{1/3}s^{-1}$ (n=0.029), and the average river width is b=120 m (Hager et al., 1986).

In August 1954, one of the highest flood waves ever recorded has been observed in the upper Rhine valley. While the period of August 12 to 19 have been considered by (Hager et al., 1986), of which the maximum inflow peak has been $\hat{Q}^* \approx 900\ m^3s^{-1}$, the present investigation examines

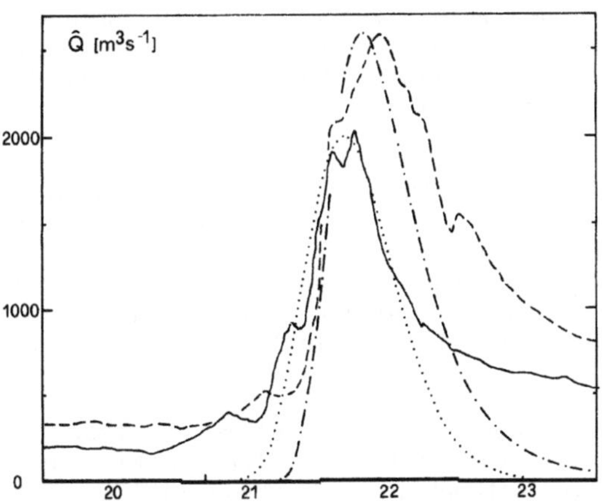

Fig. 6 : Hydrographs at the extremities of considered reach, upper Rhine valley; Felsberg,(——) inflow recorded, (•••) inflow adapted by Eq. (6); Schmitter, (---) outflow recorded; (-•-) outflow computed.

the period August 20 to 23, 1954. Figure 6 shows the inflow hydrograph at Felsberg as full curve; it is characterized by a base flow of $\hat{Q}_0 \approx$ 200 m^3s^{-1}, augmenting rapidly on August 21st, to reach peak discharge $\hat{Q}^*=2'040\ m^3s^{-1}$ on August 22nd at 06.00 a.m. Note that there is a first peak at 02.00 a.m.

The **substitute inflow hydrograph** (x=0) is shown in figure 6 as dotted curve, characterized by $\hat{Q}^*=2'000\ m^3s^{-1}$, t*=24 h=86'400 s and r=10. The temporal origin t=0 has been adjusted such that the peak inflow hydrograph zone is well represented. In particular, the inflow hydrographs as observed and modeled are nearly identical for $\hat{Q}>800\ m^3s^{-1}$. Regarding the diffusive flow pattern of the flood wave, this part of the hydrograph is significant. However, predictions regarding the flood volume and the pre- and post-history of it must be overlooked.

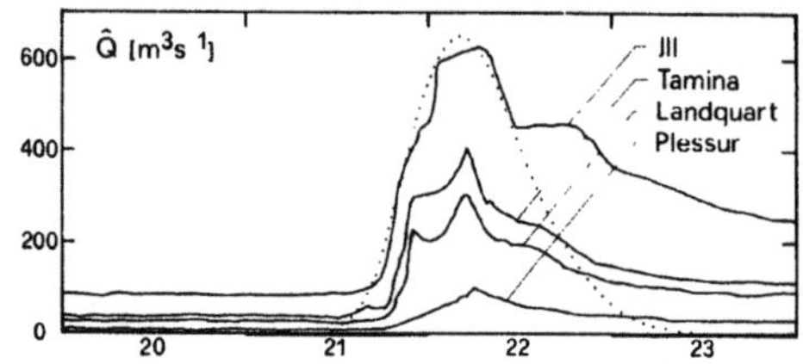

Fig. 7 : (——) Comulative hydrographs of laterals running into upper Rhine and (•••) substitute lateral inflow hydrograph, Eq. (7).

Figure 7 shows the main laterals running into the upper Rhine, namely the Plessur, the Landquart, the Tamina and the Ill. The lateral inflow hydrographs are given as comulative plot. It is seen that the maximum lateral inflow discharge is 650 m^3s^{-1}; the lateral (concentrated) hydrograph with t*=24 h and r=10 fits well with the observation except for the falling hydrograph portion. This is mainly because of a second discharge peak of the Ill river in the evening of August 22nd. However, having the simple model in mind, substitute and observed hydrographs fairly collapse.

With $\hat{Q}^*=2'000\ m^3s^{-1}$ and b=120 m, the peak inflow discharge per unit width becomes $q^*=2'000/120=16.7\ m^2s^{-1}$, whence $h^*=(q^*/(KS_o^{1/2}))^{3/5}=4.20$ m. The non-dimensional river reach is then with Eq. $(8)_1$ $\Delta X=4.2\cdot 85'000/(86'400\cdot 16.7)=0.247$ and the diffusion parameter becomes $\sigma=4.2^{1/3}/(86'400\cdot 34\cdot 0.002^{3/2})=0.0062$ from Eq. $(8)_4$. Finally, with $p^*=650/(120\cdot 85'000)=6.37\cdot 10^{-5}ms^{-1}$, the lateral inflow coefficient is $\omega=86'400\cdot 6.37\cdot 10^{-5}/4.2=1.30$.

The first obvious result from these values is that diffusive effects play an insignificant role ($\sigma \ll 1$, X<1). The kinematic wave approximation thus could be applied without loss of substantial accuracy for the flood wave considered. Detailed investigation of the flood wave given by σ= 0.0062 and ω=1.30 indicate that ζ=0.565, whence Q_{max}=1.295 and $\hat{Q}_{max}$= $1.29\cdot 2'000=2'590\ m^3s^{-1}$. Observed was the peak discharge $\hat{Q}_{max}=2'575 m^3s^{-1}$.

The time for the wave peak propagation is t=4.5 h (observed) thereby accounting for the second peak of the observed inflow hydrograph, see figure 6 and t=3.6 h (computed). The peak velocities are $V^*(x=0)$= 3.97 ms^{-1} and $V^*(x=85\ km)=4.39\ ms^{-1}$, providing an average of V^*= 4.18 ms^{-1}. The wave peak propagation velocities are $C_{obs}=85'000/(4.5\cdot 3'600)=5.25\ ms^{-1}$ and $C_{com}=85'000/(3.6\cdot 3'600)=6.55\ ms^{-1}$. The velocity ratio as computed is $(C/V)_{com}=6.55/4.18=1.57$, which is near to the value for infinitesimal wave propagation (C/V)=5/3.

CONCLUSIONS

Using a simplified model river, the discharge characteristics of diffusive flood waves in large rivers are analysed, thereby accounting for an uniformly distributed lateral inflow. The governing, non-linear convection-diffusion equation for discharge is solved by an implicit numerical scheme accounting for realistic initial and boundary conditions. Typical solutions are presented graphically; maximum discharge and corresponding time to peak is discussed in detail. Owing to the generalized approach, these solutions may be used immediately to investigate the diffusive flow pattern of large flood waves by an elementary procedure. The method presented is particularly suited to preliminary design analysis. Furthermore, questions relating to the discharge characteristics of frequency and risk analyses can be studied in a straightforward approach. Finally, simple criteria allow the qualitative prediction of the significance of flood wave diffusion.

ACKNOWLEDGEMENTS

The authors would like to thank Prof. J. Rappaz, Université de Neuchâtel, Switzerland, for his stimulating discussions.

REFERENCES

Borah, D.K., Prasad, S.N., Alonso, C.V., 1980, "Kinematic wave routing incorporating shock fitting", **Water Resources Research,** Vol. 16, 3, 529-541.

Cunge, J.A., 1975, "Two-dimensional Modeling of Flood Plains", in "Unsteady Flow in Open Channel", Vol. 2, ed. Mahmood, K., and Yevjevich, V., Water Resources Publications, Fort Collins, Col.

Hager, W.H., 1984, "A simplified hydrological rainfall-runoff model", **Journal of Hydrology,** Vol. 74, 151-170.

Hager, W.H., Hager, K., Rappaz, J., Caussignac, P., 1986, "Non-linear diffusive flood waves", **Acta Mechanica,** 60, 181-198.

Henderson, F.M., 1966, "Open Channel Flow", MacMillan, New York, N.Y.

Jansen, P.P., 1979, Principles of River Engineering", Pitman, London.

Lapidus, L., Pinder, G.F., 1982, "Numerical solution of partial differential equations", John Wiley & Sons, New York.

Liggett, J.A., 1975, "Basic Equations of unsteady flow", in "Unsteady Flow in Open Channel", Vol. 1, ed. Mahmood, K., and Yevjevich, V., Water Resources Publications, Fort Collins, Col.

Lighthill, M.H., Whitham, G.B., 1955, "On kinematic waves", Part I, **Proc. Royal Society,** A, Vol. 229, 281-316.

Raudkivi, A.J., 1979, "Hydrology", Pergamon Press, Oxford, New York,

Raviart, P.A., Thomas, J.M., 1983, "Introduction à l'analyse numérique des équations aux dérivées partielles", Masson, Paris.

Richtmyer, R.D., Morton, K.W., 1967, "Difference methods for initial-value problems", Interscience Publ., New York, Sydney, 2nd edition.

Rouse, H., Ince, S., 1957, "History of Hydraulics", Iowa Institute Hydraulic Research, Iowa State University, Iowa.

Whitham, G.B., 1974, "On linear and non-linear waves", Wiley Interscience, New York.

Woolhiser, D.A., Liggett, J.A., 1967, "Unsteady, one-dimensional flow over a plane - the rising hydrograph", **Water Resources Research,** Vol. 3, 753-771.

A RAINFALL RUNOFF MODEL FOR AGRICULTURAL DRAINAGE IN THE EXPERIMENTAL STATION OF THE THREE-RIVER PLAIN IN HEILONGJIANG PROVINCE

Yang Peishu Associate Research Fellow General Engineer
Lu Yubang Associate Engineer
Water Conservancy Research Institute of Heilongjiang Province
PRC

INTRODUCTION

The Three-River Plain Region has an immensely wide area, amounting to more than one hundred thousand square kilometers. It is one of the important commercial grain production bases in the country. The region consists of a broad marshland of low level. Although now most part of the area has been cultivated after regulation and drainage for years the drainage system in the whole region is still imperfect and the frequent occurance of drought and flood is always threatening the region very seriously. So we have many problems to be researched in the construction of water conservancy in this region.

Basing on our research work of the rainfall runoff model in the Three-River Plain Region, we applied the results of hydrologic research to the field drainage in this region and tried to solve the computation problem of drainage discharge after raining so as to provide grounds necessary for the reclamation project to be designed there.

MODELING

We modified slightly the Xinan River model, analyzed the conditions and mechanism of the Xinan River model, and applied it to the computation of the rainfall runoff of the Raoli River e.g., the single-acting hydroulic connection between tension water and free water in the surface course in the Three Water Resourse Model was modified to be a double-acting hydraulic connection; the subsurface flow to the modified model was designed to be very little, even to a negligible extent because of a water-tight layer with a small co-officient of permeability at a very low depth.

The characteristics of the Experimental Drainage Fields are that their size is small; their surface drainage speed is fast; and their network of drainage is perfect, etc., After a number of observations and trial computations we realized deeply that the fields were plain; the channels in the fields were shallow, usually 0.8-1.0m the runoff of subsurface flow hardly assembled into the total runoff, and there was only a small amount of flow of seepage in soil through the very narrow strips on each side of a channel,

V. P. Singh (ed.), Flood Hydrology, 97–105.
© 1987 by D. Reidel Publishing Company.

and most part of the flow in soil which was difficult to discharge mostly evapotranspirated from soil, so the model was further modified, i.e., in soil the outflow factor KI was designed to be very small and the subsidense factor rather large.

I. Structure of the Model (see Fig. 1)

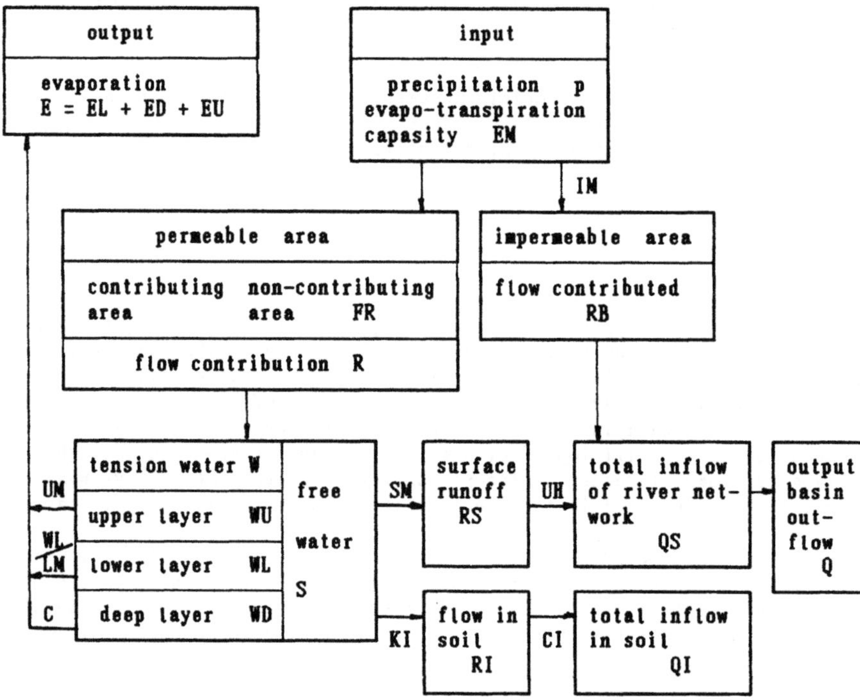

Fig. 1 A blook diagram of the model structure

The mathematical model consists of three parts, a model of flow contributed, a model of collective flow and a model of evapo-transpiration.

II. Characteristics of Flowing, Constitution of Water Resourse and Computation of Flowing

1. Characteristics of Flowing

In the years when several observations of runoff contributed were made in the experimental fields, it was discovered that most of the rúnoffs occurred in the rainy period, July and August. Before the runoff was contributed, the soil moisture basically reached the field water holding, and the underground water rose up approximately to the surface. Therefore, if there was a heavier raining, the soil moisture would quickly reach the field water holding and formed the flowing full stored. Besides, in pre-rainy period the soil moisture content was very sensitive to the runoff contribution.

2. Constitution of the water resourse

In the prosess of runoff formation, the total runoff generated by a single rainfall is not very large owing to the part played by the conditions of the lower surface and the influenses of the ridge direction and the boundary conditions. In order to speed up the discharge of the surface runoff, particularly we dug some small temporary ditches perpendicular to the drainage channels and the ridge direction. Because the coefficient of permeability of the bottom course of ploughland was very small at such a shallow depth underground in the experimental plot we held that in the runoff there was scarcely any replenishment from underground water, so the underground runoff could be neglected in computation . The total runoff whoch was obtained in the experimental plot included two parts, i.e. the surface runoff and the flow in soil was considered to be of a minor amount.

3. Computation of the runoff contributed

The flowing full stored makes it a controlling condition of the runoff contributed that the water storage of the air belt in the soil layer should be satisfied. If in the situation that all the flowing in the experimental plot is full stored (We take each area controlled by a drainage channel to be a small w. shed) , its flowing can be expressed by the storage equation,

$$R = P - (W_m - W) - E$$

In the equation, $(W_m - W)$ represents the water shortage in the air belt. Because the distribution of water shortage in air belt all over the w. shed is not uniform, and this distribution can be expressed by a distribution curve of storage capacity of water-shed, the storage capacity curve adopts a parabola of B degree, i.e., $f/F = 1 - (1 - W_m'/MM)^B$. (where B and W_m are coefficients to be determinated.)

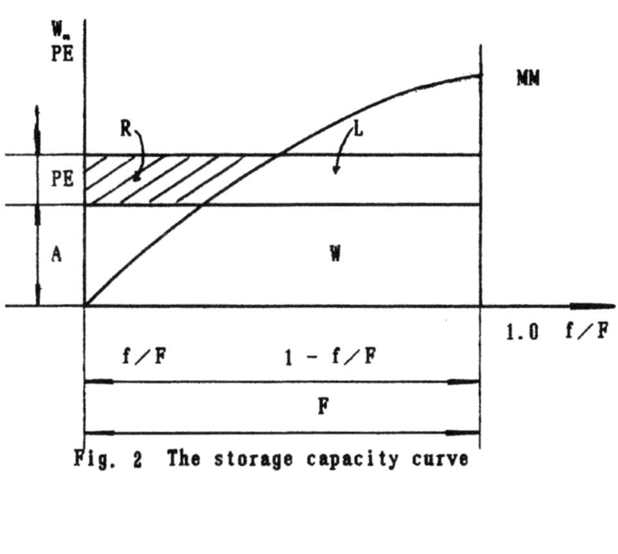

Fig. 2 The storage capacity curve

The mathematical expression of rainfall runoff relationship is as follows,

Given that the average storage capacity in the air belt in a w. shed is W_m (W_m is an srea between the lower part of the curve and the absciassa axis.)

$$W_m = \int_0^{MM} (1-f/F)\, dW'_m$$

After integration we obtain,

$$W_m = \frac{1}{1+B} MM$$

$$MM = (1+B)\, W_m$$

$$W = \int_0^{A} (1 - f/F)\, d'_{Wm}$$

After integration we obtain,

$$A = MMX\ [\ 1 - (1 - W/W_m)^{1/(1+B)}\ ]$$

Runoff,

$$R = \int_A^{P-E+A} f/F dW_m$$

After integration we obtain

$$R = P - E - W_m\ [\ (\ 1 - A/MM)^{1+B} - (\ 1 - \frac{A+PE}{MM}\ )^{1+B}]$$

In the above equation, the conditions are,

When $P - E + A < MM$, the previous equation is adopted;

When $P - E + A > MM$, $R = P - (W_m - W) - E$.

In the above equation, there are two parameters to be determinated and optimized.

4. The computing model of vapor-transpiration

In a water shed the factors which effect on evapor-transpiration are various, such as temperature, wind speed, humidity, plant cover, soil water content, etc.. Therefore the evapo-transpiretion capacity E_m in the water shed was first determined, and then the relationship between E_m and the actual value was established. Considering that the vertical distribution of the soil water content was not uniform, we divided the model into three layers of evapo-transpiration, i.e., the upper layer of the evapo-transoiration which evapourates according to its evapo-transpiration capacity, $E_u = E_m$; lower layer of evapo-transpiration which evapourates in proportion to the soil water content, $EL = (EL - EU) \times WL / LM$, and the deep layer of evapo-transpiration, $Ed = C \times EM$.

5. Physical significance of the parameters

In the experimental fields, there were many control sections of flow measurement, and their areas were various. Each of them was taken to be an individual small water shed in computation. The related parameters ans their physical sighificance in the computation of the runoff contributed and the collective flow are listed in Table 1.

The physical significance and unit of parameters and their corresponding symbols are as follows,

M, Calculating the numer of time length;

N, Number of unit graph time length;
IM, Ratio between an impermeable area and the whole area of a water shed;
WM, Average tension water capacity in a water shed (mm) ;
UM, Tension water capacity of upper layer in a water shed (mm) ;
LM, Tension water capacity of lower layer in a water shed (mm) ;
B, Index of storage capacity curve;
C, Conversion factor of evaporation in the deep layer;
WC, Control value of free water replenishment (mm) ;
U, Conversion factor of unit;
K, Conversion factor of evaporation;
KI, Outflow factor in soil;
SM, Storage capacity of free water;
CI, Subsidence factor of flow in soil;
F, Area of water shed.

Parameters TABLE 1

Experi. sield	Baoqing T.R. Experimental Station 1983				Fr.Farm Ex.F 1983	
Flood No.	717-1	717-2	717-3	722-3	7-22	7-30
Symbol	Parameter					
M	43	43	40	27	30	30
N	17	17	10	10	12	12
IM	0.01				0.02	
WM	120				120	
UM	15				15	
LM	65				65	
B	0.2				0.2	
C	0.111				0.111	
WC	40				40	
U	0.026	0.028	0.004	0.004	0.03	0.03
K	0.62				0.62	
KI	0.0005				0.005	
SM	20				16	
CI	0.95				0.90	
F	0.094	0.101	0.015	0.015	0.01	0.01

III. Computating of Collective Flow

In computation of time length it was given that $\Delta t = 1$ hr.. In computation of runoff contributed and collective flow and the division of water resources, the parameters were rated. The results were listed in Table 1.

1. Division of water resources

The water resources were divided according to the structure of a free reser-

voir.
This reservoir was of linear,

S, The storage depth of free water;
SM, Its maximum.

when $S + PE < SM$,

$RI = S \times KI \times FR$

$RS = 0.$

when $S + PE > SM$,

$RI = SM \times KI \times FR$

$RS = (S = PE - SM) \times FR.$

According to the model of flowing stored, i.e., after the flowing capacity R was found from the storage curve of water shed, the flowing entered first into the free reservoir. If the storage capacity of free water smaller than the maximum of SM, the outflow in soil was found according to the relationship (KI) , and then the total inflow in soil was found accorage to the subsidence factor of flow in soil (CI) . If the storage capacity of free water was larger than the maximum of SM, the excess part became the surface runoff, in addition to that formed in the area of runoff contributed. The surface runoff and the total inflow in soil made up the total channel inflow, and the flow process at the outlet section was obtained through unit graph routing.

The free reservoir could merely occur in the area of runoff contributed. Its bottom width FR was variable, and the runoff R, which had entered into the runoff area P - E, and meanwhile it was the storage depth added in the free reservoir.

2. Computation of the surface runoff process

In computation of surface runoff, the following formula was used,

$$QS_t = QS_{t-1} + RS \times UH \times U$$

where

QS_t, Runoff in soil in the time t;
QS_{t-1}, Surface runoff in the time t - 1;
RS, Surface runoff in the time length;
UH, Unit graph (coefficient of collective flow) ;
U, Unit conversion factor.

According to the above formula, the surface process in the water shed (the experimental district) can be found through the unit graph routing by linearly repeated addition.

3. Computation process of flow in soil

In the computation process of flow in soil, the following formula was adopted,

$$QI_t = QI_{t-1} \times CI + RI \times (1 - CI) \times U$$

where

RI, Runoff in soil in time length;
CI, Subsidence factor in soil;
QI_{t-1}, Runoff in soil in a t-1 time length;
QI_t, Runoff in soil in a t time length.

After the above formula was calculated by linearly repeated addition, the process of outflow in soil was obtained. The total outflow process Q was equal to the sum of the surface runoff process QS and the flowing process in soil QI. Through these steps, the runoff process of the analog computation could be obtained.

GENERAL DESCRIPTION OF THE EXPERIMENTAL DRAINAGE FIELDS

1. Friendship Farm Experimental Drainage Field

The experimental drainage field lies to the east by north of the farm office at a distance of 5 km. Except a small part, most of its topography is low, and its height is between 63.5 - 65 m with a steady slope, therefore the conditions of gravity drainage are very poor. At present, the surface runoff is discharged eastward from No. 2 drainage station of the farm (control area, 7,000 hactares) into the ground drain collector, and then flows into the Piaofa River. After preliminary regulation, the drainage conditions have been improved. The soil here was of light marsh land, a kind of heavy loam, but now becomes black calcareous earth of heavy loam after cultivation and ripening. The thickness of its black earth layer is 0.7 - 1.0 m. The cultivated layer is always in the state of water saturation in rainy years owing to its dense stickness and low permeability. Its underground water storage is very rich, extending upwards from the water surface to the ground at a distance of about 0.4 -1.0 m in low land and about 2.0 m in high land in the flood period of rainy years. The water level in low land almost reaches earth surfac, and the perched water of upper layer in soil even joins the surface water.

2. Baoqing Three-River Experimental Drainage Station

Its experimental field is situated to the northwest of the town of Baoqing at a distance of 4 km. The height of its topograpgy is about 75.5 m above the sea level. In the experimental district rainfall is discharged into the Raoli River through the general drain collector of West Dihezi. Its topography is high in the north and low in south. At the northeast corner of the district a drainage pumping plant has been established.

The underground water of the experimental feild is very rich (output of single well, 1,000 - 3.000 m^3/day) . In the flood period the underground water reaches approximately the earth surface at a distance of 50 - 100 cm. After raining, the water flows into low land with a steady slope and then forms a great amount of perched water in soil due to its unsmooth flowing. Its soil is of black earth of meadow land with dense stickness and poor permeability.

ACCURACY OF THE RESULTS OBTAINED

About the accuracy of 6 different peak values and peak values and peak hours in the three experimental drainage districts, a comparason between the curve of the process and that of actual measurement process is listed in Table 2 and Table 3.

In the analogy of the ten runoff processes, three of two had their relative errors within 5%, two within 5 - 10% and two within more than 10%; and for six times the errors of the peak hours were zero. Therefore the analog accuracy looks satisfactory.

A Statistical List of the Rainfall Runoff Anallog Accuracy in Friendship Farm Experimental Drainge Field

TABLE 2

Seri. No.	Date	Peak value (m^3/s)				Peak haur error ※
		Act.m. flow	Calculated flow	Absolute error	Relative error	
83722-1	JULY 22 (1983)	0.00177	0.00207	0.0003	16.9%	0
83730-1	JULY 30 (1983)	0.00433	0.00476	0.00043	9.9%	0

※ unit, Number of time length

A Statistical List of the Rainfall Runoff Anallog Accuracy in Baoqing T.R. Experimental Drainage Field

TABLE 3

Seri. No.	Data	Peak value (m^3/s)				Peak haur error ※
		Act.M. flow	Calculated flow	Absolute error	Relative error	
83717-1	JULY 17 (1983)	0.00435	0.00459	0.0024	5.5%	0
83717-2	JULY 17 (1983)	0.0103	0.0104	0.0001	1.0%	0
83717-3	JULY 17 (1983)	0.0130	0.0133	0.0003	2.0%	0
83722-3	JULY 22 (1983)	0.000822	0.000697	0.000120	14.6%	0

※ unit, Number of time length

AN ANALYSIS ON THE RESULTS OBTAINED

Generally, it is satisfactory that the analog model has attained so high an accuracy although the areas and flows in our experimental districts are comparatively small. In the analog process the rising limbs were coincides accurately, except that in the process of actual measurements, twice in the Friendship Farm and once in the Baoqing Station, water dropped faster in the recession limb. According to the analysis afterwards, this was because a measuring weir was used in the process of actual measurements so as to make sure that a free outflow would be generated after the weir had been established. So the height of the weir mouth must be higher than that of the channel bottom which caused water to drop faster in recession limb. Besides, in Baoqing Station, there are two experimental districts with similar areas. After a raining, their peak values of the flow process were different greatly. One was twice as many as the other. This was because the drainage of subsidary drain in one of the districts was included.

APPLICATIONS OF GEOMORPHOLOGIC THEORY TO UNGAUGED WATERSHEDS IN SINAI

Dr. Mohamed Nasr Allam
Assistant Professor, Irrigation and Hydraulics Department, Faculty of Engineering, Cairo University, Giza, Egypt
(Now at Faculty of Meteorology, Environment & Arid Land Agriculture, King Abdulaziz University, Jeddah, Saudi Arabia)

ABSTRACT. Recently hydrologic models based on the geomorphologic theory for the prediction of discharge hydrographs of ungauged watersheds have been developed. The paper presents a review on these models and on their mathematical development. One of these models, which accounts for the infiltration in watersheds, is applied for Elmaghara basin in the northern part of Sinai. A geomorphoclimatic model for the prediction of surface runoff volumes is also presented and applied for the same basin. The model has a simple structure and is applicable for large scale basins. It is designed in such a way that a physically based expression for iniltration can be considered. The effects of infiltration on the surface runoff of the basin are illustrated.

INTRODUCTION

The Government of Egypt is carrying out an extensive development program for improving the economic and social conditions of Sinai. A number of industries have already commenced there and others are planned for the near future. Large areas on the order of 500,000 feddans (210,000 hectars) at the various parts of Sinai are planned for agricultural development. Because of this development program, the population is expected to increase from 200,000 citizens at present to about 1,000,000 by the year 2000. Water represents a major input factor for this large development program. Water is necessary for agriculture, industrial, and municipal uses. The water sources of Sinai consist of two main components: surface water runoff of the wadis (watersheds),and groundwater in deep and shallow alluvial aquifers. In this paper, only the surface water component is emphasized.

Most of the wadis in Sinai are ungauged. They lack the hydrologic data required for the traditional runoff estimation techniques, which are based on regression analysis of the historic rainfall and runoff records. The linear systems theory is commonly used in runoff prediction, particularly for ungauged watersheds. In this theory, transformation of a rainfall to runoff is assumed linear. Given the Instantaneous Unit Hydrograph (IUH), then for a rainfall event, the corresponding

V. P. Singh (ed.), Flood Hydrology, 107–116.
© 1987 by D. Reidel Publishing Company.

runoff discharge can be determined via the convolution transformation. The main issue then is the estimation of the IUH of the concerned watershed.

Recently, a methodology for estimating the IUH based on the watershed geomorphology has been developed (Rodriguez-Iturbe and Valdes, 1979). The IUH is interpreted as the probability density function (Pdf) of the travel time that a drop of water, landing anywhere in the watershed, takes to reach the outlet (Gupta, Waymire and Wang, 1980). It is assumed that the Pdf of the time of travel in the watershed streams is exponential. The resultant IUH is called the Geomorphologic IUH (GIUH). Kirshen and Bras (1982) were worried about the assumption of the exponential travel time distribution and proposed instead a time distribution based upon the linearized equations of motion. The resultant GIUH was found to be different from that of Rodriguez-Iturbe and Valdes. In both approaches however, no consideration was given to the infiltration losses in the watersheds. Diaz-Granados, Bras, and Valdes (1983) modified the above two approaches to account for the infiltration losses with a simple linear function of the surface runoff. After this modification, they applied them for various watersheds in U.S.A. and Puerto Rico. Both approaches yield similar GIUHs and discharge hydrographs. However, the approach based on the exponential distribution assumption for the time of travel has much simpler mathematical derivation and utilization than the other approach.

This paper addresses two main issues: (1) prediction of surface runoff discharge hydrographs of ungauged watersheds using the GIUH after the infiltration modification of Diaz-Granados, Bras and Valdes (1983), and (2) prediction of surface runoff volumes of ungauged watersheds via a geomorphologic model too, accounting for infiltration as a linear function of surface runoff.

Elmaghara basin in the northern part of Sinai, is selected here as a case study. At Elmaghara, coal mining industries started in 1983, and are planned to be expanded in the near future. In addition, near Elmaghara about 2,000 hectars are in suitable conditions for agriculture. ·ediction of surface runoff yield and hydrographs will aid the policy makers in deciding on the size of intended development and on the appropriate water management policy.

A REVIEW OF THE MATHEMATICAL FORMULATION OF THE DISCHARGE HYDROGRAPH PREDICTION MODEL

According to the probabilistic interpretation of the GIUH, stated earlier in the paper, and according to the fact that number of the paths which a water drop, landing anywhere in a watershed, may follow to reach the outlet is finite, the GIUH (h(t)) can be expressed as:

$$h(t) = f_{TB}(t) = \sum_{s \in S} f_{Ts}(t)\ P(s) \qquad (1)$$

where TB is the travel time to the outlet of a watershed, P(s) is the probability that a water drop will follow path s, Ts is the travel time through path s, S is the set of all possible paths, and $f_{TB}(t)$, and $f_{Ts}(t)$ are the Pdfs of TB and Ts, respectively.

Assuming that the time a drop spends as overland flow is negligible then the travel time in a particular path is equal to the sum of travel times in the streams of this path. Assuming that the travelling in each stream is an independent random variable, then the Pdf of Ts is the convolution of the Pdfs of the travelling time in the streams of this path.

The infiltration losses in the streams are accounted for in the probability distribution of the time of travel of runoff in these streams. This distribution ($f_{T(i)}(t)$) has an exponential form as

$$f_{T(i)}(t) = \mu_i \, e^{-\lambda_i t} \qquad i<\Omega \text{ ; order of basin}$$

$$f_{T(\Omega)}(t) = \mu_\Omega^{*^2} \, t \, e^{-\lambda^{*^2} t} \tag{2}$$

where $f_{T(i)}(t)$ is the probability distribution of the time of travel of runoff in the streams of order i, $\mu_i = C_i/\bar{L}_i$ is the dynamic velocity of the flood wave in the streams of order i and equals $\bar{V}_i + \sqrt{g\bar{Y}_i}$; $\bar{L}_i$, $\bar{V}_i$, and $\bar{Y}_i$, are the average length, flow velocity and flow depth of the streams of order i; $\lambda_i = C_i/\bar{L}_i(1-I_i)$, I_i is the infiltrated percentage of the flow in the streams of order i, and $\mu_\Omega^{*^2} = 4\mu_\Omega\lambda_\Omega$. These values of λ_i and μ_i ($i = 1,..,\Omega$) make the area under $f_{T(i)}(t)$ equals to $(1-I_i)$.

On the other hand, the probability P(s) can be determined as

$$P(s) = \Theta_i \, P_{ij} \, ... \, P_{\ell\Omega} \qquad s=1,...,S \tag{3}$$

where Θ_i is the probability that a water drop falls in an area draining to a stream of order i and equals the total area draining directly into streams of order i divided by the total basin area and P_{ij} is the transition probability from streams of order i to streams of order j which is equal to number of streams of order i draining into streams of order j divided by the total number of streams of order i.

As shown in equations (1) through (3), the GIUH is defined in terms of geomorphologic parameters and runoff velocities in the watershed streams. Having determined the GIUH, the runoff discharge can then be estimated via the well known convolution transformation.

COMPUTATIONS OF THE SURFACE RUNOFF VOLUME

Given the discharge hydrograph, the surface runoff volume can easily be computed. It equals the area under the hydrograph. In the real life however, there are many situations in which hydrologists are not concerned with discharge hydrographs but only with surface water potential. In this case it will be a cumbersome to go through all these geomorphologic computations of discharge hydrograph to determine the runoff volume. Accordingly a mathematical model is developed for surface runoff volume computations.

The infiltration losses in a watershed equal the sum of these losses in all possible paths of surface runoff to the outlet multiplied by the probabilities that surface runoff will follow these paths. By representing the infiltration as a percentage of the surface runoff as done by Diaz-Granados, Bras, and Valdes (1983), then the surface runoff yield (Y) of a catchment of area A_Ω due to a storm of intensity i and duration t_r, can be computed as:

$$Y = A_\Omega \; i \; t_r \sum_{s \in S} P(s) \; (1-I_s) \tag{4}$$

where, I_s is the infiltration as a percentage of the surface runoff during its travelling through path s and can be determined from:

$$(1-I_s) = (1-I_i) \; (1-I_j) \; \ldots \; (1-I_\Omega) \tag{5}$$

By substituting with the average rainfall intensity and duration in equation (4) the average runoff volume per storm can be determined. Multiplication of this runoff volume by the average number of storms per year will result in the annual surface runoff yield.

THE CASE STUDY

Elmaghara basin is at about 60 kilometers south to the Mediterranean sea coast (Figure (1)). About 1500 of Sinai inhabitants are living at Elmaghara on farming and sheepherding. The basin has a mountainous nature with maximum and minimum elevations of 700m, and 400m above S.W.L., respectively. It has an area of about 40 Km^2 divided among five wadis; Wadi El Samar, Wadi Abu Trafia, Wadi Elmaghara, Wadi Aum Kutifat and Wadi El Kahlia. The surface runoffs of these wadis are naturally collected in a main stream which carries the water to the outlet of the basin as shown in Figure (2). The mountains of Elmaghara are of limestones with numerous cracks. In the second and third order streams, the limestones are topped with an alluvial layer. The depth of the alluvial layer reaches to more than hundred meters in the third order streams.

Of each wadi, three streams of each order are surveyed. An average cross-sectional area, an average slope and a reference flow depth for each stream order of each wadi are determined. The reference velocity is assumed here equal to the uniform flow velocity and is computed using Manning formula.

Aerial photographs for the basin were made available and are analyzed using a stereoscope. Size of the drainage areas of the different streams of each wadi, and the length of these streams are measured;and the Horton's ratios (1945) are determined.

A meteorologic station at Elmaghara was installed at the end of year 1983. Before 1983, no rainfall records for Elmaghara basin were taken. The nearest meteorologic stations from Elmaghara are at Hasana and at El-Arish, which are 45 Kms and 65 Kms to the southeast and northeast of Elmaghara, respectively. At these stations no records on rain-

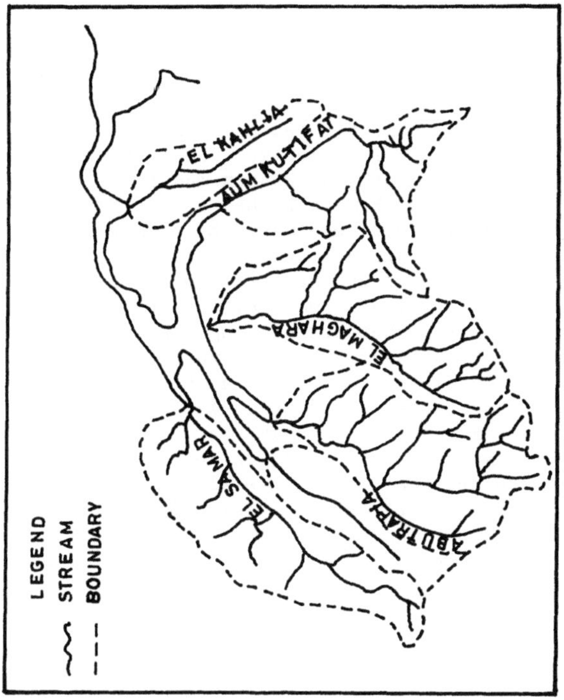

FIG.(2): EL MAGHARA BASIN.

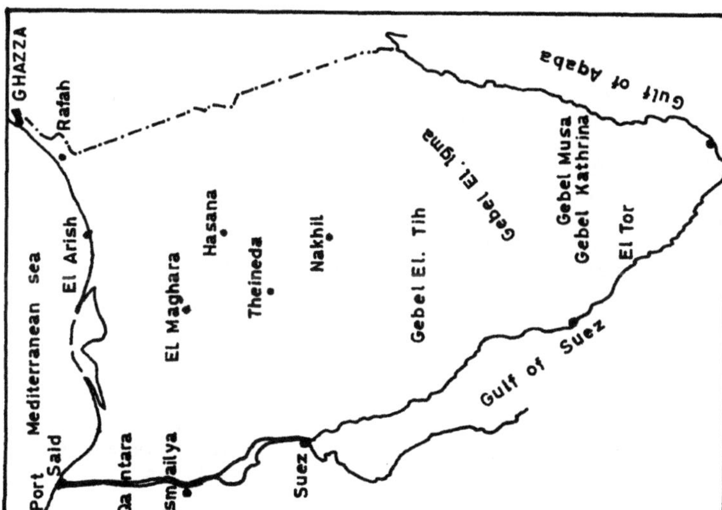

FIG.(1) THE PENINSULA OF SINAI.

fall intensity or duration are found. The available records are only on rainfall depth. The annual rainfall depth at Hasana ranges between 0. and 60 mm/year. The mean and standard deviation of the rainfall depth are found 18 mm/year and 16 mm/year, respectively. On the other hand, the rainfall depth at El-Arish varies from 0. to 200 mm/year. The mean and standard deviation of the rainfall depth at El-Arish are computed as 72 mm/year, and 43 mm/year, respectively. The topography of Elmaghara is different from the topography of Hasana and El-Arish. While Hasana and El-Arish are flat areas, Elmaghara has a mountainous nature. It is a difficult task therefore to estimate the rainfall depth at Elmaghara using the available rainfall records at Hasana and El-Arish. Investigations on flood damages at the three locations, Elmaghara, Hasana, and El-Arish and personal communications with the inhabitants of these locations as well as of surrounding areas indicate that the annual rainfall depth at Elmaghara is usually much higher than that of Hasana and close to the annual rainfall depth at El-Arish. An annual isohyetal map of Sinai (Sanyu Consultants Inc., 1984) indicates that the annual rainfall depth at Elmaghara is 45 mm/year. This map is based, particularly for the northern part of Sinai, on records taken at flat areas. It is believed therefore that the isohyetal map is underestimating the annual rainfall depth at Elmaghara. This 45 mm/year however, is considered here as a lower limit for the rainfall depth at Elmaghara with the rainfall depth at El-Arish (72 mm/year) as an upper limit.

SOLUTION OF THE CASE STUDY

The GIUHs and discharge hydrographs of the five wadis of Elmaghara are determined using different combinations of the infiltration losses in the different order streams (Allam, 1984). For illustration, the discharge hydrographs of Aum Kutifat and El Kahlia, using different combinations of the infiltration losses, are presented in figures (3) through (6). The rainfall depth is taken equal to 300 mm, which is about the average rainfall depth per storm at El-Arish,with different durations of 0.5, 1, 1.5 and 2 hours. The figures show that an increase in the infiltration losses, causes a decrease in the peak discharge as well as the area under the discharge hydrographs. They also show that for a constant storm depth, an increase in storm duration results in a decrease in the peak discharge and in an increase in runoff duration.

Application of the surface runoff prediction model as presented in equation (4) requires selection of the appropriate values for I_i $(i = 1,\ldots,\Omega)$. Based on a linearized form of Burkham expression (1970) for infiltration $f_i(x,t)$ in a channel

$$f_i(x,t) = K\, q(x,t) \tag{6}$$

where K is the infiltration coefficient and $q(x,t)$ is the discharge per unit width of the channel at distance x and time t, Diaz-Granados, Bras and Valdes (1983) derived a relationship between K_i and I_i as:

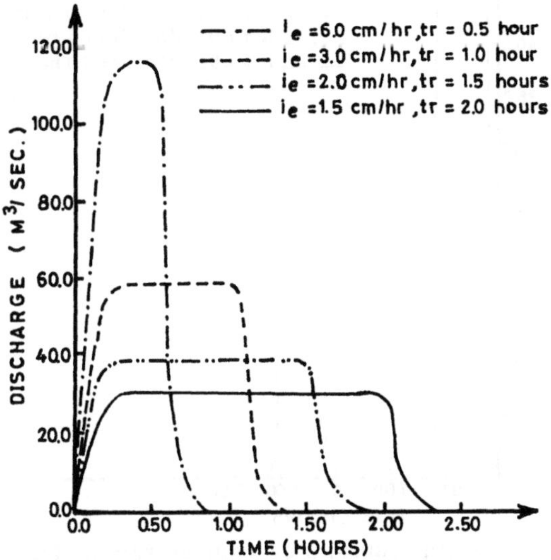

FIG.(3): DISCHARGE HYDROGRAPHS OF WADI AUM KUTIFAT (I_1 = 0, I_2 = 0 and I_3 = 0)

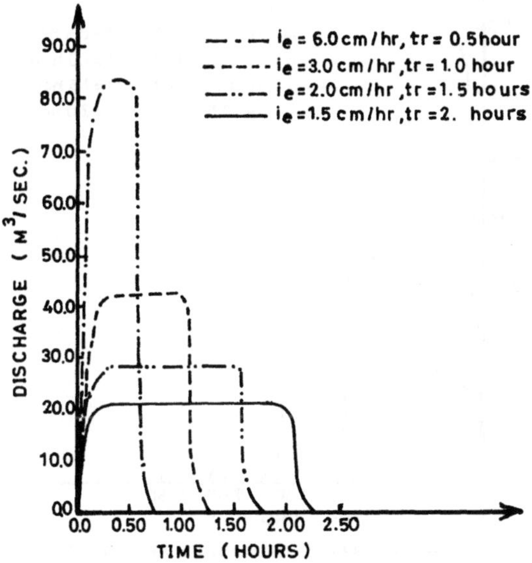

FIG.(4): DISCHARGE HYDROGRAPHS OF WADI AUM KUTIFAT (I_1 = 10, I_2 = 15 and I_3 = 20)

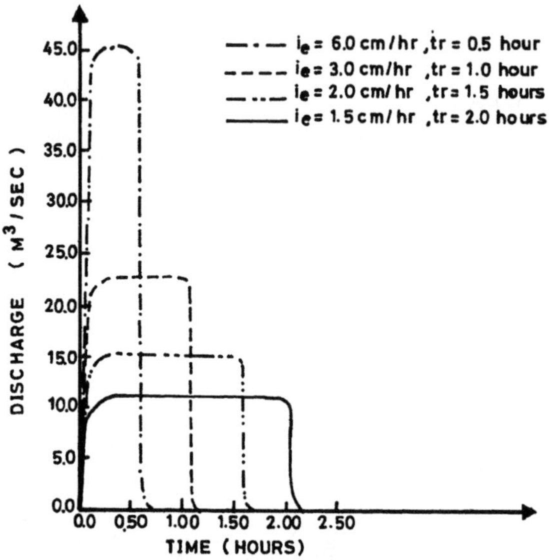

FIG.(5): DISCHARGE HYDROGRAPHS OF WADI EL KAHLIA (I_1 = 10, and I_2 = 15)

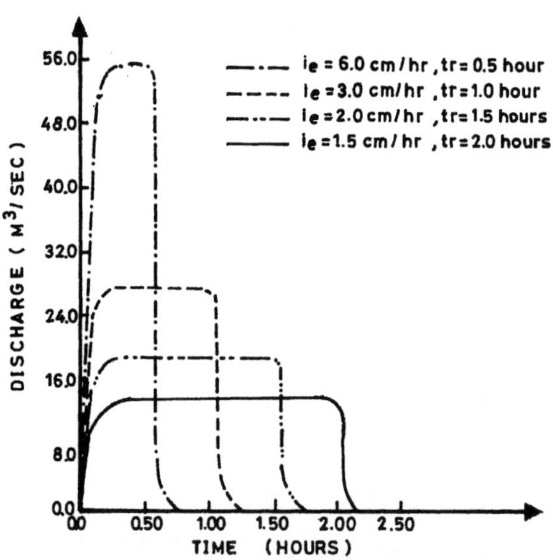

FIG. (6): DISCHARGE HYDROGRAPHS OF WADI EL KAHLIA (I_1 = 0, and I_2 = 0)

$$I_i = 1 - e^{-\bar{L}_i \bar{K}_i} \qquad (7)$$

where $\bar{L}_i$ and $\bar{K}_i$ are the average length and infiltration coefficient of the streams of order i. Then with knowledge on $\bar{K}_i$, the infiltrated percentage of the runoff can be determined. The soil of Elmaghara in the second and third order streams may be classified as sandy loam, and K probably about 0.2 per Km of the stream length. Corresponding to this value of K, $\bar{I}_1$, $\bar{I}_2$ and $\bar{I}_3$ for the five wadis are computed. The runoff coefficient of the basin is found .50. When increasing K to 0.5, the runoff coefficient decreases to .14. The latter runoff coefficient, which is reasonable for this arid region, will result in an annual runoff volume of $.25 \times 10^6$ to $.40 \times 10^6 m^3$ corresponding to the rainfall depths 45 and 72 mm, respectively.

DISCUSSION

The surface runoff coefficient as estimated in this study is relatively high. In this arid region of the world, the runoff coefficient is usually within the range of .0 - .05. The main reason behind this overestimation problem of the runoff coefficient is the negligence of the infiltration process in the over-land flow phase. Incorporation of this infiltration component in the surface runoff prediction model may be performed as follows.

Consider a physically based infiltration expression (Horton 1935, Philip 1969, and others) for overland flow phase. Consider a storm with intensity i and duration t_r. Suppose that the resultant surface runoff depth equals R. Assuming that the infiltration process is homogeneous within the drainage areas of the streams of the same order, then the surface runoff yield (Y) of a catchment can be computed as:

$$Y = \sum_{i=1}^{\Omega} \sum_{s=1}^{S} A_i R_i P(s_i) (1-I_{i,s}) \qquad (8)$$

where A_i is the size of all areas which drain directly to streams of order i, equal to $\Theta_i A$, (L^2); R_i is the surface runoff depth at area A_i, (L); $P(s_i)$ is the probability that the runoff in stream i will follow path s to reach the outlet, equal to $P_{ij} \ldots P_{\ell\Omega}$; and $I_{i,s}$ is infiltration as a percentage of the surface runoff yield of area A_i through path s, equal to $(1-I_i)(1-I_j) \ldots (1-I_\Omega)$.

This runoff prediction model accounts for the infiltration losses in the drainage areas as well as in the streams. In addition, it allows classification of large scale basins into sub-basins with homogeneous soil and rainfall characteristics.

The discharge hydrograph prediction model can also be modified to account for the infiltration process in the overland flow phase. This can be done by considering the effective instead of the gross rainfall intensity in the convolution transformation.

SUMMARY AND CONCLUSIONS

Two rainfall-runoff models, based on the geomorphologic theory, are presented. The first one is for the prediction of surface runoff discharge hydrograph. The second model predicts the surface runoff volume. The models are applied for five ungauged wadis in Sinai. The discharge hydrographs of the five wadis are obtained. Effects of infiltration on the shape of the hydrographs, peak discharges and time to peaks are illustrated. The annual runoff volumes of these wadis are determined under different configurations for infiltration. The surface runoff potential of Elmaghara is found within the range of $.25\times10^6$ to $.4\times10^6 m^3$ per year, which probably is not enough for the intended development.

Modification of both models through incorporation of a physically based expression for infiltration in the overland flow phase is recommended . This modification should significantly improve the runoff prediction capability of these models.

REFERENCES

Allam, M.N. 1984. 'A geomorphologic rainfall-runoff model incorporating infiltration losses: Case study of Elmaghara basin in Sinai'. Technical Report, Institute of Research for Water Resources, Ministry of Irrigation, Cairo, Egypt.

Burkham, D.E. 1970. 'Depletion of stream flow by infiltration in the main channels of the Tucson basin, southwestern Arizona'. Water Supply Paper, No. 1939-B, U.S. Geological Survey.

Diaz-Granados,M., Bras,R.L., and Valdes,J.B. 1983. 'Incorporation of channel losses in the GIUH'. Technical Report # 293, Ralph M. Parsons Lab Massachusetts Institute of Technology, Cambridge, Massachusetts.

Gupta,V.M., Waymire,E., and Wang,C.T. 1980. A representation of an IUH from geomorphology. Water Resources Research,Vol.16,No.5,pp.855-862.

Horton,R.E. 1935. 'Surface runof phenomena: Part I, Analysis of the hydrograph'. Horton Hydrology Lab Publ. 101. Edwards Bros,Inc. Ann Arbor

Horton,R.E. 1945. Erosional development of streams and their drainage basin, hydrophysical approach to quantitative morphology. Bulletin of the Geological Society of America, Vol. 56, pp. 275-370.

Kirshen,D.M., and Bras, R.L. 1982. 'The linear channel and its effect on the GIUH'. Technical Report #227, Ralph M. Parsons Lab., Massachusetts Institute of Technology, Cambridge, Massachusetts.

Philip, J.R. 1969 .'The theory of infiltration'. In advances of hydroscience, Vol. 5, edited by V.T. Chow, Academic, New York.

Rodriguez-Iturbe,I., and Valdes,J.B. 1979. The geomorphologic structure of hydrologic response. Water Resources Research, Vol. 15, No. 5, pp. 1409-1420.

Sanyu Consultants Inc. 1984. 'Development strategy for water resources and agriculture in Wadi El-Arish basin in Sinai', The Arab Republic of Egypt. Preliminary Report, Nagoya, Japan.

STATISTICAL METHODS OF DETERMINING TYPICAL WINTER AND SUMMER HYDROGRAPHS FOR UNGAUGED WATERSHEDS

Andrzej Ciepielowski
Warsaw Agricultural University
Dept. of Land Reclamation
Warsaw, 166 Nowoursynowska St., Poland

ABSTRACT. The present paper gives regression equations used for determining parameters of typical winter and summer flood hydrographs. They were calculated basing on hydrologic and physiographic data from lowland and highland watersheds of medium size which are situated in the Vistula basin, applying stepwise regression method. An attempt was made to utilize the obtained equations in small watersheds which differ in their character from those used for determining the equations. The results of the experiment show the possibility of applying these equations for estimative calculations in small ungauged watersheds, allowing for a possibility of an error of the order of several per cent. The problem dealt with in the paper should interest hydrologists who possess basic hydrologic-historical data: water stages and flows, and who have no opportunities of obtaining more accurate data from experimental watersheds.

1. INTRODUCTION

In many European countries typical flood hydrographs are traditionally applied in engineering practice for determining the capacity of water overflowing embankments, designing cross sections of outlets, estimating polder capacity and deciding upon water economy in small reservoirs utilized for agricultural purposes, etc.

The author suggests that the well-known and so far applied methods of Reitz and Kreps, of the Soviet hydrologists and others, should be replaced by a method consisting in identifying parameters of density function of Pearson distribution type III and IV. In the hierarchy of accuracy typical (hypothetical) hydrographs methods should be classified lower than the methods based on the deterministic, physical-mathematical models of water circulation in the watershed, which rely on the knowledge of hydrologic process genesis. However, they have some advantages over the latter ones: they do not require expensive empirical research, they are simple to use and they can utilize abundant historical data. In the countries in which modelling of this kind is more advanced than in Poland, typical hydrograph methods can be used for estimative com-

V. P. Singh (ed.), Flood Hydrology, 117–124.
© 1987 by D. Reidel Publishing Company.

putations made by other methods. The adopted typical hydrograph method does not result in an observed hydrograph but in an empirical-conventional one of a definite probability of occurrence.

The present paper aims at clarifying problems which limit the application of a typical hydrograph method in various conditions. The limitations result mainly from the lack of understanding of the character of relations among flood hydrograph parameters, watershed characterictics, etc.

2. METHODOLOGY

The study discussed in the paper made use of statistical and probability methods.

For describing typical flood hydrographs the modified Strupczewski method was used, based on the equations of density function of Pearson distribution type III and IV. The basis of this method is an equation of a typical flood hydrograph which takes the form:

$$Q_t = Q_{\max_{xp\%}} \left\{ a\left(\frac{t}{t_s}\right)^m \cdot e^{\frac{m}{n}\left[1-\left(\frac{t}{t_s}\right)^n\right]} + b \right\} \quad \text{for } t \in \{0 \to \infty\} \tag{1}$$

where

- Q_t - discharge in moment t $(m^3 \cdot s^{-1})$
- $Q_{\max_{xp\%}}$ - maximum annual or seasonal discharge with definite probability of occurrence
- a, b - equation parameters shown in Fig. 1
- m, n - parameters of hydrograph shape
- t - time elapsing from the moment accepted as the beginning of flood hydrograph (h)
- t_s - time peak calculated as a mean value of the examined natural hydrographs
- e - natural logarithm base

Parameters m, n are determined from the function:

$$f_o(m,n) = \frac{n^{\frac{m+1}{n}-1}}{m^{\frac{m+1}{n}}} e^{\frac{m}{n}} \Gamma\left(\frac{m+1}{n}\right) = \frac{\bar{V}_b}{a} \qquad f_1(m,n) = \left(\frac{n}{m}\right)^{\frac{1}{n}} \frac{\Gamma\left(\frac{m+2}{n}\right)}{\Gamma\left(\frac{m+1}{n}\right)} = m_1$$

where

- $\bar{V}_b$ - volume of averaged natural hydrograph in dimensionless scale
- m_1 - initial moment equals $m_1 = \frac{\Sigma\, Pi\, Ci}{\Sigma\, Pi}$; $i = 1, 2, \ldots, n$

where

- Pi - quantities equal to areas of part diagram
- Ci - limbs of quantity Pi (Fig. 1)

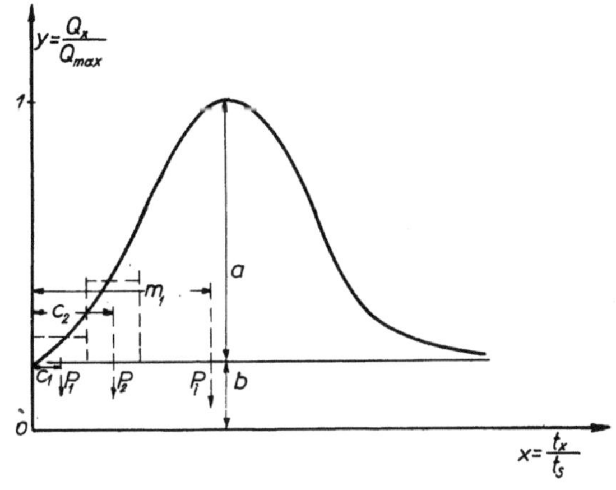

Figure 1. Diagram of denominations for equation (1)

The assumed method was adapted to determining typical hydrographs for winter and summer period separately. Besides, new criteria were introduced for selecting data. Basing on annual hydrographs from the years 1951-1975 hydrographs of summer floods, caused by rains, and of winter floods, caused by snowmelt, were chosen. Their peaks were constituted by maximum flows in particular seasons, and their bases by mean flow for summer floods - summer mean LSQ, and for winter period - annual mean SQ. That is, a straight horizontal line was assumed for the basis of a flood hydrograph, at the same height for a given hydrograph type. The hydrographs whose maximum flows from a given season constituted the peaks of complex hydrographs were neglected. Moreover, very small hydrographs were eliminated when flood values were smaller than mean flow values assumed as a basis, as well as hydrographs with a peak smaller than Q_{max} 90%. The first condition results from the error in flow estimation whereas the second one from statistical prerequisites of series of uniformity.

The basis of the assumed method for determining typical hydrographs in ungauged watersheds were regression equations, combining typical flood parameters: a, b, t_s, f_0 and f_1, occurring in equation (1) with watershed characteristics. Basing on these equations and substituting the value of a proper watershed characteristics particular equation parameters of typical hydrographs were computed. The computations were verified on dependent and independent data.

After substituting maximum flow of definite probability, calculated by one of the recommended methods, it was possible to determine a typical flood hydrograph for a given season in the required profile.

The Efroymson's version of the stepwise regression method which was developed at the University of Michigan was applied for establishing regression equations and the OSIRIS III computation programme was used. The computations were executed in the computation centre at SGGW-AR by means of a RIAD 1035 computer.

The normality condition of variables which is required when testing regression equations, was checked by λ-Kolmogorov test. When there were no reasons for assuming a series as normal, a transformed function was adopted and behaviour of the transformed variable distribution was examined from the point of view of its agreement with the normal distribution.

Apart from typical for OSIRIS III programme equations of partial and multiple correlation coefficients, equations of test characteristics *F* and *t*, standard deviation and errors, the following equations were used:

- Standard deviation relative individual error

$$S_o = \sqrt{\frac{1}{N-1} \sum_{i=1}^{N} (\delta_i - \bar{\delta})^2} \tag{2}$$

- Standard error of the mean

$$S_{\bar{x}} = \frac{S_o}{\sqrt{N}} \tag{3}$$

where

$\bar{\delta}$ - mean value of relative error

δ_i - relative error.

The differences occurring between the values of typical hydrograph parameters determined on hydrometric data, and the data obtained from the suggested regression equations, were checked. As well as defining the differences between parameters, the agreement was made by means of three criteria proposed by P.B.S. Sarma and others, 1973: correlation coefficient (*R*), special correlation coefficient (*RS*) and integral square error (*CBK*). The obtained values were compared with the evaluation scale of agreement proposed by those authors. Variation coefficient Y of residual error given in F. Neaf's 1982 paper, was also used.

3. RAW DATA

The basis for selection of watersheds for which typical hydrographs were determined, was constituted by three criteria: physiographical, hydrologic and statistical ones. A multifactoral analysis made for particular criteria led to the selection of 20 watersheds of various regional characteristics, situated in the central and northern part of the Vistula basin but having a uniform set of fundamental observed data.

The physiographical criterion requires that the watersheds should be small or of medium size and lowland in majority.

In Poland we do not have satisfactory, from the point of view of statistics, observed material for small lowland watersheds and that is why the data from IMGW (Institute for Meteorology and Water Economy) network were used for the basic analysis whereas the small watershed data from the observation network of departamental institutes of IBL (Institute for Forestry Studies) represented independent observed material, used for the verification of computations.

When selecting watershed area a condition was made that within a particular area range at least two watersheds of similar size should differ in character. An attempt was also made to differentiate them in respect of the number of lakes, swamps and woodlands. When applying a hydrologic criterion the most frequently occurring in lowlands and highlands river regimes were taken into consideration. The adoption of the first three criteria resulted in varied observed material due to which it was possible to make multicriterion and multifactoral analysis.

The last criterion, a statistical one, concerned satisfying the condition of homogeneity of hydrologic data and their sufficient number. A 25 year period was assumed for all watersheds (1951-1975). We made use of the daily hydrographs based on the observed water stage and discharge measurements made systematically on 20 gauged profiles controlled by the IMGW in these years (500 graphs).

4. STUDY RESULTS

Regression equations combining typical flood hydrograph parameters with watershed characteristics in the way previously discussed were established by means of the stepwise regression method calculated by OSIRIS III programme. 25 watershed characteristics were used. Some of these characteristics concerned the same phenomena, e.g. the number of lakes in a watershed was characterized by four values, therefore apart from eliminating variables which are automatically utilized by the programme, it was necessary to eliminate the characteristics of the same genetic sense.

The left-out characteristics were those which had been introduced first and whose partial correlation coefficient was the highest. The results of computations show that the confidence test results t-Student for the variables taking part in regression are in all cases greater than the critical value. Also the values of F-test which allow to estimate the confidence of the adopted regression equation are greater than the critical value. Multiple correlation coefficients are greater than the critical values (table I).

Computation results were verified on raw and independent material. In the first stage the values of particular parameters of typical flood hydrographs obtained from empirical dependence and calculated from observed data were compared. Relative errors in particular gauge profiles oscillated within considerable range. Parameter a and values of functions $f_0(m,n)$ and $f_1(m,n)$ were burdened with the smallest relative errors whereas parameter t_s - with the biggest ones.

The mean relative errors calculated on data from all research watersheds in the most unfavourable case only slightly exceeded 30%. The largest standard deviation of individual relative error exceeded 32% in only one case. The value of these errors was undoubtedly affected by the varied character of watersheds. The errors would have been smaller if the watersheds which considerably departed in their character from others had been neglected, for example lake watersheds. However, such procedure would have been inconsistent with the assumed concept of studies.

In the second stage, flood hydrographs in dimensionless system were compared. The computations showed considerable agreement of hydrographs. According to the adopted interpretation this agreement was on average very good and in some cases excellent agreement prevailed. From the verification of empirical equations made on the observed material, which was the basis for model identification, it is evident that the obtained equations can be applied in practice in watersheds of similar hydrologic and physiographical characteristics.

In most general terms, the computed regression equations concern lowland watersheds of medium size (A = 257-1534 km^2). However, we should take into account the possibility of the greatest errors in case of culmination time identification t_s.

Next step was the verification of the method on independent material. It should be noted that not only data from various periods of observation were assumed for the verification but also observations from watersheds with different characteristics. The verification was executed on the data from 5 small lowland watersheds situated in the Bug and the Narew basin and considerably different in physiographical, orographical and hydrologic respect. The areas of watersheds ranged between 57.2 and 121 km^2, the degree of afforestation was 7.1-92%, permeability was 9.1-86.5%. The climatic variation, however, was small, including the mean annual precipitation: 596-640 mm. The highest winter and summer flood hydrographs were chosen from the period 1966-1979 and making use of the adopted methodology typical flood hydrographs were established.

In the further course of studies typical flood equations were established from regression dependence between equation parameters and watershed characteristics given in table I. With this objective in view characteristics of watersheds included in regression equations were identified.

After substituting watershed characteristics into suitable regression equations typical flood hydrograph parameters, later called empirical, were obtained. The degree of agreement of these hydrographs with the parameters of hydrographs established from the observed data was checked. The greatest discrepancies were observed among $a_{(Z)}$, $b_{(Z)}$, $t_{s(L)}$, and $t_{s(Z)}$ though in particular cases very good agreement occurred. The question arises what kind of agreement can exist between the shape of empirical flood hydrographs and that of hydrographs established basing on observations with considerably large changeability of particular typical flood hydrograph parameters. R, RS, CBK and Y were used as measures of agreement. The computations indicate that the poorest agreement expressed by RS, CBK (i.e. good), occurred in case of winter hydrographs in the watersheds in which simple observed hydrographs were not sharply outlined and the numbers of single or complex hydrographs giving the maximum season flows were similar.

5. CONCLUSIONS

The computations show that in case of watersheds smaller than 257 km^2 but not smaller than 50 km^2 and differing in distribution and the values of physiographic and hydrologic characteristics the application of re-

Table I. Linear dependence $y = a + bx_1 + cx_2$ between parameters of typical flood hydrograph equations (winter - Z, summer - L) and watershed characteristics

Variables			Values of equation parameters							r	r^2	F	$S_{y \cdot x}$	Equation number
dependent	independent		a	b			c							
y	x_1	x_2		value	$S_{(b)}$	t	value	$S_{(c)}$	t					
WINTER HYDROGRAPHS														
$Q_{(Z)}$	L+1	δ_1	-2.80	-0.0016	0.0008	2.12	3.75	0.788	4.76	0.853	0.727	26.3	0.051	1
$t_{s(Z)}$	1+LAKE	Ir_2	-226	383	69.1	5.54	-48.8	21.0	2.33	0.811	0.657	19.2	43.9	2
$f_{o(Z)}$	1+LAKE	Ir_3	0.417	1.05	0.180	5.85	-0.177	0.0469	3.76	0.873	0.762	31.4	0.112	3
$f_{1(Z)}$	Nr reg	-	1.08	0.0085	0.0026	3.30	-	-	-	0.586	0.343	10.9	0.097	4
SUMMER HYDROGRAPHS														
$Q_{(L)}$	P	1+LAKE	0.937	0.0010	0.0003	3.34	-0.838	0.116	7.25	0.879	0.772	33.2	0.0736	5
$t_{s(L)}$	f_b	-	63.8	8.03	1.68	4.78	-	-	-	0.732	0.535	22.9	35,4	6
$f_{o(L)}$	Wd	Fc	1.69	-0.0030	0.0006	4.79	-0.398	0.181	2.20	0.738	0.545	12.4	0.139	7
$f_{1(L)}$	Wd	Fc	1.53	-0.0014	0.0004	3.73	-0.251	0.110	2.29	0.659	0.434	8.28	0.084	8

For equations (4) and (6) $F_\alpha = 4.41$, $t_\alpha = 2.10$, $r_\alpha = 0.444$, $n_{o/k} = 18/1$ whereas for remaining equations $F_\alpha = 3.59$, $t_\alpha = 2.11$, $r_\alpha = 0,545$, $n_{o/k} = 17/2$. In all cases $\alpha = 0.05$.

gression equations given in table I can be recommended only for estimative computations, reckoning with the possibility of making an error of the order of several percent.

REFERENCES

Ciepielowski A., 1984. The Interrelations between Flood Wave Parameters: Proceedings of the 5th International Conference on Water Resources. Planning and Management "Water in the Year 2000" Athens, Greece.

Ciepielowski A., Kucharska K., Tyszka J., 1984. Role lasów w wyrównaniu odpływu z małych zlewni w północno-wschodniej Polsce. *Gospodarka Wodna* nr 4.

Czarnecka H., 1983. Ustalanie charakterystyk fizycznogeograficznych dla wybranych zlewni rzecznych. Z. H. IMGW Warszawa (manuskrypt).

Draper N.R., Smith H., 1973. Applied Regression Analysis. John Wiley and Sons, Inc., New York, London, Sydney.

COSSTROJ SSSR-1972. Ukazanija po opredleniju rasčetnych gidrołogičeskich charakteristik SN 435-72, Gidrometeoizdat, Leningrad.

Naef-1982. Can We Model the Rainfall-Runoff Process Today. *Przegląd Geofizyczny* z. 1-2.

Sarma P.B.S., Delleur I.W., Rao A.R,, 1973. Comparison of Rainfall-Runoff Models for Urban Areas, *Journal of Hydrology*, z. 18.

Sokołov A.A., Rantz S.E., Roche M., 1978. Flood Flow Computation. Methods Compiled from World Experience. UNESCO.

Strzepczewski W., 1964. Równanie fali powodziowej. Wiadomości Służby Meteorologicznej i Hydrologicznej, WKiŁ, z. 57, Warszawa.

APPENDIX – SYMBOLS

The following symbols are used in this paper:

- F_c – stream frequency ($1/km^2$)
- Ir_2 – river slope – obtained mean longitudinal profile method (%)
- Ir_3 – river slope – obtained S(1085) method (%)
- L+l – river length together with dry valley (km)
- 1+LAKE – number of lakes in watershed index
- fb – swamp percentage in watershed area (%)
- P – average annual rainfall (mm)
- Wd – elevation of cross section closing watershed (m asl)
- δ_1 – reduction coefficient of number of lakes (acc. to GOCCTROJ SSSR)
- Nr reg.– type of region according to Kondracki's division (code)

ASPECTS OF FLOOD LEVEL COMPUTATIONS

H.J.M. Ogink, J.G. Grijsen, J.H.A. Wijbenga
Delft Hydraulics,
P.O.Box 152,
8300 AD Emmeloord,
The Netherlands

ABSTRACT. Flood level computations discussed in the paper refer to determination of water levels, discharge distribution and flow velocities in single and compound channels for the design of civil engineering works.
Following hydrological and hydraulic aspects are discussed:
i external boundary conditions
ii internal boundary conditions
iii computational models.

External boundary conditions refer to either extreme upstream discharges, or downstream water levels or to their joint occurrence depending on the location of the channel section. The application of one- or two-dimensional statistical models require homogeneous series of sufficient length. For various reasons, series are often inhomogeneous and too short. A procedure is discussed to eliminate inhomogeneities and to fill in missing data. It includes the sequential and efficient use of hydrologic and channel routing models. The application of this technique on a drainage area of 1500 km^2 in northern part of the Netherlands is dealt with. The results have led to considerable savings on civil engineering works.

The internal boundary conditions comprise besides channel geometry and structure conditions also the hydraulic roughness of the cross profile, i.e. main channel and flood plains. Special attention is focussed on the variation of the roughness of a sandy river bed during the passage of a flood wave. Empirical relations between bedform dimensions and hydraulic roughness based on recent investigations are dealt with. The effect of retarded adaptation of bedform dimensions on the development of the roughness is indicated. Its consequence for the hydrological boundary conditions for these type of computations is discussed. It is concluded that some commonly used statistical techniques oversimplify the boundary conditions for proper design.

Finally one- and two-dimensional numerical flood routing models, their limits and preference for application are discussed. Examples of their application for simulation of dam-break waves on the Rio Parana and for flood level computations for design of embankments along the Rhine branches in the Netherlands are presented.

V. P. Singh (ed.), Flood Hydrology, 125–141.
© 1987 by D. Reidel Publishing Company.

1. INTRODUCTION

Flood level computations aim first of all at the determination of water levels in a river section, or generally in an open water course, under extreme boundary conditions for the design of dikes (height), hydraulic structures, pumping stations, etc. Besides water levels, the flow pattern and discharge distribution in the considered section are often also of importance in order to quantify the effects of changes in the river profiles and to determine the extent of compensating measures.

The computations are carried out by means of one-dimensional (1D) and two-dimensional (2D) numerical models based on the laws of preservation of mass and momentum. The results of the computations depend on both internal and external boundary conditions. Internal boundary conditions are, for example, the geometry of the river or open water course, the hydraulic roughness of the bottom, hydraulic conditions due to structures, bifurcations, etc. The external boundary conditions refer to downstream stages and upstream and/or lateral in- and outflows. Various aspects of boundary conditions and computational models are discussed below.

2. EXTERNAL BOUNDARY CONDITIONS

2.1. General

To determine the boundary conditions for flood level computations in a river section a distinction has to be made between the upper river and the lower river. In the latter case the stages are not solely determined by the upstream (and lateral) inflow as in the upper river but are also subject to backwater effects from the downstream boundary. So depending on the location of the river section the flood levels are a function of the upstream discharge or of a combination of the upstream discharge and the downstream water level. The boundary conditions for the design situation are generally obtained by extrapolation of historical data by means of 1D and 2D statistical models, depending on the sphere of influence and the quantities of concern, often the maxima.

Besides the maximum other important aspects are a.o. the duration and shape of the flood wave, which determine the attenuation of the wave and the development of the hydraulic roughness and the changes in the bottom topography during the passage of the wave. The damping of a flood wave per unit length is roughly inversely proportional to the square of the wave period; so a short steep wave attenuates more than a very long wave. The ratio of conveyance width and storage width also strongly influences the damping.

As shown in Section 3 for the determination of the external boundary conditions the hydrological conditions outside the section and the hydraulic and morphological conditions inside the section are of importance and should therefore not be considered in isolation. So any automatism or standard procedure does not exist.

2.2. Creation of homogeneous series

The use of statistical models for the external boundaries requires homogeneous data sets of sufficient length. Series are often statistically inhomogeneous due to:

a. changes in the runoff characteristics of a catchment in the course of time as a result of:
   - urbanization with increase of impervious areas
   - improvement of drainage
   - deforestation
   - construction of embankments, causing reduction of storage.

   These changes will generally lead to higher flood waves of shorter duration.
b. systematic and random measuring errors.

Also, if the series are too short additional information is required to fill in the gaps.

A number of techniques is available to extend water level and discharge series and to eliminate inhomogeneities. One of these makes use of a rainfall-runoff model, assuming that rainfall (and evaporation) data are available and are unaffected by the changes in the catchment. The procedure for the generation of a series of homogeneous runoff extremes then could be:

a. generate a discharge series with the aid of a rainfall-runoff model for the land phase of the runoff process, calibrated to the present situation, to which a simple model for channel routing is connected (simple to keep computational costs low) → a long discharge series;
b. select from this series the annual maximum flood waves or all flood waves which exceed a certain threshold, and
c. route these extremes with a detailed and physically adequate open channel model through the main stems of the drainage system with the land phase output as upstream and lateral input. This step is carried out to represent wave damping and propagation properly.

The resulting homogeneous series of extremes form the input of the statistical models used for purpose of extrapolation towards the design conditions.

2.3. Example

An example of the procedure outlined above is a recent study on the drainage requirements of East Groningen and North-east Drente, an area of about 1500 km^2 in the northeastern part of the Netherlands (Figure 1), which drains to the sea. Excess water is discharged either by gravity via sluices at low tide or artificially via pumping stations if necessary. The main drainage network is shown in Figure 2; the major courses are known as the Eemskanaal and the Dollardboezem. This drainage system receives its water partly via pumping stations from polder areas and partly through natural drainage.

The objective of the study was to determine the required capacity of the sluices and pumping station complex of the Dollardboezem. This capacity depends on the joint occurrence of excessive rainfall and high stages at sea, making gravity drainage impossible, and on the available

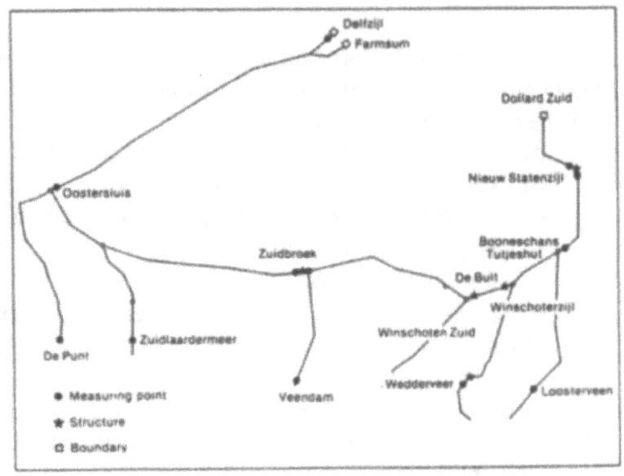

Fig. 2 Drainage network Eemskanaal and Dollardboezem

Fig. 1 Drainage area of Eemskanaal and Dollardboezem and Dutch main rivers

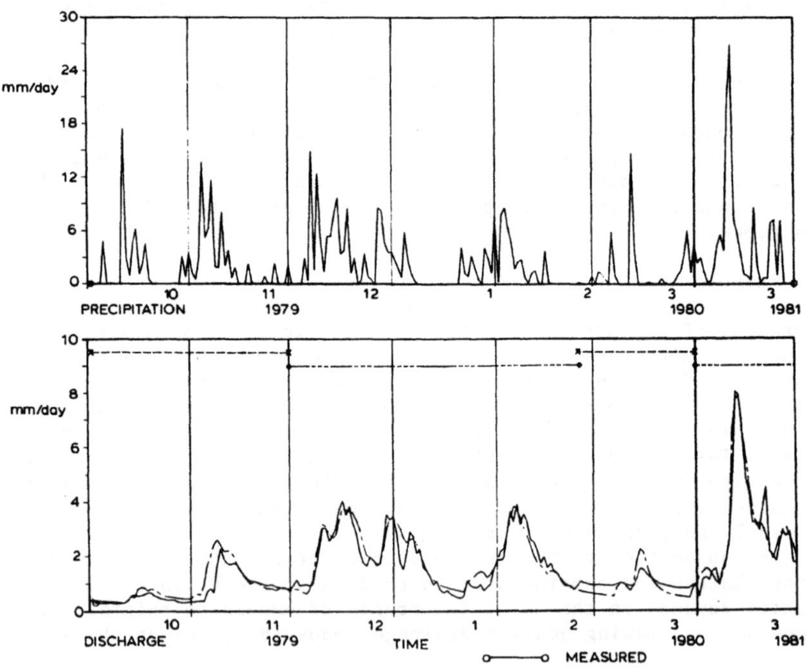

Fig. 3 Calibration of rainfall-runoff model

storage capacity as the drainage system. In this study sequential use was made of hydrological, hydraulic and statistical models. The inflow to the main drainage system was determined with the DHL-version of the Sacramento rainfall-runoff model. The model is a system of interconnected tension and free water storages for the upper and lower zone pertaining to a simple channel routing component. The parameters of the model have some physical meaning, which facilitates the calibration. Generally, when using this model the drainage area is subdivided into a number of segments; for each segment the drainage characteristics and rainfall are assumed to be uniform in space. In this particular case one set of parameters was sufficient to represent the drainage area characteristics. The result of a calibration is presented in Figure 3. It shows series of rainfall and measured and computed channel inflows. From this figure it is seen that the model is able to simulate the runoff properly. The applicability of the model was confirmed by verification runs. With this rainfall-runoff model 50 winter-half years in the period 1891-1983 were simulated. Only those winters were chosen which were expected to create extreme water levels in the drainage channels. Next, out of these 50 winter-half years a number of periods was selected, each covering a number of days with high inflows to the channel system in joint occurrence with high water levels at sea. The inflows were routed through the main channel network using the 1D mathematical hydraulic model NETFLOW, see Section 4. This resulted into a series of 18 most extreme water levels out of a period of 92 years, which were subsequently analyzed statistically using a truncated extreme value distribution. Furthermore, NETFLOW computations were carried out for possible future situations. This study showed that in the past design conditions were estimated too pessimistic. The computations led to considerable savings on investments in civil engineering works; an earlier planned pumping station (estimated cost Dfl. 50,000,000.-) was shown to be superfluous due to the improved understanding of the occurrence of extremes.

3. INTERNAL BOUNDARY CONDITIONS

3.1. General

Besides an adequate representation of the geometry of a water course an accurate prediction of the hydraulic roughness is of high importance for flood level computations. The aspects of hydraulic roughness discussed in this section refer in particular to the Rhine branches in the Netherlands (Figure 1) but do hold for other rivers as well. With respect to flow resistance a distinction is made between

- roughness of the flood plains with grass, trees, bushes, elevated roads, etc.
- roughness of the groyne section in the main channel, and
- roughness of the alluvial river bed.

The roughness is represented by a roughness length k_s, which is a measure for the bottom unevenness. For a hydraulically rough bottom the relation between the Chezy coefficient C $[m^{\frac{1}{2}}.s^{-1}]$, the hydraulic radius R [m] and the roughness length k_s [m] is given by the formulae of White-Colebrook or Strickler-Williamson, respectively:

$$C = 18 \log (12\ R/k_s) \approx 26.4\ (R/k_s)^{1/6} \qquad (1)$$

Combining the latter formula with the expression for uniform depth h_e, assuming $R \approx h_e$, it follows that h_e is proportional to $k_s^{1/10}$. This implies that a doubling of the roughness length does increase the uniform depth with 7%; for a flow depth of about 10 m it means an increase of h_e of 0.7 m. So the roughness length has to be determined with care.

3.2. Flood plain and groyne section roughness

The roughness of the flood plain varies seasonally. In the growing season the roughness length is higher than during winter time for obvious reasons. Measurements in the flood plains of the Rhine branches indicate k_s-values of 0.2 to 0.3 m for grass. During the passage of a flood k_s appeared to decrease from about 0.35 to about 0.20 m, probably due to bending of the grasses (Klaassen and van Urk, 1985); the roughness length did not increase again with decreasing velocities , possibly as a result of the bent grasses silted up, as shown in Figure 4. For the determination of the flow resistance of drowned hedges, the hedges can be schematized as a combination of a system of submerged openings and a short crested weir (Klaassen and van der Zwaard, 1984), where for the discharge coefficient of the latter a value of 1.0 can be used.

The roughness of the groyne section is determined by the form resistance of the groyne and the bottom roughness in between the groynes.

3.3. Alluvial river bed roughness

An interesting phenomenon is the roughness of the alluvial river bed consisting of grain and form resistance of ripples and/or dunes, whereby the form resistance very often dominates the grain resistance. An example of a dune-covered river bottom is shown in Figure 5, where the height of the bedforms varies from about 0.1 m to more than 1 m. The roughness length can be expressed as a function of the characteristic grain size D_{90}, the height H, the lenght L and the shape of the bedform. For the roughness of bedforms van Rijn (1985) drew up the following empirical relation valid for $k_s/h_w < 0.1$, see Figure 6:

$$k_s = 3\ D_{90} + 1.1\ H\ (1 - \exp(-25\ H/L)) \qquad (2)$$

From Equation (2) it is seen that for steep bedforms, $H/L > 0.1$, the roughness length is approximately equal to the height of the bedform. For smaller steepnesses $H/L < 0.1$, it follows that $k_s < H$. The above equation applies to bedforms with steep leeward faces where flow separation takes place behind the dune top. For smaller lee-side slopes the roughness is less than according to Equation (2), see Figure 7 (Ogink, 1984). Reduced lee-side slopes are often characteristic for large dunes where smaller dunes are superimposed on. For these composite dunes the hydraulic roughness can be predicted by the following expressions:

$$f_c = f' + \beta_1''\ f_d'' + \Sigma \left(f_{ri}''\ L_{ri}\ u_i^2/(L_d\ u^2)\right) \qquad (3)$$

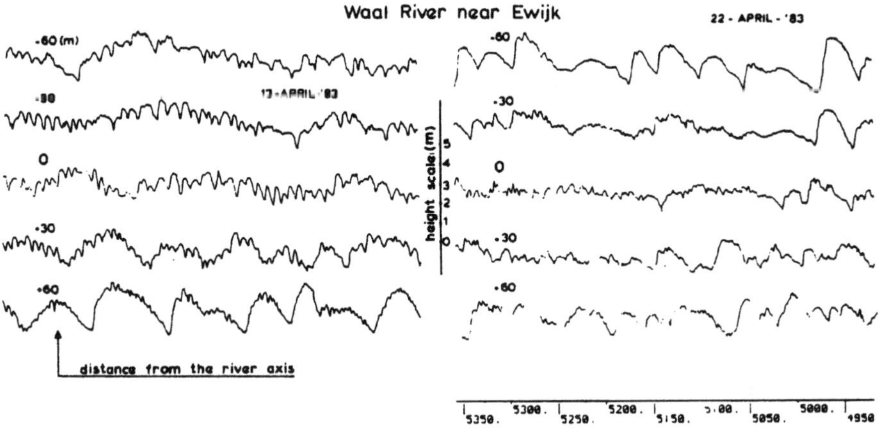

Fig. 5 Dune pattern on river bottom

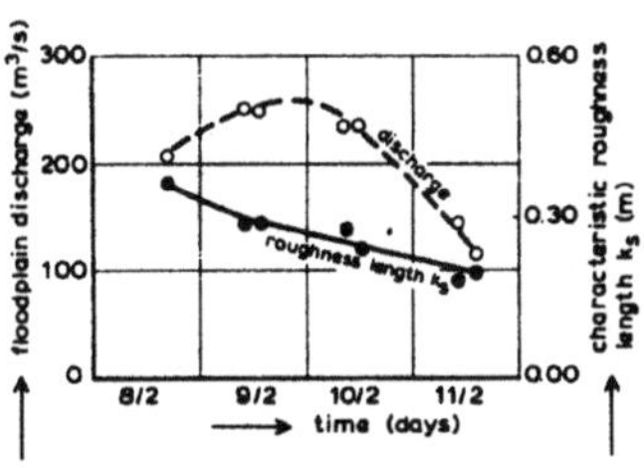

Fig. 4 Variation in flood plain discharge and roughness length k_s in Rhine branch for 1980

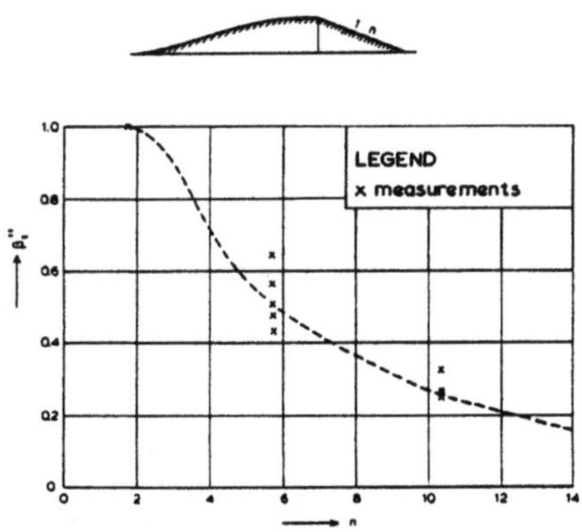

Fig. 7 Reduction of form roughness as a function of lee-side slope (Ogink, 1984)

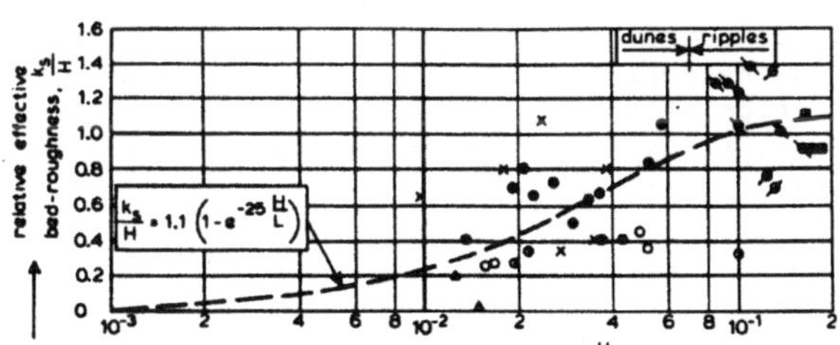

Fig. 6 Hydraulic resistance of dunes k_s/H as function of dune steepness H/L (van Rijn, 1984)

where:

f = flow resistance related to Chezy coefficient by: $f = 8g/C^2$
f_c = flow resistance of composite bedforms
f' = particle roughness
f''_d = form roughness of large dunes
f''_{ri} = form roughness of i-th superimposed dune or ripple
β''_i = correction factor for lee-side slope
L_d = length of large dunes
L_{ri} = length of i-th superimposed dune or ripple
u_i = flow velocity above i-th dune
u = average flow velocity

To estimate the roughness of the river bed for extreme high flows properly, it is of importance to predict the dimensions of the bedforms accurately. This, however, is not an easy task as the dimensions vary with changing flow conditions. In the development of bedforms a number of stages can be distinguished. For this van Rijn's transport stage parameter T (van Rijn, 1984) will be used, which is defined by:

$$T = \left[(u'_*)^2 - (u_{*,cr})^2\right] / (u_{*,cr})^2 \qquad (4)$$

where:

u'_* = $(g^{0.5}/C')u$ = bed shear velocity related to grains
C' = $18 \log (12R_b/3D_{90})$ = Chezy coefficient related to grains
R_b = hydraulic radius of bed
$u_{*,cr}$ = critical bed shear velocity according to Shields

The following stages are distinguished:

a. $T < 0$: plane bed without sediment transport
b. $0 < T < 5$: predominantly bed load transport with initiation and growth of bedforms; height and steepness of the bedforms increase with T
c. $5 < T < 25$: high sediment transport rates with stretching of bedforms; the dune length increases with T but the relative dune height, H/h_w, remains approximately constant so the dune steepness decreases with T. This development tends approximately towards a plane bed for $T \approx 25$
d. $T > 25$: upper flow regime with generation of anti-dunes.

When a flow process evolves from stage a to c, the hydraulic roughness will first increase, upto early stage c, after which it will decrease due to decrease of the bedform steepness.

Based on a large number of flume and prototype data van Rijn (1984) developed the following empirical relationships for dune height and dune steepness as a function of the characteristic particle size D_{50}, the flow depth h_w and the transport stage parameter T for steady flow (equilibrium conditions):

$$H/h_w = 0.11 \ (D_{50}/h_w)^{0.3} \ (1 - \exp(-0.5T)) \ (25-T) \qquad 0 < T < 25 \qquad (5)$$

$$H/L = 0.015 \ (D_{50}/h_w)^{0.3} \ (1 - \exp(-0.5T)) \ (25-T) \qquad 0 < T < 25 \qquad (6)$$

Recent tests in the DHL sand flume (Termes, 1986) indicate that Equation (5) should not be used for $T > 8$ since the predicted decrease

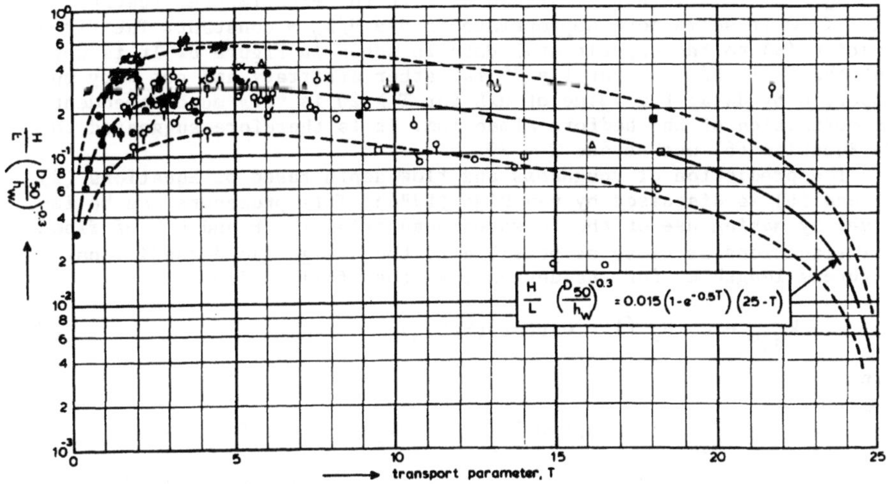

Fig. 8 Bedform steepness H/L as a function of particle size (D_{50}), flow depth (h_w) and transport stage (T) (van Rijn, 1984)

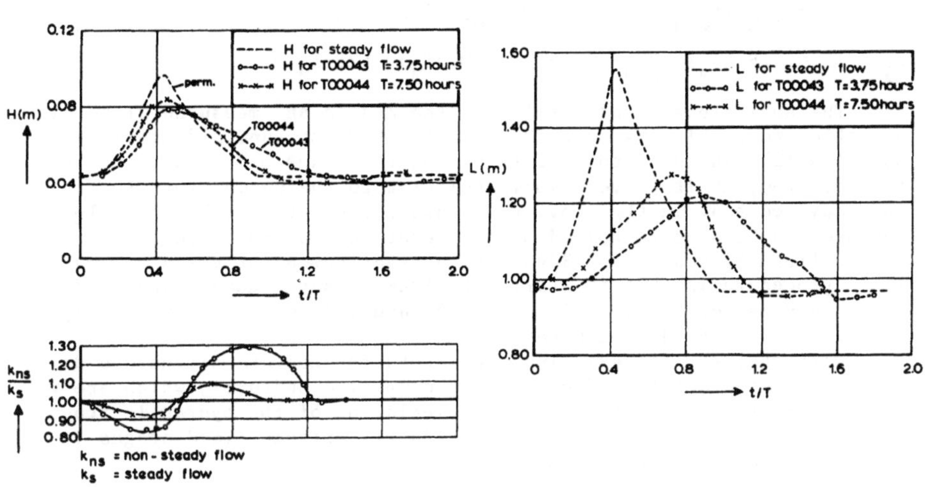

Fig. 9 Changes in bedform dimensions and roughness during the passage of flood wave in the DHL sand flume (Wijbenga, 1983)

of H/h_w for $T > 8$ is not in agreement with the experimental results. For $8 < T < 20$ it seems more appropriate to use H/h_w = constant. The fit of Equation (6) to the experimental data is shown in Figure 8. It is noted that the limit $H/L = 0$ for $T = 25$ is rather arbitrary and not very much supported by data; the value of H/L for $T > 20$ is very much dependent on the definition of the bedform dimension. It is therefore suggested to use Equation (6) not beyond $T = 20$.

The prediction of roughness and flow depth based on Equations (2), (5) and (6) is discussed by van Rijn (1984). This procedure can be improved by making use of the measured quantities k_s, H and H/L of flood flow situations as close as possible to the design flood conditions. The roughness length k_s for the design case then follows from:

$$k_{s,E} = c_1 H_E \left(1 - \exp(-25(H/L)_E)\right) \tag{7}$$

with:

$$H_E = c_2 h_{w,E}^{0.7} \left(1 - \exp(-0.5T_E)\right) (25-T_E) \qquad \text{for } 0 < T < 8$$

$$H_E = (h_{w,E}/h_{w,M}) H_M \qquad \text{for } 8 < T < 20$$

$$(H/L)_E = c_3 h_{w,E}^{-0.3} \left(1 - \exp(-0.5T_E)\right) (25-T_E) \qquad \text{for } 0 < T < 20$$

$$c_1 = k_{s,M}/\left(H_M\left(1 - \exp(-25(H/L)_M\right)\right)$$

$$c_2 = H_M/\left(h_{w,M}^{0.7} \left(1 - \exp(-0.5T_M)\right) (25-T_M)\right)$$

$$c_3 = (H/L)_M/\left(h_{w,M}^{-0.3} \left(1 - \exp(-0.5T_M)\right) (25-T_M)\right)$$

The subscripts E and M refer to the design and measured flood flow situation respectively. Note that the first right hand side term in Equation (2), which is only of value for the plane bed condition, has been omitted, as it is very small relative to the second term for a dune-covered bed. In the above equations the quantities $h_{w,E}$ and T_E are unknown and have to be solved by an iterative procedure similar to the one outlined by van Rijn (1984). The attractiveness of the above outlines procedure is that it starts off from measured quantities, and it makes only use of the formulae (2), (5) and (6) in a relative sense; the coefficients have been eliminated.

Nevertheless, the existing predictors for bedform dimensions do not yet have a high accuracy (for example composite bedforms, which are very common, are not included) so the extrapolation of the hydraulic roughness from the measured situation towards the extreme flood stages includes considerable uncertainties.

Furthermore, a complicating factor is the retarded adaptation of the bedforms to the flow condition, as shown in Figure 9 (Wijbenga, 1983). Here the development of the bedforms for flood waves of different duration is shown and is compared with the dimensions the bedforms would have had in a sequence of steady flow conditions. Different roughness values in the considered cases result. The situations refer to dune developments under stage b, i.e. growing height and steepness. From Figure 9 it is observed that the roughness length for the rising limb of the flood wave is smaller than the corresponding equilibrium roughness, while after the passage of the top the reverse is true. Note also the differences in k_s between the steep and the flat flood wave. For T > 8 the picture may differ from the above, since a retarded adaptation will lead here to higher roughness values for steep flood waves.

These observations indicate that the extreme values should be interpretted with care when used for extrapolation towards the design condition. Furthermore, it clearly shows that isolating the external boundary conditions from the internal ones oversimplifies reality.

Extensive research programs on the above subjects are presently carried out at the DHL (Klaassen et al., 1986) to eliminate the uncertainties in the prediction of bedforms and hydraulic roughness.

4. COMPUTATIONAL MODELS

4.1. Choice of model

For flood level computations use is made of:

a. 1-dimensional mathematical hydraulic network models, which describes cross-sectional averaged flow (1D model)
b. 2-dimensional mathematical hydraulic models, which schematize the water motion to a depth-averaged flow in two mutually perpendicular horizontal directions (2DH model).

The choice among these two type of computational models depends on:

- the desired information: type, detail, accuracy
- the available information: if little data is available, then it is not appropriate to apply computational models which require and produce many details (apparent accuracies)
- available budget: a 2DH model approach is more expensive than a 1D network model computation.

In general 2DH models for flood level computations are only used for steady states. In case the routing of flood waves have to be computed, the use of a 1D network model is more appropriate.

Combinations of these two types of models in studies become of interest, in case locally detailed information is required not only about stages but also about velocities and flow distribution.

4.2. 1D hydraulic network model

The basis of DHL's 1D hydraulic network model NETFLOW and its micro computer version WAFLOW, (Grijsen and Meijer, 1979), is formed by the following two equations (see also definition sketch Figure 10):

$$\partial A/\partial t + \partial Q/\partial x + q = 0 \qquad (8)$$

$$\partial Q/\partial t + \partial(\alpha Q^2/A_s)/\partial x + g\,A_s\,\partial z_w/\partial x + g\,A_s Q|Q|/(C^2 R A_s) - B_s\,\tau_s/\rho + g\,A_s\,\eta = 0 \qquad (9)$$

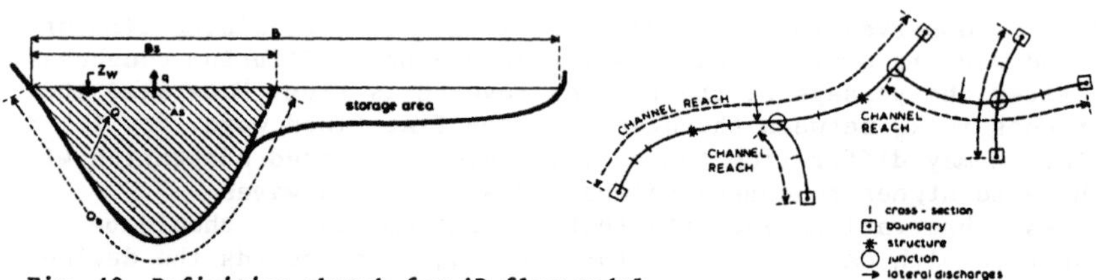

Fig. 10 Definition sketch for 1D flow model

Fig. 11 Parana and Paraguay Rivers

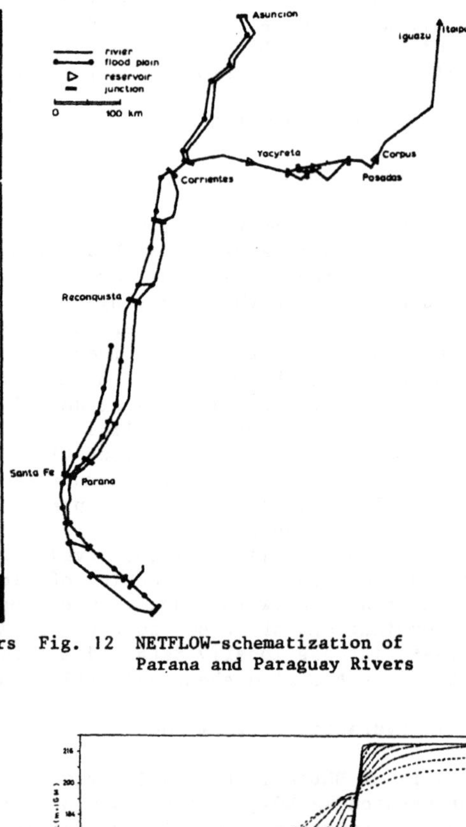

Fig. 12 NETFLOW-schematization of Parana and Paraguay Rivers

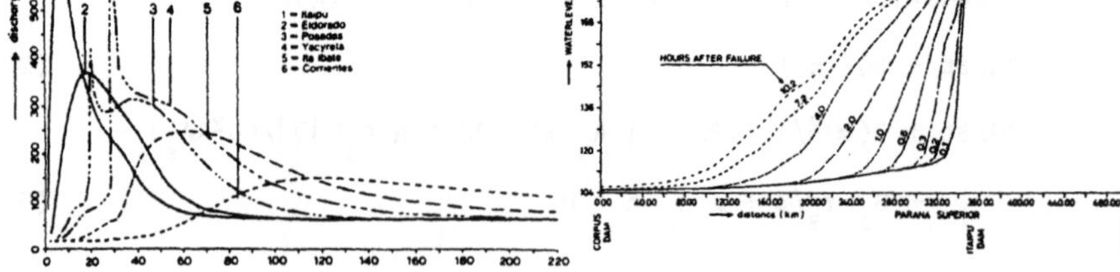

Fig. 13 Dam-break waves in Parana River

with:
A = total cross-sectional area
A_s = conveying cross-sectional area
Q = discharge
q = lateral outflow per unit length
x = location
t = time
z_w = height of water level above reference level
R = hydraulic radius
g = acceleration due to gravity
B_s = width of conveying cross-section at the water line
τ_s = wind shear stress
ρ = density of water
η = local energy head loss per unit length
α = momentum correction factor

In a network model the water courses are schematized in a system of branches and nodes. Since cross-sectional averaged flow is considered, detailed information about the velocity field cannot be obtained. In case the hydraulic roughness and/or the depth in the cross-section varies strongly, the resulting non-uniformity of the flow can be taken into account by means of the Engelund algorithm. For this the cross-section is divided into a number of parallel sections, which characteristics (Chezy coefficient, conveying cross-sectional profile and hydraulic radius) are accounted for separately via the algorithm. Computationally it means the determination of an adjusted value for C^2R in Equation (9). The 1D hydraulic models are very suitable for flow computations in (complex) channel networks ranging from polder networks to river systems.

As an example the NETFLOW application in the Parana-Paraguay flood protection and management study is presented. The project covered a 1700 km section of the Parana River in Argentina from Iguazu at the Brasil border to the river mouth at Buenos Aires and the Paraguay River downstream of Asunción (see Figure 11). In its middle and downstream part the Parana River has 15 to 60 km wide flood plains. In order to obtain a proper understanding of the flooding mechanism a hydraulic model was developed based on NETFLOW. The schematization of the rivers in a system of branches and nodes is presented in Figure 12. The network consists mainly of two parallel branches which are interconnected at a number of locations. At those connections in- and outflow conditions have been defined in order to simulate the complex interaction between the main channel and the flood plain properly. With this model the routing of extra-ordinary flood waves and dam-break waves were computed to study their damping and propagation (see Figure 13). Furthermore, with this model effects of construction of reservoirs, embankments and diversions on the attenuation of flood waves were investigated.

4.3. 2DH hydraulic models

The 1D network models fail to provide detailed information regarding the flow field. In the past in those cases use was made of hydraulic scale models or of stream tube model computations, where the latter refers to a quasi-two dimensional approach by schematizing the river with flood plains into a number of parallel stream channels interconnected at the nodes (see Jansen, 1979).

The disadvantage of scale model studies is their time consumption and therefore high costs. Scale effects also reduce their attractiveness. The neglectance of a number of terms in the momentum equation limits the accuracy of the stream tube method. This neglectance is compensated by adjusting the Chezy coefficient such that the computed water level fits the measured one. As a result the Chezy coefficient gets an empirical character which hampers the extrapolation to design conditions. Furthermore, just like a scale model the stream tube method is very time consuming as the schematization is stage dependent, requiring adaptation of the schematization for different flow conditions.

These drawbacks are removed by making use of 2D depth-averaged flow models which are based on the following equations of continuity and momentum:

$$\partial z_w/\partial t + \partial(h_w u)/\partial x + \partial(h_w v)/\partial y = 0 \tag{10}$$

$$\partial u/\partial t + u\ \partial u/\partial x + v\ \partial u/\partial y + g\ \partial z_w/\partial x - f_C\ v + \tau_{bx}/(\rho h_w) + \left(\partial(h_w \tau_{xx})/\partial x + \partial(h_w \tau_{xy})\right)/(\rho h_w) = 0 \tag{11}$$

$$\partial v/\partial t + u\ \partial v/\partial x + v\ \partial v/\partial y + g\ \partial z_w/\partial y + f_C\ u + \tau_{by}/(\rho h_w) + \left(\partial(h_w \tau_{yx})/\partial x + \partial(h_w \tau_{yy})\right)/(\rho h_w) = 0 \tag{12}$$

with:

x, y = horizontal coordinates
u, v = depth-averaged flow velocities in x- and y-direction
f_C = Coriolis parameter
τ_{bx}, τ_{by} = bottom shear stress in x- and y-direction: $\tau_{bx} = g\ u\ (u^2 + v^2)^{\frac{1}{2}}/C_x^2$
τ_{xx}, τ_{xy}, etc = effective stresses in vertical planes

(For simplicity reasons in the momentum equations wind shear stress, external forces, etc. have been omitted.) Note that the Coriolis and diffusion term do not appear in the 1D momentum equation.

The meaning of the last term in the momentum equations can be illustrated by considering the case of steady uniform flow in a prismatic channel parallel to the x-axis. Then Equation (11) simplifies to the following expression:

$$\tau_{bx} = \rho\ g\ h_w\ i_b + \partial(h_w \tau_{by})/\partial y \tag{13}$$

with:

τ_{xy} = $\rho\ \nu_e\ \partial u/\partial y$
i_b = $-\partial z_w/\partial x$
ν_e = effective viscosity coefficient $[m^2.s^{-1}]$

Equation (13) shows that positive gradients of τ_{xy} increase the bottom shear stress, while negative gradients reduce τ_{bx}. The gradient of τ_{xy} is positive for a convex shaped velocity profile u(y), a situation which occurs in the flood plain close to the main channel. In the

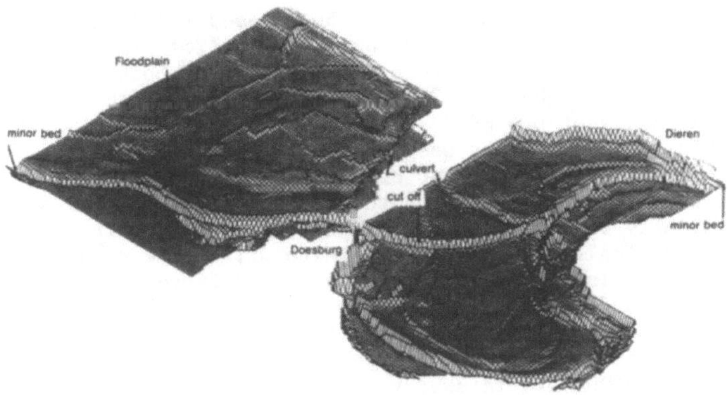

Fig. 14 WAQUA-schematization of IJssel River and flood plain near Doesburg

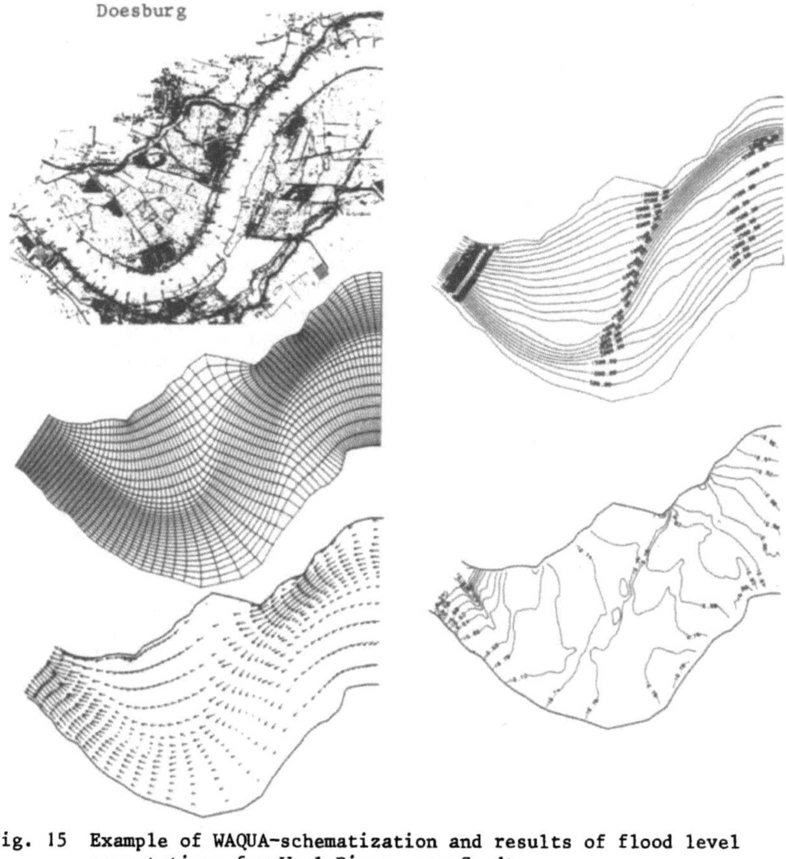

Fig. 15 Example of WAQUA-schematization and results of flood level computations for Waal River near Gendt

main stream the velocity profile near the flood plain is concave, so there τ_{bx} is less than the gravity component in flow direction. Hence the diffusion term affects the velocity pattern and bottom shear stresses (Vreugdenhil and Wijbenga, 1983). For the Rhine branches, where an effective viscosity of 0.5 $m^2.s^{-1}$ was introduced in the computations (Ogink, 1985), this component was shown to affect water levels and discharge distribution only to minor extent.

The Equations (10)...(12) are the basis of the 2DH flow model WAQUA, originally developed by Leendertse (1967, 1974). The equations are solved numerically by means of a finite difference method defined on a staggered grid. In the present versions the difference scheme of Stelling (1983) is used which provides high computational stability and accuracy. The computational grid adjusts itself automatically to drying and wetting areas, hence a flow independent schematization is obtained. Two versions of WAQUA are in use, viz. a version using a rectangular grid and one with a curvi-linear grid (Wijbenga, 1985a, b). In the rectangular grid version the smallest detail, generally the flow in the main channel, determines the grid size; at least 3 to 5 grid points in the channel are required in transversal direction to represent the flow accurately. In case of a small river with a wide flood plain, like in the example of the IJssel River near Doesburg in the Netherlands (see Figure 14), this is very unfortunate; the model gives extremely detailed information in the flood plain where it is not required. A solution would be the application of nested models with only locally a small grid size. Another attractive solution is the use of a curvi-linear grid (Wijbenga, 1985a, b), which allows a considerably more flexible schematization as the grid size can be adapted locally according to the requirements. This approach saves memory space, so larger river sections can be covered in one run. Figure 15 shows results of computations carried out for the Waal River near Gendt. It gives an example of the way flood level computations for the determination of the design height of the river dikes have been carried out for the Rhine branches recently.

5. ACKNOWLEDGEMENTS

The authors are indebted to the Dutch Public Works Department for its kind permission to publish some of the figures in this paper.

6. REFERENCES

Grijsen, J.G. and Meijer, Th.J.G.D., 1979. 'On the modelling of flood flow in large river systems with flood plains'. Proc. XVIII IAHR Congress, Cagliari, Italy (also DHL publication No. 227, 1980).

Jansen, P.Ph., Bendegom, L. van, Berg, J. van den , Vries, M. de and Zanen, A. (1979). Principles of River Engineering. The non-tidal alluvial River. Pitman Book Ltd., London.

Klaassen, G.J. and Zwaard, J.J. van der, 1974. 'Roughness coefficients of vegetated flood plains'. Journal of Hydraulic Research, Vol. 12, No. 1.

Klaassen, G.J. and Urk, A. van, 1985. 'Resistance to flow of flood plains with grasses and hedges'. Proc. XXI IAHR Congress, Melbourne, Australia.

Klaassen, G.J., Ogink, H.J.M. and Rijn, L.C. van, 1986. 'DHL-research on bedforms, resistance to flow and sediment transport'. 3rd International Symposium on River Mechanics, Jackson, Mississippi, U.S.A.

Leendertse, J.J., 1967. 'Aspects of a computational model for long-period water wave propagation'. Rand Corporation, Santa Monica, Memorandum RM-5294-PR.

Leendertse, J.J. and Gritton, E.C., 1971-1974. 'A water quality simulation model for well-mixed estuaries and coastal seas'. Volumes I-IV, Rand Corporation, Santa Monica.

Ogink, H.J.M., 1984. 'Hydraulic roughness of the Rhine and Waal Rivers'. Delft Hydraulics Laboratory, Report R 2017 (in Dutch).

Ogink, H.J.M., 1985. 'On the effective viscosity in 2D depth-averaged flow models'. Proc. XXI IAHR Congress, Melbourne Australia.

Rijn, L.C. van, 1984. 'Sediment transport. Part III: Bedforms and alluvial roughness'. Proc. ASCE Journal of Hydraulic Engineering, Vol. **110**, No. 12 (also DHL publication No. 334, 1985).

Stelling, G.S., 1983, 'On the construction of computational methods for shallow water flow problems'. Dissertation, Delft University of Technology, Delft.

Termes, A.P.P., 1986. 'Dimensions of bedforms for steady flow conditions and high rates of sediment transport'. Delft Hydraulics Laboratory, Report M 2130 (in Dutch with English summary).

Vreugdenhil, C.B. and Wijbenga, J.H.A., 1982. 'Computation of flow patterns in rivers'. Proc. ASCE Journal of the Hydraulics Division, Vol. **108**, HY11 (also DHL publication No. 298, 1983).

Wijbenga, J.H.A., 1983. 'Flow resistance and bedform dimensions for varying flow conditions'. Results of flume experiments with flood waves. Applied Research Rivers (TOW), Delft Hydraulics Laboratory, Information R 657-XLI.

Wijbenga, J.H.A., 1985a. 'Determination of flow patterns in rivers with curvi-linear coordinates'. Proc. XXI IAHR Congress, Melbourne, Australia.

Wijbenga, J.H.A., 1985b. 'Steady depth-averaged flow calculations on curvi-linear grids'. 2nd International Conference on the Hydraulics of Floods & Flood Channels, Camebridge, England.

A SHOT-NOISE MODEL OF STREAMFLOW

P. Todorovic, Professor
Dept. of Statistics
University of Kentucky
Lexington, KY 40506

D. A. Woolhiser
Hydraulic Engineer
ARS, USDA
Aridland Watershed Management Research Unit
2000 E. Allen Road
Tucson AZ 85719

ABSTRACT. A version of the classical shot-noise process is adopted in this paper as a streamflow model. We discuss some of its basic properties which may have a potential interest in applications. Among the features investigated here are its covariance function, the duration of its excursions above a certain threshold level, x_0, and its extreme values. We obtained the distribution function of its maximum value in an arbitrary time interval [o,t] in a simple form. Our principal concern here is the physical base for application of the shot-noise process as a streamflow model. We discuss, in some detail, the climatic conditions and geomorphological features of a watershed which suggest that the shot-noise model may be feasible. In this respect, of central importance is the set of the discontinuity points of the process, which represents the building blocks of the entire structure. Their importance stems from the fact that they represent the occurrence times of those rainfall events which are capable of generating an effective runoff. Section 3 provides a physically plausible explanation of why this set of points represents a nonhomogeneous Poisson point process.

1 INTRODUCTION

Hydrologists have long been searching for a mathematical flood frequency formulation with a strong physical base. It has been recognized that to obtain such a formulation, one must take into account certain critical aspects of the precipitation and infiltration phenomenon. In addition, it appears necessary to consider evapotranspiration and the dynamics of surface and subsurface runoff as affected by the climate, soils, and geomorphological characteristics of a basin.

V. P. Singh (ed.), Flood Hydrology, 143–163.
© 1987 by D. Reidel Publishing Company.

Although many of these processes have been studied extensively and are well understood individually, their joint effects and mutual interactions are too complex to handle without numerous simplifying assumptions concerning the nature of their relationships.

One of the first attempts to construct a mathematical model of floods which would incorporate at least some of these factors was made by Eagleson in 1971. Beginning with certain simplifying assumptions concerning the distribution of the storm duration and the precipitation intensity, and expressing the catchment dynamics with the kinematic wave (for this purpose, a simple open-book type geometry of the catchment is assumed), he obtained an analytic expression for flood peak Q_p as a function of recurrence interval and several hydrological parameters related to climate and catchment characteristics.

In 1972, Eagleson modified his earlier equation to include partial area runoff, resulting in the expression:

$$Q_p = \frac{645\ K\ A_r}{\beta} \ln[\phi_1\ \phi_2\ \theta\ T_E e^{-2\sigma} \sigma^{1-\sigma}\ \Gamma(\sigma)] + 0.074(1 - \phi_2)\phi_1\ p\ A_c \tag{1.1}$$

A_c: catchment area (square miles),
ϕ_1: fraction of the rainfall occurring as runoff,
ϕ_2: fraction of the runoff occurring as direct runoff,
θ : average annual number of rainfall events,
T_E: annual exceedance recurrence interval, years,
K : fraction of point rainfall depth occurring as areal rainfall,
p : average annual rainfall (in inches), and
β : parameter of conditional distribution of rainstorm depth given duration (hours/inch).

where:

A_r: area producing runoff;
σ : dimensionless catchment-climate parameter

$$\sigma = \{\frac{2.21\ \lambda^2\beta\ A_r}{K\alpha_c\ L_s} \cdot [1 - \frac{1830\ \alpha_s^{4/3}\ A_r^2}{\alpha_c L_s^3 (Q_p - \overline{Q}_b)^{1/3}}]\}^{1/3}$$

λ^{-1}: mean duration of storm (in hours),
α_c : overland flow parameter,
α_s : stream bed parameter,
L_s : length of idealized basin, and
$\overline{Q}_b$ = $0.074(1 - \phi_2)\phi_1\ p\ A_c$ = average annual base flow.

The formulation of the relationship between Q_p and T_E, in eq. (1.1), has a number of interesting features. It is expressed in terms of climate and catchment parameters. Given that we are able to evaluate the parameters, it can be used as a basis for flood-frequency analysis when no streamflow record is available.

In deriving this relationship, a number of expedient assumptions were made to insure mathematical tractability of the model. For instance, the runoff-producing area A_r is assumed to be independent of the rainfall excess variables. In addition, the infiltration loss is supposed to be constant for all rainfall events, and the rainfall intensity uniform. Finally, it is important to emphasize that the equation (1.1) allows Q_p to assume negative values (with a positive probability). In spite of the shortcomings and imperfections of the model, this is an appealing approach.

Recent attempts in relating the instantaneous unit hydrograph (IUH) to the geomorphologic structure of a catchment (Rodriguez-Iturbe and Valdes, 1979) may represent an improved procedure for relating the peak runoff rate to the storm characteristics while accounting for the geometry of the particular basin. This line of research is currently being actively pursued Diaz-Granados et al. (1983), (Rodriguez-Iturbe et al., 1982, Hebson and Wood (1982), Cordova and Rodriguez-Iturbe (1982), and appears promising.

The models discussed so far exhibit the non-linear character of the surface runoff response to rainfall. On the other hand, they require a considerable number of simplifying assumptions, particularly vis-a-vis the precipitation process. Furthermore, the base flow is supposed to be constant as well as the initial soil water content. Thus, interactions between base flow, soil water content, and infiltration, which may cause significant differences in runoff for otherwise identical rainfall events, are not considered.

Koch (1982) demonstrated that the streamflow process implied by these physically based models is the well-known filtered Poisson process if the hydrologic routing processes are linear, the rainfall excess counting process is Poissonian, and the random variables describing the system input are independent of the counting process. Weiss (1977) applied a special case of the filtered Poisson process, known as the shot-noise process, to modeling streamflow, but did not explicity consider the physical relationships that transform the rainfall process to the runoff process. The shot-noise model was also used as an example of the time series approach to flood estimation in the United Kingdom (Natural Environment Research Council, 1975).

In this paper, we suppose that, in some cases, the flow process can be described with sufficient accuracy by a version of a non-stationary shot-noise stochastic process. We identify physical conditions under which applications of this model seem appropriate (here we refer to the assumption that the number of local maxima of this process represents a non-homogeneous Poisson process). In addition to an analysis of various properties of this process, we also determine the distribution function of the duration of its excursion above a certain threshold level x_0. Finally, we investigate the maximum value of the process in a given interval of time and evaluate its distribution.

2. SOME REMARKS ON THE RAINFALL PROCESS

Denote by $\xi(t)$ the precipitation intensity at a given point at time t. As is very well known, $\{\xi(t);\ t \geq 0\}$ represents a non-negative, intermittent stochastic process which vanishes on random time intervals.

In Fig. 1 is a possible realization of $\xi(t)$.

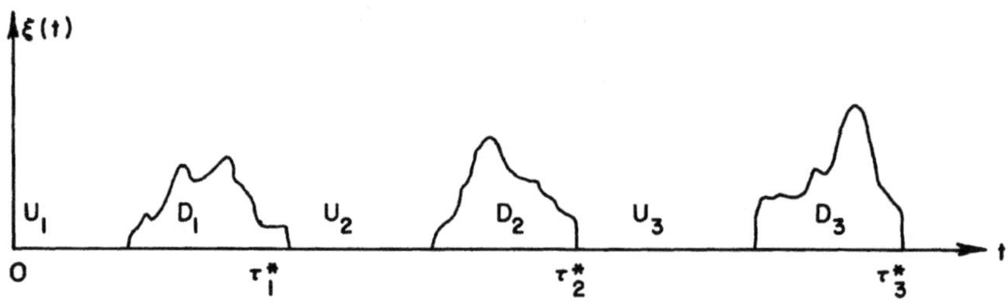

Fig. 1. A possible realization of $\xi(t)$

The sequence of termination times $\{\tau_n^*\}_1^\infty$ of rainfall events forms a point process on the half-line $R_+ = (0, \infty)$. We shall denote by $N(t)$ the counting random function of this point process, i.e.,

$$N(t) = \sum_{n=1}^{\infty} I_{\{\tau_n^* \leq t\}} \qquad N(0) \equiv 0 \tag{2.1}$$

where $I_{\{\cdot\}}$ is the indicator function.

In applications, various simplified versions of the intermittent stochastic process $\xi(t)$ are often used. For instance, it is frequently assumed that each rainfall event has a uniform precipitation intensity, say R_i. In this case, the stochastic process $\xi(t)$ has the following form:

$$\xi(t) = [\, R_{N(t)+1}] I_{\{\tau^*_{N(t)} + U_{N(t)+1} \leq t\}} \tag{2.2}$$

where $\{R_i^\infty\}_1$ is a sequence of positive random variables; R_i is the precipitation intensity of the i - th rainfall event. In this case, the intensity process is characterized by the family of random variables $\{R_i^\infty\}_1$, and the counting random function $\{N(t);\ t > 0\}$. A realization of process (2.2) is given in Fig. 2.

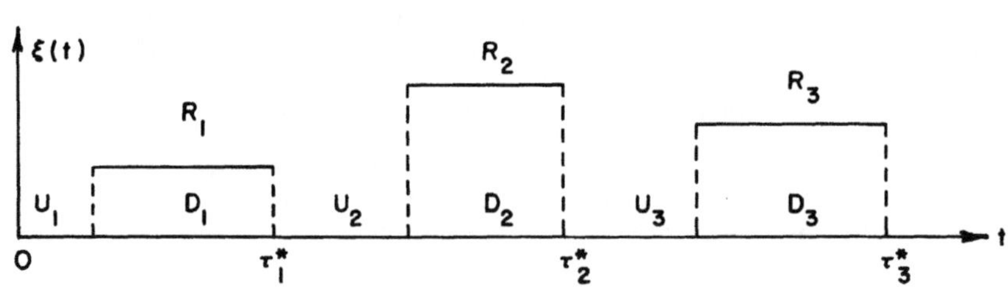

Fig. 2. A sample function of random process (2.2)

Another model of the rainfall intensity which is sometimes used is one where the precipitation intensity pattern of every rainfall event is a random isosceles triangle. In this case, one can write

$$\xi(t) = \frac{2R_{N(t)+1}}{D_{N(t)+1}} \{[\tau^*_{N(t)+1} - t] I_{\{\tau^*_{N(t)+1} - 0.5D_{N(t)+1} < t < \tau^*_{N(t)+1}\}} \tag{2.3}$$

$$+ [t - \tau^*_{N(t)} - U_{N(t)+1}] I_{\{0 < t - \tau^*_{N(t)} - U_{(t)+1} < 0.5 D_{N(t)+1}\}}\}$$

A realization of the stochastic process $\xi(t)$ in the form (2.3) is given in Fig. 3.

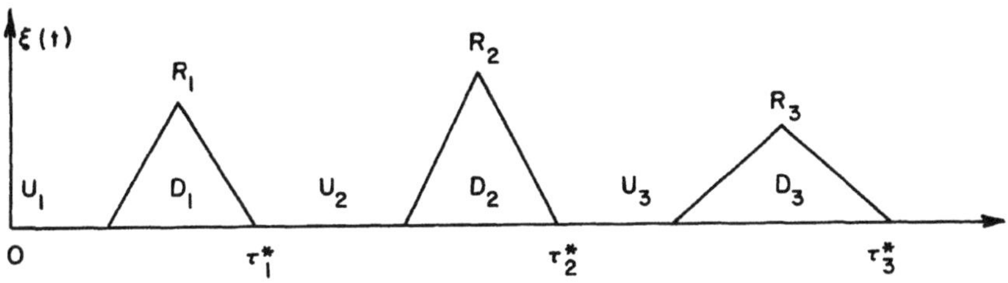

Fig. 3. A realization of the intensity process $\xi(t)$ in form (2.3)

Consider now the conversion process of rainfall into runoff. It is quite clear that, due to infiltration and other factors, the small volume rainfall events do not form runoff at all. On the other hand, of those precipitation events that can form runoff, only a very small number are capable of generating flows high enough to exceed a certain threshold level, say x_0, so that they can be qualified as floods. Of course, very much depends on such characteristics of a watershed as soil texture, the moisture content of the soil, vegetation, etc.

Several investigators have described the mechanisms of the process whereby runoff events are "selected" from rainfall events (c.f. Fogel et al., 1974; Woolhiser and Todorovic, 1974; Diaz-Granados et al., 1983; Koch, 1982). Diaz-Granados et al. (1983) utilized an expression for infiltration capacity based upon Philips' equation for unsaturated flow in porous media to determine the joint distribution function of the duration and average rate of rainfall excess. Koch (1982) used a slightly different formulation to obtain the distribution function and the first two moments of the volume of excess precipitaton. The parameters of these distributions include the parameters of the joint distribution of precipitation intensity and duration, as well as saturated hydraulic conductivity, sorptivity, and initial soil water content. It is instructive to examine the distribution functions derived by Koch (1982) using this approach, as shown in Fig. 4. In Fig. 4(a), we note that, for the climate of Denver, CO, the probability of no runoff increases as the soil texture becomes coarser until the probability of runoff is essentially zero for a sandy loam soil. The importance of initial soil water content is demonstrated in Fig. 4(b). It is significant that the initial water content has less effect on the distribution function at high levels of excess precipitation (rare events) than it does at lower levels. It should be noted that these distributions will be sensitive to the assumed rainfall intensity process. For example, the triangular representation of Fig. 3 would result in more runoff events than the constant intensity formulation of Fig. 2 for the same values of infiltration parameters.

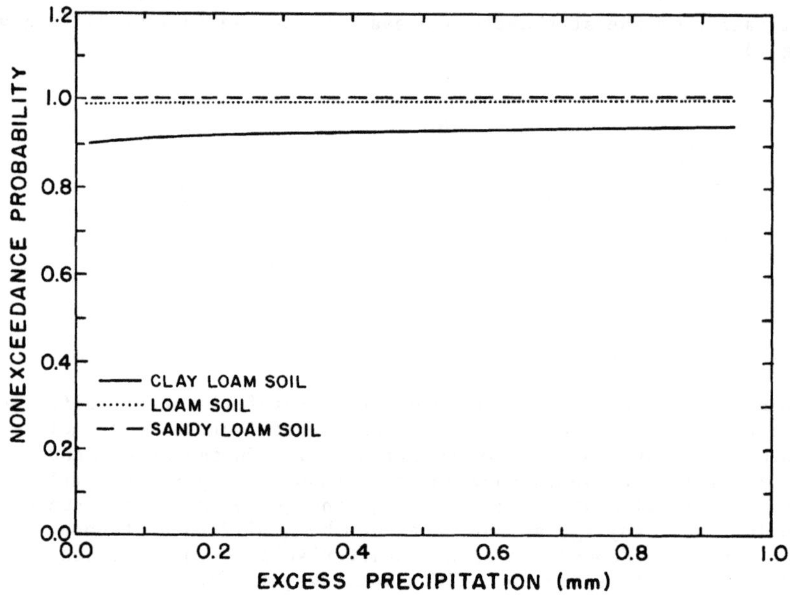

Fig. 4a. Cumulative distribution function of excess precipitation volume for selected soil types, Denver, CO (from Koch, 1982)

Denote by $N(t;x_o)$ the counting function of the "thinned" version of the point process $N(t)$. In other words, $N(t;x_o)$ is the number of those rainfall events in $(o,t]$, each of which is capable of creating runoff rates that exceed threshold level x_o. Then, under certain conditions (for instance, x_o is sufficiently high), $N(t;x_o)$ becomes roughly speaking, a Poisson point process. In what follows, we shall give a more precise formulation of this important property of a point process.

Suppose that we have the following situation: A point process $\{\tau^*_n\}_1^\infty$ on $R_+ = (0, \infty)$ with counting random function $N(t)$ undergoes independent thinning. This means that each point τ^*_n of the process is deleted with a probability q and retained with a probability $p = 1 - q$, independently, for each point. The points that are not deleted form a thinned version of the original version of the point process. The question we ask ourselves is, "What will happen to the thinned version when we let $p \to 0$?" The answer to this question is well known: When $p \to 0$, the thinned verson converges weakly to a Poisson process, weak convergence, meaning here convergence of all finite dimensional distributions of the thinned process.

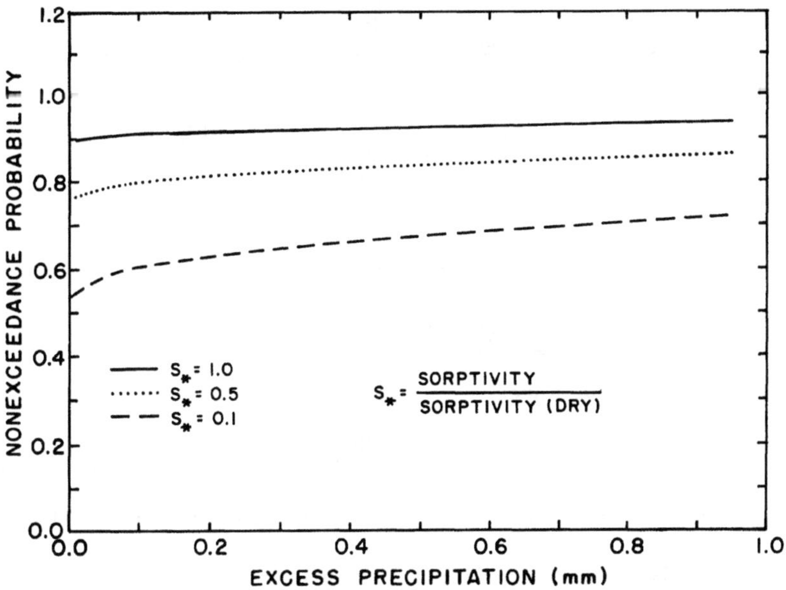

Fig. 4b. Cumulative distribution function of excess precipitation volume for different water contents, clay loam soil, Denver, CO climate (from Koch, 1982)

We now give a proof of this assertion: Suppose that

$$N(\frac{t}{p})\ p \to \Lambda(t) = \int_0^t \lambda(s)ds \tag{2.4}$$

as $p \to 0$, then

$$N_p(t) \overset{w}{\to} \eta(t)$$

where $N_p(t)$ is the thinned version of the point process $N(t)$ after a scale contraction by a factor p^{-1} and $\{\eta(t);t > 0\}$ is a Poisson process with the mean function $\Lambda(t)$, i.e.,

$$P\{\eta(t) = n\} = e^{-\Lambda(t)}\frac{[\Lambda(t)]^n}{n!} \tag{2.5}$$

and $\eta(t)$ has independent increments.

To make the derivation simple and brief, we shall use the Laplace-Stieltjes transform. Consider a thinned version $N_p(t)$ of the point process $N(t)$ and write:

$$E\{e^{-\Theta N_p(t)}\} = \int E\{e^{-\Theta N_p(t)}\ |N(\frac{t}{p})\}dP \tag{2.6}$$

The thinning mechanism implies that the conditional distribution of $N_p(t)$, given $N(\frac{t}{p})$, is binomial with parameter p. Thus,

$$E\{e^{-\theta N_p(t)} | N(\tfrac{t}{p})\} = \sum_{k=0}^{N(\frac{t}{p})} e^{-\theta k} P\{N_p(t) = k | N(\tfrac{t}{p})\}$$

$$= \sum_{k=0}^{N(\frac{t}{p})} e^{-\theta k} \binom{N(\frac{t}{p})}{k} p^k (1-p)^{N(\frac{t}{p})-k}$$

$$= [(1-p) + e^{-\theta} p]^{N(\frac{t}{p})}$$

$$= e^{N(\frac{t}{p}) \ln[1 - p(1 - e^{-\theta})]}$$

From this and (2.6) we have:

$$E\{e^{-\theta N_p(t)}\} = E\{e^{N(\frac{t}{p}) \ln[1 - p(1 - e^{-\theta})]}\} \tag{2.7}$$

On the other hand, for any $0 < \varepsilon < \frac{1}{2}$, we have the following inequalities:

$$p(1 - e^{-\theta}) < -\ln[1 - p(1 - e^{-\theta})] < p(1 - e^{-\theta})(1 + \varepsilon) \tag{2.8}$$

if $p < \varepsilon$.

As $p \to 0$, one can choose $\varepsilon > 0$ arbitrary small. From (2.6) and (2.7), we deduce that

$$\lim_{p \to 0} E\{e^{-\theta N_p(t)}\} = \lim_{p \to 0} E\{e^{-N(\frac{t}{p}) p(1 - e^{-\theta})}\}$$

$$= e^{-(1 - e^{-\theta}) \Lambda(t)}$$

due to (2.4). Thus,

$$\lim_{p \to 0} E\{e^{-\theta N_p(t)}\} = e^{-(1 - e^{-\theta}) \Lambda(t)} \tag{2.9}$$

Since the right hand side of (2.9) is the Laplace-Stieltjes transform of a Poisson distribution with mean value $\Lambda(t)$, the assertion follows (after some additional technicalities which will not be presented here).

The result obtained is not new. For more information on independent thinning, see Belyaev (1963), Kallenberg (1975) and Westcott (1976).

Corollary 2.1

Consider again the point process of rainfall $\{\tau_n^*\}_1^\infty$ with counting random function $N(t)$ and its thinned version $N(t;x_o)$. According to its definition, $N(t;x_o)$ represents the number of those rainfall events in $(o,t]$ which are capable of generating enough runoff so that the flow will exceed a certain threshold level x_o. Let p_{x_o} be the probability of a rainfall event exceeding the threshold x_o; it is clear then that

$$p_{x_0} \to 0 \quad \text{as } x_0 \to \infty$$

Thus, $N(t;\, x_0)$ converges weakly to a Poisson process as $x_0 \to \infty$. In practical terms, it means that for a sufficiently high threshold x_0, the number of those precipitation events that will exceed this level has a Poisson distribution. This theoretical result explains, among other things, why the number of exceedances in a partial duration series frequently follows a Poisson distribution.

3. SHOT-NOISE PROCESS

In this section, we discuss conditions under which the use of the shot-noise process as a streamflow model seems appropriate. We begin our presentation with a formal definition of a version of this process. We then investigate some of its fundamental properties which are needed in the remainder of this paper. To be more specific, we attempt to formulate a set of physically based criteria which may be of help in deciding if the geomorphological and climatic characteristics of a particular watershed warrant the uses of the process as a streamflow model.

Let $\{\tau_n\}_1^\infty$ be a Poisson point process on $R_+ = (0, \infty)$, with counting random function $\{\eta(t);\ t > 0\}$ and denote by

$$\Lambda(t) = E\{\eta(t)\} \tag{3.1}$$

Throughout this paper we shall assume that $\Lambda(t) < \infty$ whenever $t < \infty$. Let $\{V_n\}_0^\infty$ be a sequence of positive random variables. The stochastic process $\{Y(t);t>0\}$ defined as

$$Y(t) = \sum_{k=0}^{\eta(t)} V_k e^{-\gamma(t - \tau_k)} \qquad (\tau_0 \equiv 0) \tag{3.2}$$

$$= \int_0^t V_u\, e^{-\gamma(t-u)} d\eta(u) \qquad V_{u_j} \equiv V_j$$

where the parameter, $\gamma > 0$, is called the shot noise process.

According to its definition, $Y(t)$ has discontinuities at the points $\tau_1, \tau_2, \ldots$ and an exponential recession between any two consecutive jumps. In Fig. 5, a realization of $\{Y(t);\ t > 0\}$ is given.

It seems that a version of the shot noise process as a streamflow model was first proposed by Bernier (1970). Subsequently, Weiss published two papers on the same subject (1973 and 1977). Here, we consider a non-stationary version of $Y(t)$; it is expected that this modified version will reflect, in a certain way, the effects of seasonal variations.

Let us outline a set of physical conditions under which an application of the process seems plausible. To this end, consider a catchment where the prevailing geometry, significant infiltration capacity due to particular soil texture, and extensive vegetation preclude formation of significant runoff except in the case of heavy rainfall. When the occurrence probability of such an event is sufficiently small, the counting random function of these filtered rainfall events becomes "approximately Poissonian" (according to the results of the previous section on thinning of a point process). In this case, the shot-noise process seems suitable to model the streamflow dynamics.

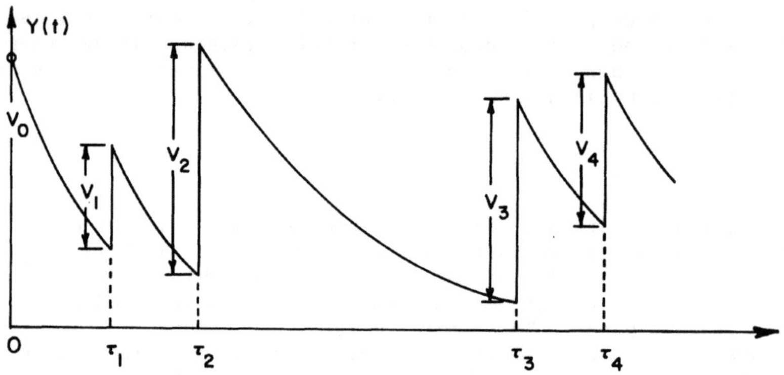

Fig. 5. A realization of the process Y(t)

Strictly speaking, realizations of a natural streamflow process are continuous functions. This, unfortunately, is not the case with sample functions of the shot-noise process (although the number of discontinuities in every finite time interval is finite). However, Y(t) should model streamflow recessions very well, since recessions have frequently been approximated with an exponential decay. In addition, the runoff component is incorporated into its structure and appears explicitly in its formulation.

Let us specify regularity conditions on the variables $\{V_i\}_0^\infty$ and $\{\eta(t);\ t > 0\}$ under which Y(t) will be used as a streamflow model.

(i) $\{V_i\}_0^\infty$ is a sequence of independent random variables such that

$$H_0(x) = P\{V_0 \le x\}\ ;\ H(x) = P\{V_k \le x\} \quad \text{if } k \ge 1 \tag{3.3}$$

(ii) $\{V_i\}_0^\infty$ and $\{\tau_i\}_1^\infty$ are independent sequences.

Before we proceed further, a few remarks on the sequence $\{\tau_n\}_1^\infty$ would be in order. Since every process with independent increments is a Markov process, the Poisson process $\{\eta(t);\ t \ge 0\}$ is also a (homogeneous) Markov process. From this it follows that $\{\tau_n\}_1^\infty$ is also a homogeneous Markov process with the transition probability ($s \le t$)

$$P\{\tau_{n+1} \le t | \tau_n = s\} = 1 - e^{-[\Lambda(t) - \Lambda(s))]} \tag{3.4}$$

As a matter of fact, this result can be generalized as follows (Todorovic, 1983): Let $\{\tau_n\}_1^\infty$ be an arbitrary point process on R_+, and N(t) its counting random function. If $\{N(t);\ t \ge 0\}$ has the Markov property, so does the sequence $\{\tau_n\}_1^\infty$.

Denote by $\phi_0(\theta)$, $\phi_1(\theta)$ and $L(\theta;t)$ the Laplace-Stieltjes transform of V_0, V_1, and $Y(t)$, respectively, i.e.,

$$\phi_0(\theta) = E\{e^{-\theta V_0}\};\ \phi(\theta) = E\{e^{-\theta V_1}\};\ L(\theta;t) = E\{e^{-\theta Y(t)}\} \tag{3.5}$$

In what follows, we shall evaluate $L(\theta;t)$. To this end, the following result is needed. Consider a random variable U with values in [0,t] and the distribution function

$$P\{U \leq u\} = \frac{\Lambda(u)}{\Lambda(t)} \qquad 0 \leq U \leq t \tag{3.6}$$

Let $U_1, U_2 \ldots U_n$ be a random sample of size n from this distribution and denote by

$$U_{n1}, U_{n2}, \ldots U_{nn} \tag{3.7}$$

the order statistic from this sample ($U_{n1} \leq U_{n2} \leq \ldots \leq U_{nn}$). Write

$$h_n(u_1, u_2, \ldots, u_n)$$

for the joint probability density of the system (3.7); then, as is well known,

$$h_n(u_1, u_2, \ldots, u_n) = \frac{n!}{[\Lambda(t)]^n} \prod_{i=1}^{n} d\Lambda(u_i) \tag{3.8}$$

We can now formulate the following well known result which is needed to determine $L(\theta;t)$.

Lemma (Poisson)

Given that $\eta(t) = n$, the joint distribution of $(\tau_1, \tau_2, \ldots \tau_n)$ is that of the system (3.7). In other words,

$$P\{\tau_1 \in du_1, \tau_2 \in du_2, \ldots, \tau_n \in du_n | \eta(t) = n\} \tag{3.9}$$

$$= \frac{n!}{[\Lambda(t)]^n} \prod_{i=1}^{n} d\Lambda(u_i)$$

Using this lemma, we will proceed to evaluate the Laplace-Stieltjes transform $L(\theta;t)$. As a matter of fact, we have the following:

$$L(\theta;t) = \phi_0(\theta e^{-\gamma t}) e^{-\int_0^t [1 - \phi(\theta e^{-\gamma(t-s)})] d\Lambda(s)} \tag{3.10}$$

Denote by

$$v_0 = E\{V_0\} \qquad v = E\{V_1\} \tag{3.11}$$

Differentiating (3.10) twice with respect to θ, and keeping in mind that

$$E\{Y(t)\} = -L'(\theta;\ t)|_{\theta = 0}$$

$$E\{Y^2(t)\} = L''(\theta;\ t)|_{\theta = 0}$$

one obtains

$$E\{Y(t)\} = e^{-\gamma t}[v_0 + v\int_0^t e^{\gamma s}\, d\Lambda(s)] \tag{3.12}$$

$$E\{Y^2(t)\} = [E\{Y(t)\}]^2 + e^{-2\gamma t}[E(V_0{}^2) - v_0^2 + E(V_1{}^2)\int_0^t e^{2\gamma s}\, d\Lambda(s)]$$

In addition, one can show that for any $0 \le t_1 \le t_2 < \infty$

$$E\{Y(t_1)\, Y(t_2)\} = E\{Y^2(t_1)\} \cdot e^{-\gamma(t_2 - t_1)} \tag{3.13}$$

$$+ e^{-\gamma(t_1+t_2)}\{v_0 + v\int_0^{t_1} e^{\gamma u} d\,\Lambda(u)\}\{\int_{t_1}^{t_2} e^{\gamma u} d\,\Lambda(u)\}v$$

4. REMARKS ON EXCURSIONS OF Y(t)

Consider a certain threshold level x_0. In various problems pertinent to flood frequency analyses or planning the operation of a reservoir, we are interested in excursions of the process Y(t) above the level x_0. Of particular interest is the duration of such an excursion. In this section, we study this and related problems. For this purpose, we will have to point out some particular properties of the local maxima of Y(t).

According to its definition, the points of the local maxima of the stochastic process Y(t) are $0, \tau_1, \tau_2, \ldots$; the value of these maxima are:

$$Y_0 = Y(0) = V_0$$

$$Y_1 = Y(\tau_1) = V_0 e^{-\gamma\tau_1} + V_1$$

$$Y_2 = Y(\tau_2) = V_0 e^{-\gamma\tau_2} + V_1 e^{-\gamma(\tau_2 - \tau_1)} + V_2$$

and so on. In general, we have for every $n = 0, 1, 2, \ldots$

$$Y_n = Y(\tau_n) = \sum_{k=0}^{n} V_k e^{-\gamma(\tau_n - \tau_k)} \tag{4.1}$$

where by its definition $\tau_0 \equiv 0$

Denote by

$$\tau_1 = Z_1, \qquad \tau_n - \tau_{n-1} = Z_n \qquad n = 2, 3, \ldots \tag{4.2}$$

From (4.1), it follows then that

$$Y_n = e^{-\gamma Z_n}\, Y_{n-1} + V_n \tag{4.3}$$

Indeed, write the right hand side of (4.1) as

$$\sum_{k=0}^{n-1} V_k \exp\{-\gamma \sum_{j=k+1}^{n} Z_j\} + V_n = e^{-\gamma Z_n}(\sum_{k=0}^{n-2} V_k \exp\{-\gamma \sum_{j=k+1}^{n-1} Z_j\} + V_{n-1}) + V_n$$

$$= e^{-\gamma Z_n}\, [\sum_{k=0}^{n-1} V_k\, e^{-\gamma(\tau_{n-1} - \tau_k)}] + V_n$$

$$= e^{-\gamma Z_n}\, Y_{n-1} + V_n$$

This proves the assertion. Equation (4.3) represents a relation between two neighboring local extremes of $Y(t)$.

Denote by

$$G_n(x) = P\{Y_n \leq x\} \tag{4.4}$$

For $n = 0$, we have from (3.3) and (4.1) that

$$G_0(x) = H_0(x)$$

Taking into account (4.1), we can write

$$G_1(x) = P\{e^{-\gamma Z_1} V_0 + V_1 \leq x\}$$

$$= \int_0^x P\{e^{-\gamma Z_1} V_0 \leq x - u\} dH(u)$$

On the other hand, bearing in mind that V_0 and Z_1 are independent random variables, we have:

$$P\{e^{-\gamma Z_1} V_0 \leq x - u\} = \int_0^1 P\{V_0 \leq \frac{x-u}{s} | e^{-\gamma Z_1} = s\} dP\{e^{-\gamma Z_1} \leq s\}$$

$$= \int_0^1 H_0(\frac{x-u}{s}) \frac{1}{\gamma s} e^{-\Lambda(-\frac{1}{\gamma} \ln s)} \lambda(-\frac{1}{\gamma} \ln s) ds$$

where $d\Lambda(s)/ds = \lambda(s)$. Hence,

$$G_2(x) = \int_0^x \int_0^1 H_0(\frac{x-u}{s}) \frac{1}{\gamma s} e^{-\Lambda(-\frac{1}{\gamma} \ln s)} \lambda(-\frac{1}{\gamma} \ln s) ds\, dH(u) \tag{4.5}$$

To evaluate G_n for $n \geq 3$, we invoke the fact that $\{\tau_n\}_1^\infty$ is a homogeneous Markov process with transition probability function (3.4). From this, one can easily obtain ($s_0 = 0 \leq s_1 \leq s_2 \leq \ldots \leq s_n$)

$$P\{\tau_1 \in ds_1, \tau_2 \in ds_2, \ldots, \tau_n \in ds_n\} \tag{4.6}$$

$$= e^{-\Lambda(s_n)} \prod_{i=1}^{n} d\Lambda(s_i)$$

By virtue of this and (4.1), one can write

$$G_n(x) = P\{\sum_{k=0}^{n-1} V_k e^{-\gamma(\tau_n - \tau_k)} + V_n \leq x\}$$

$$= \int_0^\infty \int_0^{s_n} \ldots \int_0^{s_2} P\{\sum_{k=0}^{n-1} V_k e^{-\gamma(s_n - s_k)} + V_n \leq x\} e^{-\Lambda(s_n)} \prod_{i=1}^{n} d\Lambda(s_i)$$

On the other hand, if we denote by

$$H_0(x; s_n) = P\{V_0 e^{-\gamma s_n} \leq x\} \tag{4.7}$$

$$H(x; s_n - s_k) = P\{V_k e^{-\gamma(s_n - s_k)} \leq x\} \qquad k = 1, \ldots, (n-1) \tag{4.8}$$

we obtain

$$P\{\sum_{k=0}^{n-1} V_k e^{-\gamma(s_n - s_k)} + V_n\} = H_0(x; s_n) * H(x; s_n - s_1)$$

$$* \dots H(x; s_n - s_{n-1}) * H(x)$$

where $*$ stands for convolution. Hence, for $n = 3, 4, \dots$

$$G_n(x) = \int_0^\infty \int_0^{s_n} \dots \int_0^{s_2} H_0(x, s_n) * H(x, s_n - s_1) * \dots \tag{4.9}$$

$$*H(x; s_n - s_{n-1}) * H(x)\ e^{-\Lambda(s_n)} \prod_{i=1}^{n} d\Lambda(s_i)$$

Remark 4.1

It is not difficult to establish (see (4.7) and (4.8)) that

$$H_0(x; s_n) = H_0(x\, e^{\gamma s_n})$$

$$H(x; s_n - s_k) = H(x\, e^{\gamma(s_n - s_k)})$$

Consider now the following problem: Given that $Y(\tau_n) > x_0$, what can one say about the duration of the corresponding excursion of $Y(t)$ above x_0. Denote this duration by D_n (Fig. 6). It is clear that D_n is a non-negative random variable for every $n = 0, 1, \dots$. Denote by

$$\Pi_n(s; x_0) = P\{D_n \leq s | Y_n > x_0\} \tag{4.10}$$

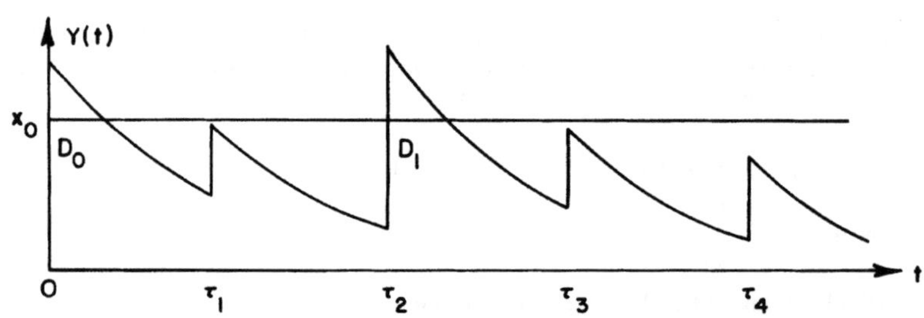

Fig. 6. Excursions of $Y(t)$ above the threshold x_0

We now embark on the problem of evaluating (4.10). Consider first the case, $n = 0$. Given that

$$Y(0) = V_0 > x_0$$

and, taking into account the definition of D_0, we have:

$$V_0\, e^{-\gamma D_0} = x_0$$

Hence

$$D_0 = \frac{1}{\gamma} \ln \frac{V_0}{x_0}$$

From this we have:

$$P\{D_0 \leq s \mid Y(0) > 0\} = P\{\frac{1}{\gamma} \ln \frac{V_0}{x_0} \leq s \mid V_0 > 0\}$$

$$= P\{V_0 \leq x_0 e^{\gamma s} \mid V_0 > x_0\}$$

$$= \frac{P\{x_0 < V_0 \leq x_0 e^{\gamma s}\}}{P\{V_0 > x_0\}} = \frac{H_0(x_0 e^{\gamma s}) - H_0(x_0)}{1 - H_0(x_0)}$$

Thus,

$$\Pi_0(s;\, x_0) = \frac{H_0(x_0 e^{\gamma s}) - H_0(x_0)}{1 - H_0(x_0)} \tag{4.11}$$

To evaluate the conditional distribution function, we proceed in the following fashion. Note first that, on the set $\{\eta(t) = n\}$

$$Y(t) = \sum_{k=0}^{n} V_k e^{-\gamma(t - \tau_k)}$$

Hence

$$Y(\tau_n + D_n) = \sum_{k=0}^{n} V_k\, e^{-\gamma(\tau_n + D_n - \tau_k)} = x_0$$

From this, it follows that

$$e^{-\gamma D_n} \sum_{k=0}^{n} V_k e^{-\gamma(\tau_n - \tau_k)} = x_0$$

or (see (4.1))

$$e^{-\gamma D_n}\, Y_n = x_0$$

Therefore,

$$D_n = \frac{1}{\gamma} \ln \frac{Y_n}{x_0} \tag{4.12}$$

By virtue of this, we have

$$P\{D_n \leq s \mid Y(\tau_n) > x_0\} = P\{\frac{1}{\gamma} \ln \frac{Y(\tau_n)}{x_0} \leq s \mid Y_n > x_0\}$$

$$= P\{Y_n \leq x_0\, e^{\gamma s} \mid Y_n > x_0\}$$

$$= \frac{P\{x_0 < Y_n \leq x_0\, e^{\gamma s}\}}{1 - P\{Y_n \leq x_0\}}$$

Hence, for every $n = 1, 2, \ldots$

$$\Pi_n(s;\, x_0) = \frac{G_n(x_0 e^{\gamma s}) - G_n(x_0)}{1 - G_n(x_0)} \tag{4.13}$$

5. EXTREME VALUES OF THE SHOT-NOISE PROCESS

Consider the stochastic process

$$\chi(t) = \sup_{0 \leq s \leq t} Y(s) \tag{5.1}$$

which represents the maximum value of Y(s) in a time interval $[o, t]$. Our principal aim in this section is to evaluate the distribution function

$$\phi(x;t) = P\{x(t) \leq x\} \tag{5.2}$$

Bearing in mind remarks of the previous section, the largest observation of the process Y(s) in [o,t] must be one of the local maxima in this interval. Thus, we have

$$\chi(t) = \sup_{0 \leq n \leq \eta(t)} Y(\tau_n) \tag{5.3}$$

On the basis of this, one can write

$$\phi(x;t) = P\{V_0 \leq x\}\ e^{-\Lambda(t)} + \sum_{n=1}^{\infty} P\{\sup_{0 \leq j \leq n} Y(\tau_j) \leq x | \eta(t) = n\} P\{\eta(t) = n\} \tag{5.4}$$

Consider now the conditional distribution under the summation sign. To evaluate it, we invoke the Poisson lemma presented in section 3 to obtain the following result:

$$P\{\sup_{o \leq j \leq n} Y(\tau_j) \leq x | \eta(t) = n\} \tag{5.5}$$
$$= P\{V_0 \leq x,\ Y_1(\tau_1) \leq x,\ .\ .\ .,\ Y_n(\tau_n) \leq x | \eta(t) = n\}$$
$$= P\{V_0 \leq x,\ Y_1(U_{n1}) \leq x,\ .\ .\ .,\ Y_n(U_{nn}) \leq x\}$$

where $U_{n1}, U_{n2}, . . ., U_{nn}$ are the order statistics with distribution (3.8). Hence,

$$P\{V_0 \leq x,\ Y_1(U_{n1}) \leq x,\ .\ .\ .,\ Y_n(U_{nn}) \leq x\} \tag{5.6}$$
$$= \int_o^t \int_{u_1}^t \cdots \int_{u_{n-1}}^t P\{V_0 \leq x,\ Y_1(u_1) \leq x, . . . ,Y_n(u_n) \leq x\} \frac{n!}{[\Lambda(t)]^n} \prod_{j=1}^n d\,\Lambda(u_j)$$

where

$$\begin{aligned} Y_1(u_1) &= V_0\ e^{-\gamma u_1} + V_1 \\ Y_2(u_2) &= V_0\ e^{-\gamma u_2} + V_1\ e^{-\gamma(u_2 - u_1)} \\ &- \\ Y_n(u_n) &= \sum_{k=o}^{n} V_k e^{-\gamma(u_n - u_k)} \qquad (u_0 \equiv 0) \end{aligned} \tag{5.7}$$

We now attempt to evaluate the distribution under integral (5.6). Taking into account assumptions (i) and (ii) from section 3, and the system of functions (5.6), we obtain for n = 1:

$$P\{V_0 \leq x,\ Y_1(u_1) \leq x\} = P\{V_0 \leq x,\ V_0\ e^{-\gamma u_1} + V_1 \leq x\}$$
$$= \int_o^x P\{v_0\ e^{-\gamma u_1} + V_1 \leq x\} dH_0(v_0)$$
$$= \int_o^x H(x - v_0\ e^{-\gamma u_1}) dH_0(v_0)$$

For n = 2, we have:

$$P\{V_0 \le x,\ V_0\, e^{-\gamma u_1} + V_1 \le x,\ V_0\, e^{-\gamma u_2} + V_1\, e^{-\gamma(u_2 - u_1)} + V_2 \le x\}$$

$$= \int_0^x P\{V_1 \le x - v_0\, e^{-\gamma u_1},\ V_1\, e^{-\gamma(u_2 - u_1)} + V_2 \le x - v_0\, e^{-\gamma u_2}\}\, dH_0(v_0)$$

$$= \int_0^x \int_0^{x - v_0 e^{-\gamma u_1}} P\{V_2 \le x - v_0\, e^{-\gamma u_2} - v_1\, e^{-\gamma(u_2 - u_1)}\}\, dH_0(v_0)\, dH(v_1)$$

$$\int_0^x \int_0^{x - v_0 e^{-\gamma u_1}} H(x - e^{-\gamma u_2}[v_0 + v_1\, e^{\gamma u_1}])\, dH_0(v_0)\, dH(v_1)$$

In general,

$$P\{V_0 \le x,\ Y_1(u_1) \le x,\ \ldots,\ Y_n(u_n) \le x\}$$

$$= \int_0^x \int_0^{x - v_0 e^{-\gamma u_1}} \int_0^{x - e^{-\gamma u_2}(v_0 + v_1 e^{\gamma u_1})} \cdots \int_0^{x - e^{-\gamma u_{n-1}}(v_0 + \sum_{j=1}^{n} v_j e^{\gamma u_j})}$$

$$\cdot\ H\Big(x - e^{-\gamma u_n}[v_0 + \sum_{j=1}^{n-1} v_j\, e^{\gamma u_j}]\Big)\, dH_0(v_0) \prod_{k=1}^{n} dH(v_k) \qquad (5.8)$$

Denote the last integral by

$$T_n(x,\ u_1,\ \ldots,\ u_n)$$

then the conditional distribution (5.5) can be written as

$$P\{\sup_{0 \le j \le n} Y(\tau_j) \le x \mid \eta(t) = n\}$$

$$= \frac{n!}{[\Lambda(t)]^n} \int_0^t \int_{u_1}^t \cdots \int_{u_{n-1}}^t T_n(x,\ u_1,\ \ldots\ u_n) \prod_{j=1}^{n} d\Lambda(u_i)$$

From this we obtain:

$$\phi(x;t) = e^{-\Lambda(t)}\Big\{H_0(x) + \sum_{n=1}^{\infty}\Big(\int_0^t \int_{u_1}^t \cdots \int_{u_{n-1}}^t T_n(x,\ u_1,\ \ldots\ u_n) \prod_{j=1}^{n} d\Lambda(u_j)\Big)\Big\} \qquad (5.9)$$

Form (5.9) of the distribution function $\phi(x;t)$ is rather cumbersome and difficult to evaluate effectively for practical purposes. This, of course, raises the question of determining a suitable approximation or examining another approach which may lead to different and, hopefully, simplified analytical expressions for $\phi(x;t)$.

In the sequel, we will outline briefly another approach which, under certain conditions, may prove to be quite useful. To this end, consider the bivariate sequence of random variables (see Fig. 5).

$$\{(\tau_n,\ Y_n)\}_1^{\infty}$$

This sequence represents a bivariate point process. We shall denote by $N^*(t,\Gamma)$ its counting random function, i.e.,

$$N^*(t,\Gamma) = \sum_{n=1}^{\infty} I_{\{\tau_n \leq t,\ Y_n \varepsilon\ \Gamma\}} \tag{5.10}$$

where Γ is a Borel subset of $R_+ = (0, \infty)$. The process $N^*(t,\Gamma)$ represents the number of points (τ_n, Y_n) in $[0, t] \times \Gamma$. The usefulness of $N^*(t,\Gamma)$ stems from the fact that

$$P\{\chi(t) \leq x\} = P\{N^*(t, \Gamma_x) = 0\} \tag{5.11}$$

where $\Gamma_x = (x, \infty)$.

We shall attempt to evaluate the probability $P\{N^*(t,\Gamma_x) = 0\}$ under the assumption that the following limit exists, and that

$$\lim_{\Delta t \to 0} \frac{P\{N^*(t + \Delta t, \Gamma_x) - N^*(t, \Gamma_x) = 1 | N^*(t, \Gamma_x) = 0\}}{\Delta t} \tag{5.12}$$

$$= \lambda(t, \Gamma_x)$$

To this end, consider

$$P\{N^*(t + \Delta t, \Gamma_x) = 0\} - P\{N^*(t, \Gamma_x) = 0\}$$

$$= P\{N^*(t, \Gamma_x) = 0,\ N^*(t + \Delta t, \Gamma_x) - N^*(t, \Gamma_x) = 0\}$$

$$- P\{N^*(t, \Gamma_x) = 0,\ N^*(t + \Delta t, \Gamma_x) - N^*(t, \Gamma_x) = 0\}$$

$$- P\{N^*(t, \Gamma_x) = 0,\ N^*(t + \Delta t, \Gamma_x) - N^*(t, \Gamma_x) = 1\} + o(\Delta t)$$

as $\Delta t \to 0$. From this, we obtain that

$$P\{N^*(t + \Delta t, \Gamma_x) = 0\} - P\{N^*(t, \Gamma_x) = 0\}$$

$$= P\{N^*(t + \Delta t; \Gamma_x) - N^*(t, \Gamma_x) = 1 | N^*(t, \Gamma_x) = 0\}\ P\{N^*(t, \Gamma_x) = 0\} + o(\Delta t)$$

Hence, by virtue of (5.12), we have:

$$\frac{\partial P\{N^*(t, \Gamma_x) = 0\}}{\partial t} = -\lambda(t, \Gamma_x)\ P\{N^*(t, \Gamma_x) = 0\} \tag{5.13}$$

A general solution of this differential equation is

$$P\{N^*(t, \Gamma_x) = 0\} = U(x) e^{-\int_o^t \lambda(s, \Gamma_x) ds} \tag{5.14}$$

From this and (5.11) we obtain:

$$\phi(x;t) = U(x) e^{-\int_o^t \lambda(s, \Gamma_x) ds} \tag{5.15}$$

To determine the unknown function $U(x)$, consider equation (5.9); for $t = 0$, we clearly have that

$$\phi(x;\ 0) = H_0(x) \tag{5.16}$$

On the other hand, for $t = 0$, we obtain from equation (5.15) that

$$\phi(x;\ 0) = U(x)$$

Taking into account this and relation (5.16), we finally have that:

$$\phi(x;\ t) = H_0(x) e^{-\int_o^t \lambda(s, \Gamma_x) ds} \tag{5.17}$$

6. GENERAL REMARKS

In this paper, we are concerned with a version of the classical shot-noise stochastic process as a streamflow model. We discuss some of its basic properties which may have potential interest in hydrological applications. Among the features that we investigate here are its covariance function and the duration of its excursions above a certain threshold level x_0. We have determined this function as well as the distributions of duration of its excursions. However, of particular interest in this paper are the extreme values of this process. We have obtained the distribution function of its maximum value in an arbitrary time interval [o,t] in a relatively simple form.

Our principal concern here is the physical basis for application of the process as a streamflow model. In section 3, we focus our attention on the physical conditions of a watershed (concerned mainly with its climate and geomorphological features) which stipulates a feasible use of the process as a model. In this respect, of central importance appears to be the set of the discontinuity points of the shot-noise process, which represents the building blocks of its entire structure. Their importance is further enhanced by the fact that they represent the occurrence times of those rainfall events which are sufficiently voluminous to generate an effective runoff. Section 3 provides a physically plausible explanation of why this set of points represents a Poisson point process.

Finally, we would like to make a few remarks concerning the distribution function $\phi(x;t)$ of the maximum value of the process in a time interval [o,t]. This distribution is clearly determined by two functions, $H_0(x)$, which is defined by the first equation in (3.3), and the $\lambda(t,\Gamma_x)$. According to its definition (5.12), one may easily infer that $\lambda(t,\Gamma_x)$ is endowed with the following properties: for every $t > 0$

$$\lambda(t,\Gamma_x) \to 0 \qquad \text{as } x \to \infty$$

$$\lambda(t,\Gamma_x) \to \lambda(t) \qquad \text{as } x \to 0$$

where

$$\Lambda(t) = \int_0^t \lambda(s)ds$$

In this respect, it seems feasible to assume that $\lambda(t,\Gamma_x)$ can be written as:

$$\lambda(t,\Gamma_x) = \lambda(t) \cdot \alpha(x)$$

where clearly $0 \leq \alpha(x) \leq 1$, and

$$\alpha(x) \to 1 \quad \text{as } x \to 0 \text{ while } \alpha(x) \to 0 \quad \text{as } x \to \infty$$

Since $\alpha(x)$ is strictly decreasing, it might be plausible, as the first approximation, to suppose that

$$\alpha(x) = e^{-x}$$

In this case, (5.17) becomes

$$\Phi(x;t) = H_0(x)\, e^{-\Lambda(t)} \cdot e^{-x}$$

The utility of this distribution function, and the distribution function of the duration of excursions above a threshold, remains to be demonstrated by case studies.

7. NOTATION

$\xi(t)$ = Precipitation intensity at time t.
τ_n^* = Termination time of n-th rainfall event.
R_+ = $[0,\infty)$.
$N(t)$ = Number of τ_k^* in $([o,t]$.
D_n' = Duration of n-th rainfall event.
U_n = $\tau_{n+1}^* - D_n' - \tau_n^*$.
$\eta(t)$ = Counting function of a Poisson Process.
τ_n = A point of $\eta(t)$.
Z_n = $\tau_n - \tau_{n-1}$
$\Lambda(t)$ = $E\{\eta(t)\}$.
$Y(t)$ = Shot-noise process.
V_0 = $Y(0)$
V_n = Magnitude of n-th discontinuity of $Y(t)$
$H_0(x)$ = Distribution function of V_0.
$H(x)$ = Distribution function of V_n.
$I_{\{a,A\}}$ $\begin{cases} = 1 \text{ if } a \in A \\ = 0 \text{ if } a \notin A \end{cases}$ indicator function.
$\phi_0(\theta)$ = Laplace - Stieltjes transform of V_0.
$\phi(\theta)$ = Laplace - Stieltjes transform of V_n.
$L(\theta;t)$ = Laplace - Stieltjes transform of $Y(t)$.
v_0 = $E(V_0)$.
v = $E(V_n)$.
Y_n = $Y(\tau_n)$.
$G_n(x)$ = Distribution function of Y_n.
D_n = Duration of n-th excursion of x_0 by $Y(t)$.
$\Pi_n(s;x_0)$ = Conditional distribution of D_n given that $Y_n > x_0$.
$\chi(t)$ = Maximum of the shot-noise process in $[0,t]$.
Γ_x = (x,∞).
$\phi(x;t)$ = Distribution function of $\chi(t)$.

8. ACKNOWLEDGEMENTS

This work was carried out when P. Todorovic was Visiting Professor, Department of Hydrology and Water Resources, University of Arizona, Tucson, AZ. We wish to thank J. J. Rocha for his skill and patience in typing this manuscript.

9. REFERENCES

Belyaev, K.Yu. 1963. 'Limit Theorems for Dissipative Flows'. Theory of Probability and Applications, Vol. 8, pp. 155-173, 196.

Bernier, J. 1970. 'Inventaire des Modèles de Processus Stochastiques Applicables à lar Decription des Débits Journaliers des Rivières'. Reviews of the International Statistical Institute, Vol. 38, pp. 49-61, 1970.

Cordova, J.R. and Rodriguez-Iturbe, I. 1983. 'Geomorphoclimatic Estimation of Extreme Flow Probabilities'. Journal of Hydrology, Vol. 65, pp. 159-173.

Diaz-Granados, M.A., Valdes, J.B. and Bras, R.L. 1984. 'A Physically Based Flood Frequency Distribution'. Water Resources Research, Vol. 20, pp. 995-1002.

Eagleson, P.S. 1971. 'The Stochastic Kinematic Wave'. In: V. Yevjevich (editor), Systems Approach to Hydrology, pp. 210-225, Water Resources Publications, Fort Collins, Colorado,

Eagleson, P.S. 1972. 'Dynamics of Flood Frequency'. Water Resources Research, Vol. 8, pp. 878-898.

Fogel, M.M., Duckstein, L. and Sanders, J.L. 1974. 'An Event-based Stochastic Model of Areal Rainfall and Runoff'. Proceedings of the Symposium on Statistical Hydrology, USDA ARS Miscellaneous Publication No. 1275, pp. 247-261.

Hebson, C. and Wood, E. 1982. 'A Derived Flood Frequency Distribution Using Horton Order Ratios'. Water Resources Research, Vol. 18, pp. 1509-1518.

Kallenberg, O. 1975. 'Limits of Compound and Thinned Point Processes'. Journal of Applied Probability, Vol. 8, pp. 269-278.

Koch, R.W. 1982. 'A Physical-probabilistic Approach to Stochastic Hydrology'. Ph.D. Thesis, 187 pp., Colorado State University, Fort Collins, Colorado.

Natural Environment Research Council. 1975. Flood Studies Report, Vol. 1, pp. 219-234, Hydrological Studies, Natural Environment Research Council, 27 Charing Cross Road, London,

Rodriguez-Iturbe, I. and Valdes, J.B. 1979. 'The Geomorphologic Structure of Hydrologic Response'. Water Resources Research, Vol. 15, pp. 1409-1420.

Rodriguez-Iturbe, I., Gonzalez M. and Bras, R.L. 1982. 'A Geomorphoclimatic Theory of the Instantaneous Unit Hydrograph'. Water Resources Research, Vol. 18, pp. 877-886.

Todorovic, P. 1983. 'Topics in Stochastic Point Processes with Applications In Hydrology'. Applied Probability in Biology and Engineering. An ORSA/TIMS Special Interest Meeting, University of Kentucky, Lexington, Kentucky, 18-20 July, 1983.

Weiss, G. 1973. 'Filtered Poisson Processes as Models for Daily Streamflow Data'. Ph.D. Thesis, Imperial College of Science and Technology, London, England.

Weiss, G. 1977. 'Shot Noise Models for the Generation of Synthetic Streamflow Data'. Water Resources Research, Vol. 13, pp. 101-108.

Westcott, M. 1976. 'Simple Proof of a Result on Thinned Point Processes'. Annals of Mathematical Statistics, Vol. 4, No. 1, pp. 89-90.

Woolhiser, D.A. and Todorovic, P. 1974. 'A Stochastic Model of Sediment Yield for Ephemeral Streams'. Proceedings of the Symposium on Statistical Hydrology, USDA ARS Miscellaneous Publication No. 1275, pp. 295-308.

RESPONSE CHARACTERISTICS OF TWO TROPICAL RIVER BASINS

Masahiko HASEBE, Associate Professor
Department of Civil Engineering,
University of Utsunomiya,
Ishii , Utsunomiya, 321 ,JAPAN
and
Mikio HINO, Professor
Department of Civil Engineering,
Tokyo Institute of Technology,
O-okayama, Meguroku, Tokyo, 152, JAPAN

ABSTRACT. In this paper, we extend our method (Hino and Hasebe,1979,1981) of hydrologic inverse detection originally for runoff data analysis to the hydrologic basin in tropical zone.

Applications of this method (Filter Separation Autoregressive Model Method) have been carried out on the daily hydrologic data of watershed of the Quai Yai River with drainage area of $A = 10800$ km^2 in Thailand and watershed of the Pampanga River with a drainage area of $A = 7850$ km^2 in Philippines.

1. INTRODUCTION

Generally, identification of the response characteristics of a system requires the data of both the input and the output from the rainfall - runoff system. Therefore, most of the hydrologic system analysis hitherto proposed have been developed by making use of the time series of both rainfall and runoff.

In this paper, we attempt to derive, from only daily data, not only the response characteristics of hydrologic subsystems (overland flow, interflow and groundwater flow), but also the effective daily rainfall components to the subsystems. The key to this method lies in the fact that the input (daily rainfall) is of white noise and the physical process in the hydrologic subsystems are well understood although they were only conceptual. The procedure of the present method is summarized as follows :

(1) Daily runoff time series are separated into time series of overland flow, interflow and groundwater flow through numerical filters whose cut off frequencies have been determined from the Autoregressive (AR) coefficient and / or from the coherence gap between rainfall and runoff.

V. P. Singh (ed.), Flood Hydrology, 165–174.
© 1987 by D. Reidel Publishing Company.

(2) Response characteristics of hydrologic subsystem (overland flow, interflow and groundwater flow) have been identified by fitting the AR models to the time series of filtered runoff components.

(3) Daily rainfall time series have been estimated as the random input to the fitted AR model.

2. SEPARATION OF THE RAW RUNOFF TIME SERIES INTO COMPONENT

2.1 Estimation of cut off frequencies.

The values of autoregressive coefficients may be used to estimate the separation frequency. When the runoff time series is fitted to an AR model, the absolute values of the AR coefficients generally decrease gradually. At a certain order, they decrease abruptly, showing that the primary runoff component is composed of a runoff component with shorter period than this particular order. The AR coefficients of the raw time series of the daily discharge are calculated. Thus, the separation periods (cutoff frequencies) are Tc = 10 and 4 days for the Quai Yai and Pampanga Rivers, respectively.

2.2 Numerical filtering of runoff time series.

To design a numerical filter, we utilized the response characteristics of a simulated mass - spring - dashpot system expressed by :

$$\frac{d^2y^{(1)}}{d\,t^2} + c_1\frac{dy^{(1)}}{dt} + c_0y^{(1)} = c_0y \tag{1}$$

which passes only the low frequency component, acting as a high frequency cutoff filter, whose response characteristics are given, when $c_1^2/4 - c_0 \geq 0$,

$$h(\tau) = \begin{cases} c_0\exp(-c_1\tau/2)\sinh\{(c_1^2/4-c_0)^{1/2}\tau\}/(c_1^2/4-c_0)^{1/2} & (\tau\geq0) \\ 0 & (\tau<0) \end{cases} \tag{2}$$

The frequency domain characteristics of $h(\tau)$ is written as :

$$|\chi(f)|^2 = \{(1-(f/f_c)^2)^2 + \delta^2(f/f_c)^2\}^{-1} \tag{3}$$

where

$$\left.\begin{aligned} f_c &= \sqrt{c_0}/2\pi \\ \delta &= c_1/2\pi f_c = c_1/\sqrt{c_0} \end{aligned}\right\} \tag{4}$$

The numerical filter contains two parameters c_0 and c_1 .
For a large lag time, $h(\tau)$ is approximated by :

$$h(\tau) = C\exp\{-(c_0/c_1)\tau\} \tag{5}$$

where

$$C = \tfrac{1}{2}c_0\left(\tfrac{1}{4}c_1^2 - c_0\right)^{-1/2}$$

Therefore, we obtain :

$$\left.\begin{aligned} Tc &= c_1/c_0 \\ f_c &= \delta/(2\pi Tc) \end{aligned}\right\} \qquad (6)$$

On the other hand, referring to the frequency response function , one can choose arbitrarily a value for the damping factor δ . In our case, we used a value δ = 2.1-2.5. Consequently, the values of c_0 and c_1 are determined as follows.

$$\left.\begin{aligned} c_0 &= (\delta/Tc)^2 \\ c_1 &= \delta^2/Tc \end{aligned}\right\} \qquad (7)$$

The filtered time series for the groundwater flow component $y^{(1)}$ is given.

$$y^{(1)}(t) = \alpha\Sigma w_k y(t-k) \qquad (8)$$

in which

$$w_k = \begin{cases} h(k\Delta t) & k = 0,1,2,\ldots\ldots\ldots,N \\ 0 & k = -1,-2,\ldots\ldots\ldots\ldots \end{cases} \qquad (9)$$

A weigting factor α should be chosen so as to satisfy the condition that the residual outputs

$$y^{(2)}(t) = y(t) - y^{(1)}(t) \qquad (10)$$

should not be negative.

2.3 Separation of runoff time series into components

As the cutoff parameter, c_0 and c_1 , are determined, the observed hydrologic data are separated into components by the numerical filter of eq.(9).

Fig.1 and 2 show the hydrograph of groundwater flow $y^{(1)} = y_1$ and interflow $y^{(2)} = y_2$ of the Quai Yai River and the Pampanga River.

In the case treated in this paper, the raw hydrographs of two watersheds have been separated into two components (interflow and groundwater flow).

3. HYDROLOGIC RESPONSE CHARACTERISTICS OF SUBSYSTEMS

3.1 Response function of subsystem

As mentioned already, the time series of each subsystem is calculated by using high frequency numerical filters. By assuming the rainfall to be a white noise, the separated runoff components are

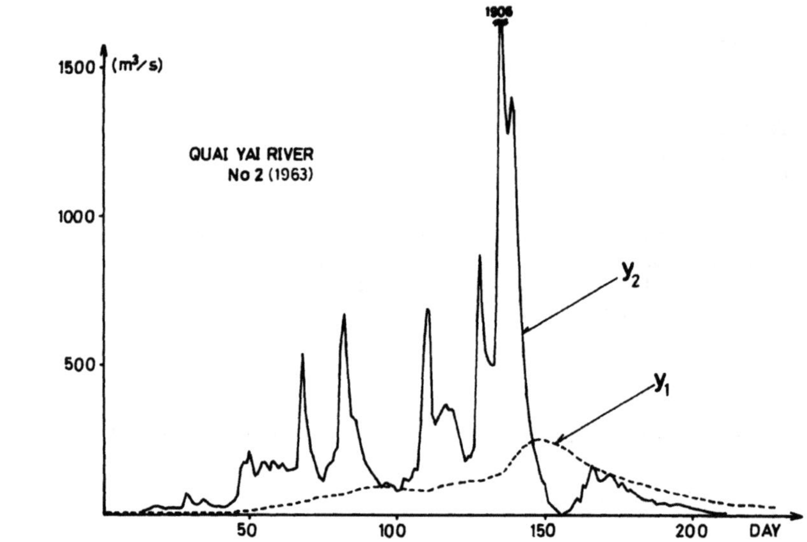

Figure 1. Time series of hydrographs of groundwater flow and interflow of the Quai Yai River (1963).

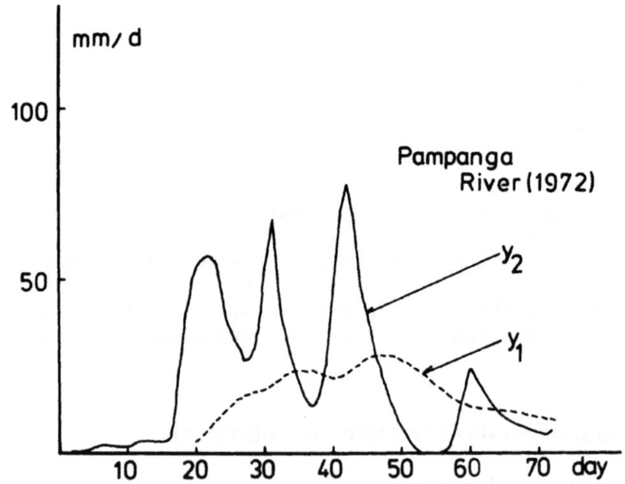

Figure 2. Time series of hydrographs of groundwater flow and interflow of the Pampanga River (1972).

fitted by the autoregressive process (AR).

$$y_t^{(\ell)} = a_1^{(\ell)} y_{t-1}^{(\ell)} + a_2^{(\ell)} y_{t-2} + \dots + a_p^{(\ell)} y_{t-p}^{(\ell)} + \varepsilon_t^{(\ell)} \tag{11}$$

where superscript (l) stands for the runoff components (i.e. l = 1 ; groundwater flow, l = 2 ; interflow and l = 3 ; overland flow), $a^{(\ell)}$ is the autoregressive coefficients and $\varepsilon^{(\ell)}$ is random noise series. A white noise ε is proportional to the daily rainfall component of the l -th subsystem. When the AR coefficients have been determined, the hydrologic response function is calculated by eq.(12)(Hino and Hasebe,1981).

$$\left.\begin{aligned} h_0^{(\ell)} &= 0 \\ h_1^{(\ell)} &= 1/\beta^{(\ell)} \\ &\vdots \\ h_m^{(\ell)} &= \Sigma h_{m-j}^{(\ell)} a_j^{(\ell)} \qquad (m>1) \end{aligned}\right\} \tag{12}$$

where $\beta^{(\ell)}$ is the conversion factor. This factor $\beta^{(\ell)}$ for daily analysis is given as follows.

$$\beta^{(\ell)} = 86.4/\{A.(1-a_1^{(\ell)}-a_2^{(\ell)}-\dots\dots-a_p^{(\ell)})\} \tag{13}$$

in which A is the drainage area expressed in unit of km^2 .

3.2 Response characteristics of the groundwater subsystem of real basins (Quai Yai and Pampanga Rivers).

The AR coefficients thus determined have been converted into the response function by eq.(12). The response functions (fourth order ;p=4) of groundwater flow of the Quai Yai and Pampanga Rivers are shown in Fig.3 and Fig.4. From these graphs, the groundwater flow of the Quai Yai River has a peak discharge at about 9 day lag and ends within 50 days. On the other hand, in the Pampanga River, the peak discharge occurs at a lag of 7-8 day and runoff ends within 50 days.

The response characteristics of the groundwater subsystem of the Quai Yai River basin is invariant regardless of the rainfall intensity distribution the same as the river basin tested before in Japan.

3.3 Response characteristics of the interflow subsystem of real basins (Quai Yai and Pampanga Rivers).

Time series of interflows are obtained by subtracting the high frequency cutoff filtered time series from the raw runoff time series. The AR coefficients (second order ;p=2) of the interflow component of the Quai Yai River and the AR coefficients (third order ;p=3) of the interflow component of the Pampanga River are estimated, respectively. The AR coefficients have been converted into response function as the same as the groundwater subsystem. The response functions of interflow subsystem are also shown in Fig.3 and Fig.4.

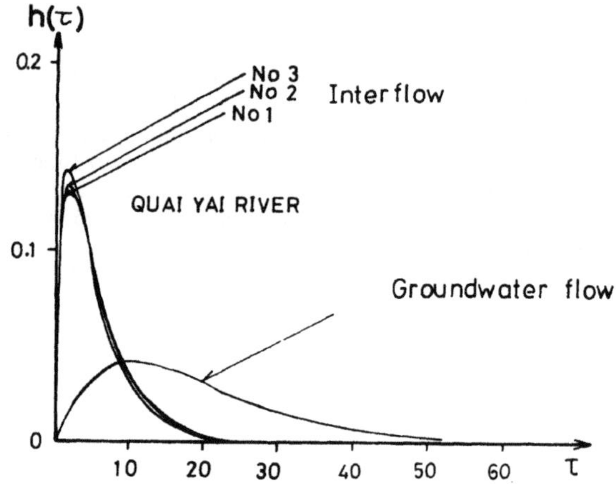

Figure 3. Response functions of groundwater subsystem and interflow subsystem of the Quai Yai River.

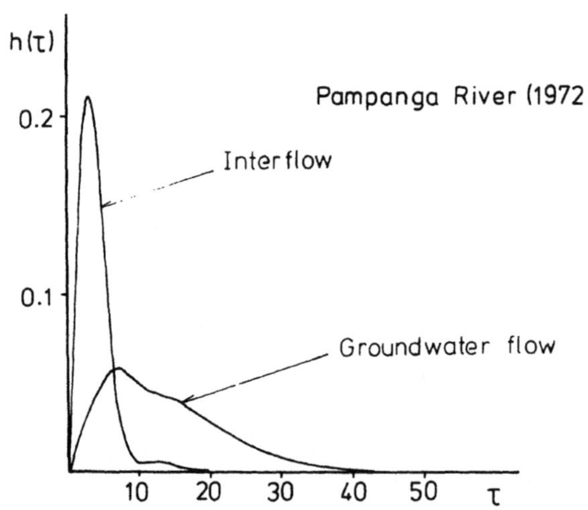

Figure 4. Response functions of groundwater subsystem and interflow subsystem of the Pampanga River.

Fig.3 indicates that the response functions of interflow for three cases of the Quai Yai River basin have peaks discharge at a lag 3 day ending within 20 day. The response characteristics of interflow subsystem in the Quai Yai River basin is also invariant from Fig.3. On the other hand, the graph of Fig.4 indicates that the response characteristics of the interflow subsystem of the Pampanga river has a peak discharge at 3 or 4 day and ends within 20 day.

4. INVERSE ESTIMATION OF EFFECTIVE RAINFALL

After the AR coefficients of subsystems, the time series of the rainfall component $\hat{x}_t^{(\ell)}$ are inversely estimated as proportional to random noise input ε_{t-1} by eq.(14), which is transformed from eq.(11).

$$\hat{x}_{t-1}^{(\ell)} = \beta^{(\ell)}(y_t^{(\ell)} - a_1^{(\ell)} y_{t-1}^{(\ell)} - \cdots\cdots\cdots - a_p^{(\ell)} y_{t-p}^{(\ell)}) \qquad (14)$$

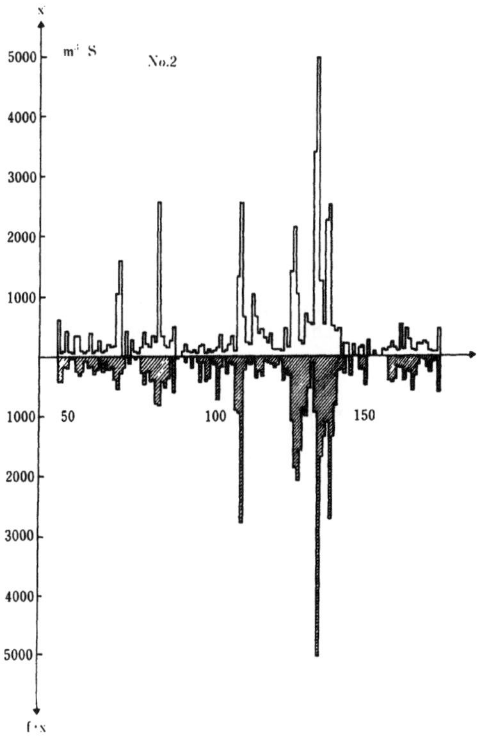

Figure 5. Comparison of time series of the inversely estimated rainfall with the observed effective rainfall (the Quai Yai River, 1963).

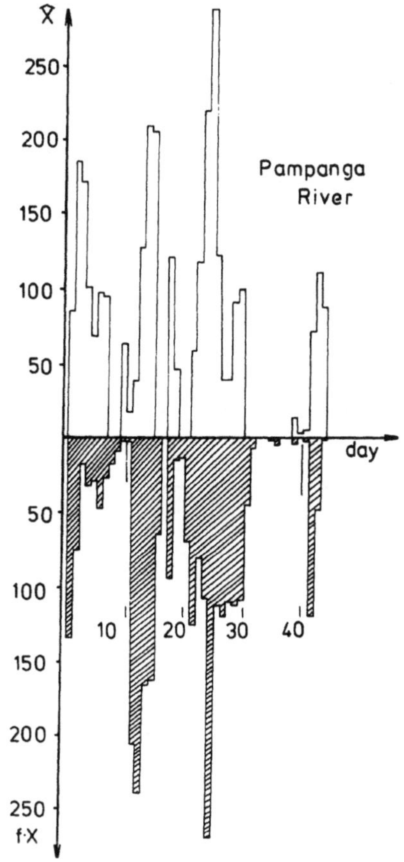

Figure.6 Comparison of the time series of the inversely estimated rainfall with the observed effective rainfall (the Pampanga River , 1972).

The total effective rainfall estimated by this present method is given by eq.(15).

$$\hat{x}_t = \hat{x}_t^{(1)} + \hat{x}_t^{(2)} \tag{15}$$

Fig.5 and Fig.6 are the comparison of the inversely estimated composite rainfall $\hat{x}_t$ with the observed effective rainfall x_t which are obtained by multiplying the precipitation X with the observed total rainfall / runoff ratio, i.e. the runoff coefficient f ($x_t = fX_t$). As shown in these figures, the time series of rainfall inversely estimated from the daily discharge compare well with the time series of effective rainfall.

In order to check the tentatively fitted model, the runoff time series is again computed from the estimated effective rainfall $\hat{x}^{(\ell)}$ and the AR coefficients. The runoff time series $\hat{y}^{(\ell)}$ is recomposed by eq.(16).

$$\hat{y}_t^{(\ell)} = a_1^{(\ell)}\hat{y}_{t-1}^{(\ell)} + a_2^{(\ell)}\hat{y}_{t-2}^{(\ell)} + \ldots\ldots + a_p^{(\ell)}\hat{y}_{t-p}^{(\ell)} + \hat{x}_{t-1}^{(\ell)}/\beta^{(\ell)} \tag{16}$$

Fig.7 and Fig.8 indicate, in the upper part, the effective rainfall $\hat{x}^{(\ell)}$ inversely estimated and, in the lower part, the observed runoff $y^{(\ell)}$ separated by eq.(9) compared with the runoff series $\hat{y}^{(\ell)}$ recomposed by eq.(16).

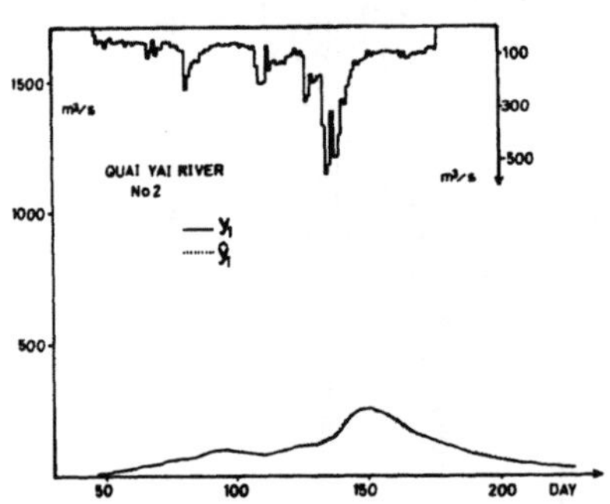

Figure 7. Time series of the inversely estimated rainfall component shown in the upper parts and those of the recomposed runoff estimation compared with the filtered observed data of groundwater flow of the Quai Yai River.

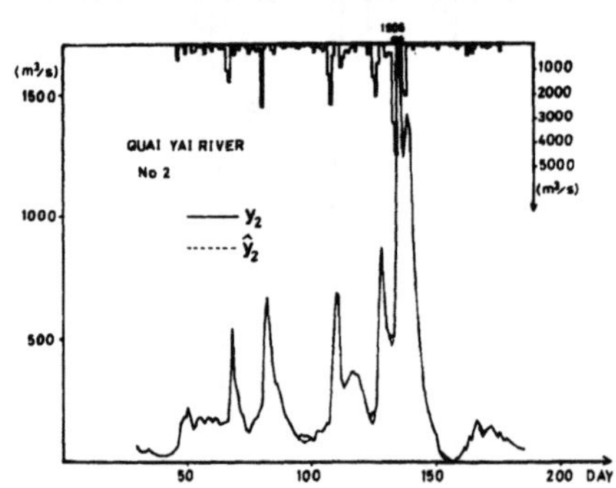

Figure 8. Time series of the inversely estimated rainfall component shown in the upper parts and those of the recomposed runoff estimation compared with the filtered observed data of interflow of the Quai Yai River.

5. NONLINEAR SEPARATION LAWS OF DAILY RAINFALL INTO COMPONENT

By plotting the daily rainfall component to a subsystem, for instance, from the groundwater rainfall component plotted against the corresponding daily total rainfalls,the separation law of daily rainfall into component is derived (Fig.9). A law of rainfall separation into components may be estimated from the method proposed in this paper, which is nonlinear as shown in Fig.9.

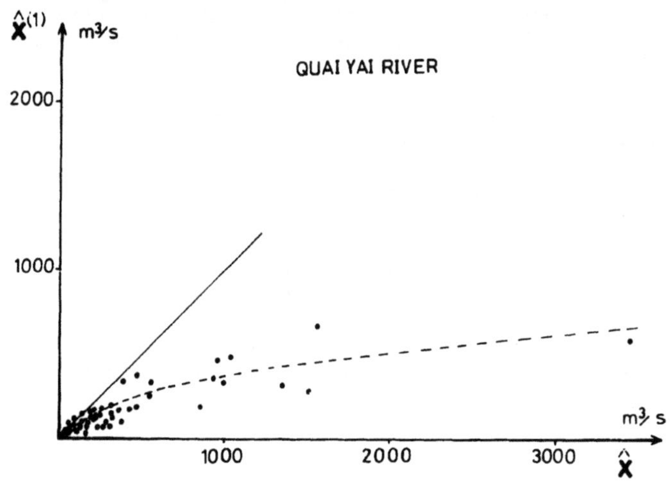

Figure 9. Separation law of rainfall.

In this respect, we assert that the nonlinearlity of hydrologic system may be caused mainly by the nonlinearity of this separation law of rainfall.

6. CONCLUSIONS

The conclusions of this analysis are as follow :

(1) Three floods of the Quai Yai River and the Pampanga River were separated into two runoff components, interflow and groundwater flow.

(2) The cutoff frequencies for separating runoff into components were 0.1 cycle / day (Quai Yai River) and 0.25 cycle / day (Pampanga River), respectively.

(3) The orders of the AR model fitted to the groundwater flow were four (p=4) both in the Quai Yai River and in the Pampanga River, respectively. While those of interflow were two (p=2 ; Quai Yai River) and three (p=3 ; Pampanga River).

(4) The response characteristics of the groundwater subsystem from the Quai Yai River basin has a peak discharge at 9 or 10 day and ends within 50 day. On the other hand, in the Pampanga River, the peak discharge occurs at a lag of 7 or 8 day and runoff ends within 50 day.

(5) The response functions of interflow for three cases of the Quai Yai River basin have been nearly identical forms with the peak discharge at a lag 3 day ending within 20 day. On the other hand, the response characteristics of the interflow subsystem of the Pampanga River has a peak discharge at 3 or 4 day and ends within 20 day.

(6) The response characteristics (response function) of the groundwater flow of the Quai Yai River basin is invariant regardless of the rainfall intensity distribution and those of the interflow subsystem in this river basin are also invariant.

(7) Time series of the daily rainfalls inversely estimated from the daily runoff compare relatively well with the effective precipitation time series of the observed data.

(8) A law of rainfall separation into components may be estimated from the method proposed in this paper, which is nonlinear as shown in Fig.9.

REFERENCES

Hino,M.,1970. *Runoff forecasts by linear predictive filter*.J.Hydraul.Div.,Proc.Am.Soc.Civ.Eng.,**96** (HY3):pp 681 - 702.

Hino,M and Hasebe,M.,1979. *A hydrologic inverse problem - Analysis of rainfall - runoff characteristics and separation of runoff components from the discharge sequences*. Proc. 23rd Jpn. Conf. on Hydraulics. Jpn.Soc.Civ.Eng.,pp 193 - 198.

Hino,M and Hasebe,M.,1981. *Analysis of hydrologic characteristics from runoff data - A hydrologic inverse problem*.J.Hydrol.,**49**:pp 287 - 313.

Hino,M and Hasebe,M.,1984. *Identification and prediction of nonlinear hydrologic systems by the filter separation autoregressive (AR) method : extension to hourly hydrologic data*.In:G.E.Stout and G.H.Davis (Editors),Global Water:Scince and Engineering - The Ven Te Chow Memorial Volume. J.Hydrol.,**68**:pp 181 - 210.

Hasebe,M and Hino,M.,1984. *Runoff analysis of a tropical river basin.* 4th IAHR International symposium on stochastic hydraulics, Illinois, U.S.A, pp 245 - 254.

Hino,M and Hasebe,M., 1985. *Analysis of hydrologic rainfall-runoff system.*,Morikita Publishing Co. Tokyo (in Japanese).

SYNTHETIC FLOW GENERATION WITH STOCHASTIC MODELS

Lankeswara H. Wijayaratne, Paul C. Chan
Department of Civil Engineering
New Jersey Institute of Technology
Newark, New Jersey 07102, U.S.A.

ABSTRACT. A first-order Markovian model was proposed based on the Three-parameter Log Normal distribution. This model was comparatively used with the Thomas-Fiering models, one formulated using Normal distribution and the other with Gamma distribution. Primary objectives of the study were: (1) to examine the satisfactory applicability of the proposed model, and (2) to investigate the importance of preserving the skewness in streamflow generation. In the methodology, each model was used to generate sequences of annual mean flows for six watersheds. Average estimates of the statistical parameters were computed for the generated sequences and then compared with the historic values. To determine the effect of skewness, the probable maximum annual flows were estimated for three return periods using generated sequences and the magnitudes due to different models were compared. The results of the study indicate the importance of preserving the skewness in flow generation. In addition, it reveals that the proposed model can be satisfactorily used.

INTRODUCTION

Synthetically generated flows have many uses to the water resources planner. They are of equal importance as historic flows in simulation and optimization schemes used to study several feasible alternatives of planning, design and operation of water resources projects.

The credit for initial work on streamflow generation probably goes to Hazen (1914), who obtained an extended sequence by combining the annual flows from fourteen different streams. Sudler (1927), Barnes (1954), Thomas and Fiering (1962), Matalas (1967) were among other researchers who made significant contributions to this field. Although a number of models are available, only the Thomas-Fiering model is reviewed in this paper for comparative application with the proposed model.

The primary objectives of this paper are twofold: first, to examine the satisfactory applicability of the proposed model and second, to investigate the importance of preserving the historic

V. P. Singh (ed.), Flood Hydrology, 175–185.
© 1987 by D. Reidel Publishing Company.

skewness in the generation of streamflows. The proposed model is a form of the first-order Markov model and uses the Three-parameter Log Normal distribution which preserves the mean, variance and skewness of the historic sequence. The objectives were achieved by comparative application of the proposed model with the two forms of the Thomas-Fiering model, one uses the Normal distribution, whereas the other is based on the Gamma distribution. In application, the three models were used to generate annual mean flows for six gaging stations located in different watersheds in New Jersey. The mean, variance, skewness and lag-one autocorrelation of the generated flow sequences were compared with the corresponding parameters of the historic flow sequence. Also, the probable maximum annual flows were estimated for three different return periods by using the properties of the generated flow sequences. The results due to different models were compared to illustrate the significance of skewness.

Thomas-Fiering (T-F) Model

The Thomas-Fiering model was developed in the 1960's (Thomas and Fiering, 1962) and has been widely used by hydrologists and engineers in synthetic flow generation schemes. The model assumes that the entire influence of the past on the current flow is reflected in the previous flow value. The basic model for annual flow generation can be written as:

$$X_i=\mu_x+\rho_1(x)(X_{i-1}-\mu_x)+\varepsilon_i\sigma_x[1-\rho_1^2(x)]^{\frac{1}{2}} \tag{1}$$

where X_{i-1}, X_i are the flows in (i-1)th and i th years, respectively. Parameters μ_x, σ_x and $\rho_1(x)$, respectively, represent the mean, standard deviation and lag-one autocorrelation of the historic annual flows. ε_i is a normal random variate that has zero mean and unit variance, which is independent of X_{i-1}.

Therefore, the model given by equation (1) preserves the mean, variance and the lag-one autocorrelation of the historic flows, but neglects the skewness. For convenience of reference, this model is referred to as T-F(basic) model in this paper.

In order to preserve the skewness of historic flows, it is suggested that the random variate ε_i should be generated by using the following equations (2) and (3), so that the generated flows are approximately gamma distributed (Fiering and Jackson, 1971).

$$\gamma_\varepsilon=\gamma_x[1-\rho_1^3(x)]/[1-\rho_1^2(x)]^{3/2} \tag{2}$$

$$\varepsilon_i=2/\gamma_\varepsilon[(1+\gamma_\varepsilon.t_i/6-\gamma_\varepsilon^2/36)^3-1] \tag{3}$$

The coefficient γ_ε is introduced as the skewness of the random component of each flow event. γ_x represents the skewness coefficient of the historic flows, and t_i is a random deviate from the standard normal population.

The model formulated by equations (1), (2) and (3) generates flows by preserving all mean, variance, skewness and lag-one autocorrelation of the historic flows. This model is referred to as T-F(skewed) model in this paper.

The Proposed Model

The probability density function of the Three-parameter Log Normal (LN3) distribution can be given as:

$$f(x;a,c,k)=[\sqrt{2\pi}k(x-a)]^{-1}\exp\{-\tfrac{1}{2}[\tfrac{1}{k}\ \ln(\{x-a\}/c)]^2\}$$

where, ln represents the natural logarithmic function and the parameters a,c and k are the distribution parameters. Using the method of moments, it can be shown that the following set of equations (4) thru (8) can be used to solve for these parameters (Hoshi et.al., 1984).

$$\text{Let}\quad u = 1+\gamma_x^2/2 \tag{4}$$

$$\phi = [u+(u^2-1)^{\frac{1}{2}}]^{\frac{1}{3}}+[u-(u^2-1)^{\frac{1}{2}}]^{\frac{1}{3}}-1 \tag{5}$$

$$\text{Then}\quad k = [\ln\phi]^{\frac{1}{2}} \tag{6}$$

$$c = \sigma_x/[\phi(\phi-1)]^{\frac{1}{2}} \tag{7}$$

$$a = \mu_x-c\phi^{\frac{1}{2}} \tag{8}$$

where μ_x, σ_x and γ_x represent the population estimates of mean, standard deviation and skewness coefficient respectively.

As the LN3 distribution preserves the first three statistical moments, it is coupled with the lag-one Markov process to propose the following model.

$$Y_i=\rho_1(y).Y_{i-1}+\varepsilon_i[1-\rho_1^2(y)]^{\frac{1}{2}} \tag{9}$$

$$X_i=a+c\ \exp(kY_i) \tag{10}$$

where Y_{i-1} and Y_i are the normalized flows in (i-1)th and ith periods, respectively. By knowing the observed flow value X_{i-1}, Y_{i-1} can be computed as follows:

$$Y_{i-1}=(X_{i-1}-\mu_x)/\sigma_x \tag{11}$$

$\rho_1(y)$ is the lag-one autocorrelation of the normalized historic flows.

Therefore, the proposed model should preserve the historic mean, variance and skewness in the real domain and the lag-one autocorrelation in the Normal domain. However, it is a question whether it can preserve the lag-one autocorrelation in the real domain, when the normalized flows are transformed by equation (10). This fact will be examined by the application of the model to generate synthetic flows.

A major advantage of the proposed model is the computational efficiency. In comparing with the T-F(skewed) model, the proposed model comprises of two computational steps thus allowing the computer generated normal random deviate to be directly usable in equation (9). Another advantage is that it avoids the transformation of real data into the log domain in parameter estimation. Also, it should satisfactorily perform in generating low flows, mainly because the difference between small numbers are significantly enlarged on the log scale.

APPLICATIONS

The three stochastic models were used to generate annual mean flows for six selected watersheds in New Jersey. Each gaging station has a record of over 60 years historic data as furnished by the U.S. Geological Survey. For the purpose of this study, the annual mean flows in the latest 60-year period (1926-1985) were selected and subject to the following analytical procedure.

(1) For each historic sequence, sample estimates of mean($\bar{X}$), standard deviation (S_x), coefficient of skewness (C_{sx}) and lag-one autocorrelation ($r_1(x)$) were computed. Sampling errors were minimized by using the appropriate formulae for unbiased estimates. These estimates were then used to replace the corresponding population parameters in the above derived equations. The lag-one autocorrelation in the normalized domain ($r_1(y)$), was estimated upon transformation of the real data by equation (11) and used to replace $\rho_1(y)$.

(2) The parameters of the models were computed using the above historic statistical parameters.

(3) Synthetic flow sequences were generated for each gaging station, by using the T-F(basic), T-F(skewed) and proposed models. With the application of each model, fifty sequences were generated so that each equals in length to the historic sequence. Then unbiased statistical parameters were estimated for all fifty sequences.

(4) In a comparison of these fifty sets of statistical parameters (for each model) with the corresponding estimates for the historic sequence, their average values were directly compared with the historic parameters. Also, the deviations from the historic value such as bias and variance were computed for each statistic.

(5) Using the average values of the generated statistical parameters for each model, the probable maximum annual flows were forecast at 50, 100, and 200-year return periods. For this purpose, the same distribution assumed by the model, is used for forecasting. For example, LN3 distribution was used to fit the flows generated by the proposed model, whereas Normal distribution was fitted to the flows generated by the T-F(basic) model.

RESULTS AND DISCUSSION

The results obtained in this study have been presented in the form of statistical parameters, their deviations and the probable maximum annual mean flows for each watershed (see Tables I & II). The bias and

TABLE I. Statistical Properties of Historic Flow Records (Annual Mean Flows)

Watershed No.	Stream & Location	USGS Station No.	Statistical Properties				
			Mean($\bar{x}$) (cfs)	Std. Dev. (S_x) (cfs)	Coef. skew (C_{SX})	Lag-one Auto-corr. Coef.	
						$r_1(x)$	$r_1(y)$
1	Whippany River at Morristown, NJ	01381500	53.4	16.7	0.686	0.346	0.346
2	Assunpink Creek at Trenton, NJ	01464000	129.0	39.4	0.674	0.219	0.219
3	Flat Brook near Flatbrookville, NJ	01440000	109.8	31.9	0.529	0.211	0.211
4	North Branch Raritan R. near Raritan, NJ	01400000	307.1	100.4	0.578	0.230	0.230
5	N. Branch Rancocas Creek near Pemberton, NJ	01467000	173.3	46.0	0.551	0.170	0.170
6	Rahway R. near Springfield, NJ	01394500	28.8	10.9	0.716	0.346	0.346

TABLE II. Statistical Properties of Generated Flow Sequences (Annual Mean Flows)

Watershed No.	T-F(basic) Model				T-F(skewed) Model				Proposed Model			
	$\overline{X}$ (cfs)	S_X (cfs)	C_{SX}	$r_1(x)$	$\overline{X}$ (cfs)	S_X (cfs)	C_{SX}	$r_1(x)$	$\overline{X}$ (cfs)	S_X (cfs)	C_{SX}	$r_1(x)$
1	52.9	16.6	0.117	0.350	52.9	16.8	0.782	0.347	52.9	16.7	0.767	0.340
2	128.0	39.1	0.126	0.230	128.0	39.6	0.789	0.227	128.1	39.5	0.788	0.222
3	109.0	31.7	0.127	0.223	109.0	32.0	0.646	0.221	109.0	32.0	0.649	0.217
4	304.6	99.6	0.128	0.241	304.5	100.8	0.693	0.238	304.7	100.5	0.693	0.234
5	172.2	45.6	0.129	0.184	172.2	46.1	0.673	0.181	172.3	46.1	0.677	0.178
6	28.7	10.8	0.150	0.341	28.7	11.1	0.820	0.340	28.8	11.0	0.796	0.334

TABLE III. Bias and Variance of Skewness Coefficient Estimated for the Generated Flow Sequences.

Watershed No.	T-F(basic) Model			T-F(skewed) Model			Proposed Model		
	Average Estimate	Bias	Variance	Average Estimate	Bias	Variance	Average Estimate	Bias	Variance
1	0.117	-0.569	0.434	0.782	0.096	0.218	0.767	0.081	0.214
2	0.126	-0.548	0.410	0.789	0.115	0.229	0.788	0.114	0.239
3	0.127	-0.402	0.272	0.646	0.117	0.200	0.649	0.120	0.205
4	0.128	-0.450	0.314	0.693	0.115	0.209	0.693	0.115	0.213
5	0.129	-0.422	0.289	0.673	0.122	0.209	0.677	0.126	0.218
6	0.150	-0.566	0.444	0.820	0.104	0.251	0.796	0.080	0.253

TABLE IV. Bias and Variance of Lag-one Autocorrelation Estimated for the Generated Flow Sequences.

Watershed No.	T-F(basic) Model			T-F(skewed) Model			Proposed Model		
	Average Estimate	Bias	Variance	Average Estimate	Bias	Variance	Average Estimate	Bias	Variance
1	0.350	0.004	0.006	0.347	0.001	0.007	0.340	-0.006	0.008
2	0.230	0.011	0.007	0.227	0.008	0.008	0.222	0.003	0.008
3	0.223	0.012	0.007	0.221	0.010	0.008	0.217	0.006	0.008
4	0.241	0.011	0.007	0.238	0.008	0.008	0.234	0.004	0.008
5	0.184	0.014	0.008	0.181	0.011	0.008	0.178	0.008	0.008
6	0.341	-0.005	0.012	0.340	-0.006	0.013	0.334	-0.012	0.013

TABLE V. Comparison of Estimated Probable Maximum Annual Mean Flows.

Watershed No.	50-year Probable Maximum (cfs)			100-year Probable Maximum (cfs)			200-year Probable Maximum (cfs)		
	seq.1	seq.2	seq.3	seq.1	seq.2	seq.3	seq.1	seq.2	seq.3
1	87.0	94.0	93.5	91.5	101.3	100.9	95.7	108.3	108.1
2	208.4	224.9	224.9	219.0	242.2	242.7	228.7	258.8	260.1
3	174.1	185.2	185.1	182.8	198.2	198.4	190.6	210.6	211.2
4	509.3	546.8	546.3	536.4	588.7	589.4	561.1	628.8	631.0
5	265.9	282.6	283.0	278.3	301.6	302.6	289.6	319.6	321.6
6	50.9	56.0	55.8	53.8	60.9	60.8	56.5	65.7	65.6

Note: Seq. 1, Seq. 2 and Seq. 3 are the annual flow series generated by T-F(basic), T-F(skewed) and proposed models respectively.

variance of the skewness and lag-one autocorrelation coefficients estimated for the generated flow sequences are given in Table III and Table IV respectively.

As indicated by Tables I and II, the mean and standard deviation are very closely preserved by all three models. It is also interesting to notice that the lag-one autocorrelation is the same in both the real domain and the normal domain.

According to the results in Table III, it can be said that the T-F(basic) model provides no guarantee in preserving the skewness as indicated by the large negative bias. On the other hand, the bias and variance resulting from the T-F(skewed) model and the proposed model are almost of the same order and comparatively small. The numbers in Table IV for bias and variance are very small and are of the same order for all three models. Therefore, the proposed model closely preserves the autocorrelation in the real domain.

The estimated magnitudes of probable maximum annual mean flows are listed in Table V. In making these estimates, the generated flows were fitted by the same distribution incorporated in each model. When compared these magnitudes, it is seen that the results due to the T-F (skewed) model and the proposed model are almost the same. The magnitudes resulting from the T-F(basic) model are always lower than the other two model estimates. Their differences become larger as both the mean flow and the return period increase. These differences in magnitudes, therefore, clearly indicate the significance of skewness in streamflow generation.

CONCLUSIONS

The major conclusions that can be made based on the results of the study are as follows:

1. The proposed model closely preserves the historic values of mean, variance, skewness and the lag-one autocorrelation as does by the Thomas-Fiering model based on Gamma distribution.
2. Annual mean flows generally reflect a positive skewness.
3. Neglection of skewness in streamflow generation causes a serious error when the generated sequences are used for forecasting. The error increases as the meanflow and the return period increases.

ACKNOWLEDGEMENTS

The authors wish to acknowledge Dr. H.N. Phien for personnel communication. Also, they express their sincere appreciation towards the Civil Engineering Department of the New Jersey Institute of Technology for the provision of computer facilities.

REFERENCES

Barnes, F.B., 1954, "Storage Requirements for a City Water Supply", J.

Inst. Eng., Aust., Vol. 26, pp. 198-200.
Fiering, M.B. and Jackson, B.B., 1971, Synthetic Streamflows, Water Resources Monograph 1, American Geophysical Union, Washington, D.C. pp. 49-78.
Hazen, A., "Storage to be Provided in Impounding Reservoir for Municipal Water Supply", Trans. Am. Soc. Civ. Eng., 77: pp. 1539-1640.
Hashi, K., Stedinger, J.R. and Burges, S.J., 1984, "Estimation of Log Normal Quantiles: Monte Carlo Results and First-order Approximations", J. Hydrology (Netherlands), Vol. 71, pp. 1-30.
Matalas, N.C., 1967, "Mathematical Assessment of Synthetic Hydrology", Water Resources Research, Vol. 3(4), pp. 937-945.
Sudler, C.E., 1927, "Storage Required for the Regulation of Streamflow", Trans. Am. Soc. Civ. Eng., Vol. 91, pp. 622-704.
Thomas, H.A. and Fiering, M.B., 1962, "Mathematical Synthesis for the Analysis of River Basins by Simulation", Chapter 12, Howard University Press, Cambridge, Massachusetts.

REDUCTION OF UNCERTAINTIES IN THE FLOOD ESTIMATION IN THE CZECHOSLOVAK SECTION OF DANUBE RIVER

B. Minárik and K. Martinka
Slovak Hydrometeorological Institute
Jeseniova 17
83315 Bratislava
Czechoslovakia

ABSTRACT. The problem of the real-time estimation and forecasting in the Czechoslovak section of Danube river is analysed. Because of various short- and long-term changes in watershed, complicated hydrologic and hydraulic phenomena, and incomplete information about the process the problem of the modelling is considerably uncertain. The problem of the runoff estimation and their short-term forecasting is formulated by the state space model. The adaptive filtering approach for the estimation of unknown parameters of the model is used. The algorithm is applied to the flood forecasting problem on the Danube system in Czechoslovakia.

1. INTRODUCTION

The computer dispatching system of the Hydrological Information and Forecasting Service in Bratislava, Czechoslovakia has to its disposal relatively good operational data basis of the hydrological and meteorological quantities from the Czechoslovak territory and from the Federal Republic of Germany and Austria. However, various water constructions and agricultural, forestry and other short- and long-term changes in watersheds cause some difficulties in the operational identification of hydrological processes. From the viewpoint of the short-term forecasting of hydrological quantities the problems may be divided into two groups: /i/ alternation of the character of the runoff process, /ii/ difficulties with the interpretation and verification of measured data. Numerous hydrological procedures with a fixed model structure and time-invariant parameters cannot be used for identification. Out of the identification methods originally developed for natural hydrological conditions the state space approaches appear to be the most suitable for the identification of watersheds influenced by man.

In the state approach time-variable parameters of models are manipulated whereby the real-time and updated algorithm is used to estimate unknown parameters. Real-time estimation allows to adapt the model from the originally inaccurately defined initial values

V. P. Singh (ed.), Flood Hydrology, 187–195.
© 1987 by D. Reidel Publishing Company.

closer to the real values and it also allows to get successively adapted to long-term non-stationary states. Wood and Szöllösi-Nagy /1978/ suggested an algorithm considering the structual and parameter changes due to random environmental effects. Simons et all /1978/ made use of Kalman's state approach for the prediction of water and sediment discharges in rivers, Kitanidis and Bras /1980/ for the identification of rainfall-runoff models.

Another group of treatises is devoted to the application of the state approach in modelling hydraulic problems of water resources systems/e.g. Jones and Moore, 1980/.

In watersheds essential influenced by man, however, the problems of rainfall-runoff modelling and water motion modelling through riverbeds or through artificial water schemes, respectively, mutually overlap. The unceasing rise of new water schemes, depositions and adaptations in older objects imply that data from water-stage-gaging sites do not render a sufficient picture about the real hydrological development of the given watershed. Artificial changes in the watershed materialize at a much faster pace than long-term nonstationaries. In the contribution the problem of runoff estimation from man-influenced Danube watersheds is solved by means of real-time balancing of water conditions and the adaptive filtration method is used to forecast it. The method is applied to the watershed of Danube(for forecasting purposes in the Czechoslovak section of the river) where hydrological changes are incessant and where 9 hydropower plants in Austrian part and other water resource objects have beeb built and where two water power plants Gabčíkovo and Nagymaros in Czechoslovakia-Hungary part are under construction.

Figure 1. shows the scheme of Danube river and Figure 2. shows an example of biased forecasts /even in relatively steady state of hydrological processes/ as the consequence of the activity of water plants in upper part of the river, where the power output varyies during a day.

2. RUNOFF ESTIMATION

Assume that the considered catchment area may be divided into interbasins limited by adequate specific profiles that may be independently modelled. The interbasin is generally characterized by an inflow from an interbasin q_1 of a higher position, by an intermediate inflow z and by the runoff from interbasin q_2. The system input and output are measured, from the interbasin, in contrast, the runoff measurements of some lateral rivers may, but need not, be available. Besides, precipitations measurements p may be accessible from the network which, however, does not represent sufficiently the precipitations of the entire interbasin.

Denote the inflow from a higher interbasin by $q_1(k-\tau)$ transformed by the riverbed within the time of translation τ^1 into the lower profile by symbol $q_s(k)$. The difference $q_A(k)=q_2(k) - q_s(k)$

Figure 1. Scheme of Danube river.

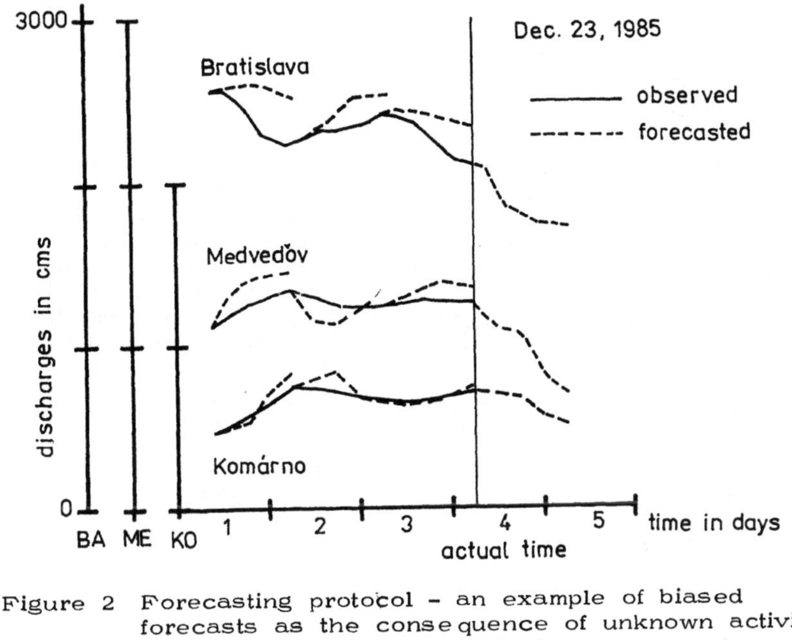

Figure 2 Forecasting protocol - an example of biased forecasts as the consequence of unknown activity of Austrian water power plants.

characterizes the contribution of the interbasin. To model the forecasting dependence use can then be made of both the directly measured data from the interbasin and of the estimation $q_A(.)$.

To derive the transformed behaviour $q_s(.)$ it is advantageous to utilize the adaptive model of the cascade of linear reservoirs /Szöllösi-Nagy, 1982/. If n is the number of hypothetic reservoirs, K their storage coefficient, ΔT the sampling interval, vector

$$x_s(.) = [x_{s1}(.), \ldots, x_{sn}(.)],$$

where $x_{si}(.)$ is the volume of an i-th reservoir, then

$$x_s(k+1) = F(k)\,x_s(k) + G(k)\,q_1(k) + w_s(k) \qquad /1/$$

where $w_s(.)$ is the system perturbation and $F(.)$ the lower tringular matrix of the dimension n x n. The output from the cascade of linear reservoirs is characterized by the equation

$$q_2(k) = H(k)\; x_s(k) + v_s(k) \qquad /2/$$

where v_s is the perturbation of the observation and $H^T=[0,\ldots, K]$

To solve the system /1/ and /2/ it is of advantage to make use of Kalman's filtration algorithm, whereby the unknown covariances Q_s, R_s of perturbations w_s, v_s may be estimated by the trial and error method. Filter characteristics are to be set at optimal values by the underlying data q_1, q_2 from periods when contribution was still negligible.

From the balance of inflow from a higher positioned interbasin and the outflow from the given interbasin as well as from data on the increment of volumes, thus it is possible to define the approximate values of the subcatchments contribution. The time sequence of contributions $q_A(k)$, k=1, 2, 3, ... may oscillate considerably. It is therefore suitable to make also use of data from tributary flows $q_B(k)$, k=1, 2, 3, ... that are accessible even if representing only a small portion of the subcatchment.

Assume that q_B represents the sum of the time synchronized measured intermediate inflows, then the forecast of interbasin contribution may be modelled by ARMAX relation

$$z(k) = H(k)\; x(k) + v(k) \qquad /3/$$

whereby

$$H(k) = [q_A(k-1), \ldots, q_A(k-n_1), \quad q_B(k-1), \ldots, q_B(k-n_2), \quad p(k-1), \ldots, p(k-n_3), \quad \overline{w}(k-1), \ldots, \overline{w}(k-n_4)] \qquad /4/$$

and $z(k)$ in Eq. /3/ is the forecast of contribution for time k, $q_A(.)$ and $q_B(.)$ are of an autoregressive character, $p(.)$ are precipitations as inputs to the system, $\bar{w}(.)$ are errors of preceding forecasts, n_1, n_2, n_3 and n_4 are the corresponding orders and $v(k)$ the perturbation of observation. About the parameters

$$x(k) = [x_1(k), x_2(k), \ldots, x_{n_1+n_2+n_3+n_4}(k)]$$

it is presumed that they are a stochastic process of the form

$$x(k+1) = A(k)\ x(k) + w(k) \qquad /5/$$

where $A(k)$ is a transtion matrix and $w(k)$ the perturbation of the system.

The recursive solution of the equation system /3/ and /5/ - the Kalman filter - requires again to know the covariance matrix Q of the system perturbation w, the covariance matrix R of the perturbation of observation v, the transition matrix A in each step, as well as the initial estimates of parameters $x(1/0)$, and of the covariance matrix of error estimation $P(1/0)$. In addition, as the first step the model structure estimation is to be carried out /3/, /5/, that is the estimation of orders n_1, n_2, n_3 and n_4 for example by means of analysing the behaviour of absolute and mean square error for various order values.

The initial estimates $x(1/0)$, $P(1/0)$ may be obtained off-line by regression analysis. Anyhow, the quantities Q, R, A, are either unknown or known only approximately such that it is necessary to estimate them inreal time by adaptive filtration methods. In view of the fast changes in the behaviour of hydrological processes, the results of adaptive filtration mostly based on the analysis of inovation or residual sequence /Martin and Stubberud, 1976/ exihibit typical delays and/or instability which depreciate hydrological forecasts. Satisfactory results of adaptive hydrological forecasts may be obtained only for advance times smaller than the runoff concentration period /see e.g. O'Connel, 1980/.

The independently working classical estimator of, e.g. Kalman type can automatically change its characteristics Q, R, or A regarding the varying hydrological situation according to the deviation of the innovation sequence which is produced by the equation

$$\nu(k) = z(k) - H(k)\ x(k/k-1) \qquad /6/$$

from the ideal white sequence. With regard to inaccuracy of modellig this tie is in practice often very free. For calibration of initial Q_o, R_o, and A_o estimates it is nearly impossible to satisfy requirements for the estimation algorithm for different hydrological situations. Let us consider only 3 types of hydrological situations - steady state, rise, decline. Then when calibrating the model for, e.g.

decline it doesn't give satisfactory results for the change on rise, etc. Besides the time loss of change indication, also the estimator's inertia has an unfavourable effect, because also after the correction of, e.g. Q to $Q+\Delta Q$ the estimator considers for the estimate for the next step the values of the state vector $x(k/k-1)$ and error covariance matrix $P(k/k-1)$ determined in the last step still on the basis of Q-value.

In the forecasting system of the Hydrological Information and Forecasting Service in Bratislava is problem of short-term forecasting handled (besides of other methods - see e.g. Kunsch and Minárik, 1984) by using a two-level procedure /Minárik, 1983/ in which use is made for estimation also of information that cannot be utilized for quantitative modelling owing to their binary character or insufficient representativeness.

On the first level of the two-level estimator there is a "fine" estimator which makes the final tuning of unknown parameters on the base of a "coarse" estimate. On the second level there is an estimator giving a coarse estimate of unknown parameters. The coarse estimator is based on a finite number of a priori models and its task is to find out which model best fits the data.

In the preparatory stage it is possible to obtain the values of unknown characteristics Q, R, and A for certain typical situations in which hydrological processes often find themselves. These values may subsequently serve as basic estimates in the coarse estimator in the real-time mode of identification. Assume that we have a coarse estimator that is based on L a priori models of uncertain characteristics of the state space representation and a fine estimator of the Kalman type. The estimates $x(k/k)$ and $x(k+1/k)$ of the fine estimator serve as results of the entire algorithm for the determination of the required estimation of the instantaneous contribution of the subcatchment $z(k/k)$ and its prediction $z(k+1/k)$.

The coarse estimator with the pre-defined Q^i, R^i, A^i values for a certain i-th type of hydrological situation tends to evaluate the real situation always as pre-defined. Corresponding a posteriori estimates $x^i(k/k)$ can be calculated with the help of the Kalman filter. During the change of real hydrological situation to j-th type, the estimates $x^j(k/k)$ are closer to the demanded values than in the case with one universal adaptive estimator, also if estimates $x^j(k/k-1)$, $P^j(k/k-1)$ were determined during the previous hydrological situation.

When operating in real time the coarse estimator produces L coarse estimates of the state vector and the error covariance matrix using the basic values Q^i, R^i, and A^i, $i = 1, \ldots, L$. Estimates $x^i(k/k)$, and $P^i(k/k)$ are time varying. The task of the decision-making mechanizm of the coarse estimator is to find out which estimate is optimal in time k. To distinguish the type of hydrological and hydraulical situation in which the process is at the moment, use can be made of addition information usually available in addition to those included in matrix H or z /Minárik, 1983/. These are various hydrometeorological data dispatched to the Control Centre, and some data (e.g. measured stages on some tributaries of low significance, and data of remote sensing and/or radar reflections), that are available in the control computer. Further they may include some features of the recent course (e.g. convexness, concaveness and the like) that cannot be utilized directly in quantitative modelling. If, for example, in the rising stage of the runoff process, after the previous increase, there arises the situation that the concave course was superseded by a convex one, runoff culmination may be expected to set in in the next steps. This fact may be similarly indicated by the transgression of culmination in some smaller tributary inflows.

These indications are not sufficiently reliable. If, however, they are used to effect changes of uncertain characteristics, that is to change from Q^i, R^i, A^i into Q^j, R^j, A^j in the coarse estimator, then the corresponding coarse estimates $x^j(k/k)$, $P^j(k/k)$ will better direct the fine estimates $x(k+1/k)$, $P(k+1/k)$ in identifying the first symptoms of culmination in observations H and z. If the mentioned insuficiently reliable indications were used directly in the basic model through H and z in the basic model of the process (e.g. Eq./3/), the resulting forecasts $z(k+1/k)$ would have essentially worse quality /Minárik, 1981/.

It has been similarly shown that although precipitations are the cause of runoff, the measurements proper of precipitations are not accurate and sufficiently representative to be included into Eq. /3/ in numerous practical cases. The utilization of precipitation measurements and other information on the process to distinguish the type of uncertainty is possible by the probability way. A posteriori distribution of different models can be updated using the Bayes rule and the most probable model can be found by maximizing the logarithm of the likelihood function /Minárik, 1983/.

The algorithm was applied to the flood forecasting problem on the Danube system in Czechoslovakia. Two examples of results of proposed algorithm in the operation on Hydrological Information and Forecasting Service, Bratislava are shown on Figure 3 and 4. Resulting protocols serve for final decision-making in the real-time operation. Figures 3 and 4 are copies of the plotter colour protocol in which are also picture results from other hydrological models (cascade of nonlinear reservoirs, impulse response model RIMO, inflow coefficient model, and Muskingum model) in reality /Minárik, 1986/.

3. CONCLUSION

A discrete time Kalman based estimator has been successfully applied to the estimation/forecasting problem in the computer dispatching system of the Hydrological Information and Forecasting Service, Bratislava for the real-time short-term forecasting of Danu-

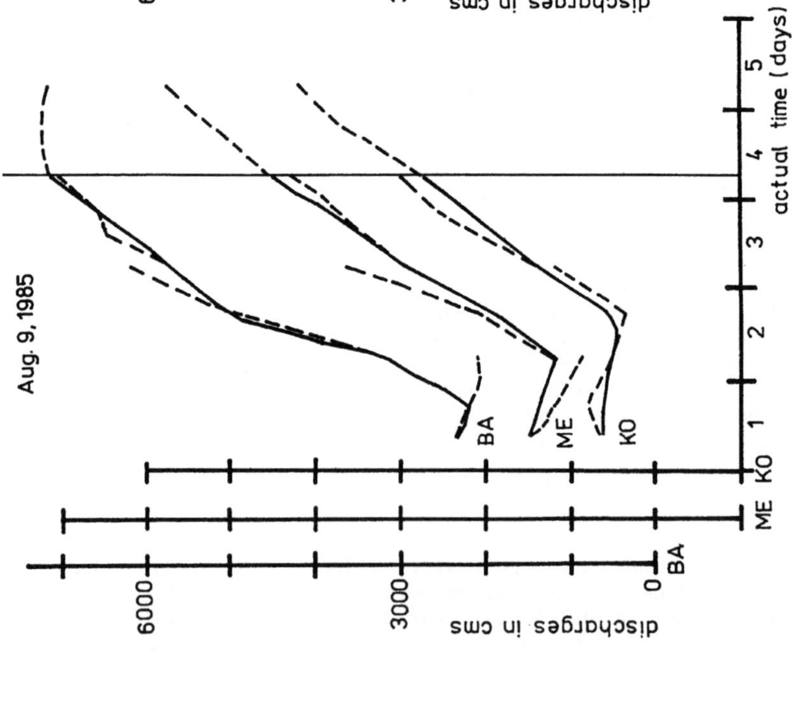

Figure 3 Forecasting protocol.

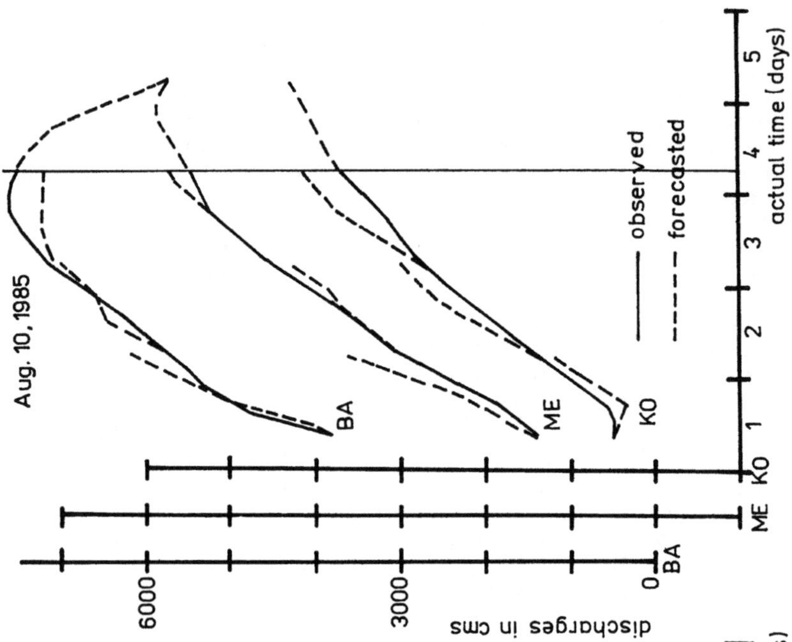

Figure 4 Forecasting protocol.

be river discharges. The proposed two-level algorithm can be used for the estimation and the forecasting of interbasin contribution as well as other variables of water resource systems.

REFERENCES

Jones, D. A., and Moore, R. J. 1980. A simple channel flow routing model for real time use, IAHS Publications, No. 129, pp. 397-408

Kitanidis, P. K. and Bras, R. L. 1980. Real-time forecasting with a conceptual hydrologic model, Water Resources Research, Vol. 16, No. 6, pp. 1025-1044

Kunsch, I. and Minárik, B. 1984. Vorhersage in Abschnitten mit bedeutenden aber nicht gemessenem Nebenfluss. Proc. Donau landes, Bratislava August 21-23

Martin, W. C. and Stubberud, A. R. 1976. The innovations processes with application to identification. In: C. T. Leondes /Editor/, Control and Dynamic Systems, Vol. 12, pp. 173-258, Academic Press, London

Minárik, B. 1981. A contribution to identification of nonstationary dynamic systems with application to water resources. Ph.D. thesis, Institute Technical Cybernetics, Bratislava /In Slovak/

Minárik, B. 1983. Two-level suboptimal filtering with application to water resources. In: G.A. Bekey and G.N. Saridis /Editors/, Identification and System Parameter Estimation, pp. 1409-1414, Pergamon Press, Oxford

Minárik, B. 1986. Contribution to the real-time flood estimation and forecasting. 2nd Scientific Assembly of the IAHS, Workshop 3: Recent Developments in Flood Routing, Budapest, Hungary, 31 pp.

O'Connel, P. E. /Editor/ 1980. Real-time hydrological forecasting and control. Institute of Hydrology, Wallingford, UK

Simons, D. B., Duong, N., and Li, R. M. 1978. An approach to short-term water and sediment discharge prediction using Kalman filter. Proc. AGU Chapman Conf., Pittsburg, University Pittsburg Press, pp. 473-481

Szöllösi-Nagy, A. 1982. The discretization of the continuous linear cascade by means of state space analysis, Journal Hydrology, Vol. 58, pp. 223-236

Wood, E. F. and Szöllösi-Nagy, A. 1978. An adaptive algorithm for analyzing short-term structural and parameter changes in hydrologic prediction models, Water Resources Researche, Vol. 14, No. 4, pp. 577-581

REAL-TIME FLOOD FORECASTING IN THE RIVER SECTION WITH UNGAGED TRIBUTARIES

Boris Minárik
Slovak Hydrometeorological Institute
Jeseniova 17
833 15 Bratislava
Czechoslovakia

ABSTRACT. In the contribution, the solution of the problem of the adaptive flood forecasting of tributaries of large-scale water resource systems is described. The water resource system is divided into interbasins limited by significant specific profiles. The basic dynamic model of interbasin behaviour is formulated by the state space model and solved by the adaptive Kalman filtration algorithm. In view of fast changes in the behaviour of hydrological processes, the results of adaptive filtration mostly based on the analysis of innovation sequence exhibit typical delays and/or instability which depreciate hydrological forecasts. In real time operation therefore the following procedure can be advantageous: /i/ to estimate the type of situation in which the hydrological process is positioned, /ii/ to use, for the given type of hydrological situation, the corresponding basic model and the values of the unknown parameters, and to let the adaptive algorithm operate around these values till the subsequent change in the type of hydrological situation sets in. The problem of the effective utilization of inaccurate and incomplete information on the behaviour of hydrological processes in ungaged interbasins is solved by the two-level estimator. The algorithm is applied to the real-time estimation and forecasting problem in the Czechoslovak section of the Danube river and other Czechoslovak rivers.

1. INTRODUCTION

The endeavour at utilizing the methods of the up-to-date estimation theory for solving the problem of real-time runoff forecasting has led in recent years to a wide development of modelling in the state space and of the filtering estimation methods. In hydrology the use of state space methods in forecasting has been documentsd since 1975 especially due to Todini, Szöllösi-Nagy and Wood /see for example the papers Todini and Bouillot, 1975, Wood and Szöllösi-Nagy, 1978/. The consideration of a stochastic character of hydrological phenomena and of the adaptive principle

V. P. Singh (ed.), Flood Hydrology, 197–210.
© 1987 by D. Reidel Publishing Company.

of estimation, low claims laid upon input data, simple calibration of models and the orientation upon a real-time mode of computer processing appear to be the main advantages of the procedure.

It appears, however, that the use of filtering methods in operational hydrology did not yield the expected results in terms of accuracy and stability of estimates.

The filtering estimation approaches hitherto known are oriented upon linear type models and only upon some special types of nonlinear models. A natural tendency to improve results lies in the endeavour at applying the approach to hydrological models of a more conceptual type. Such models would allow a more detailed description of the individual phases of the runoff process and hence more accurate forecasts. This trend in research is borne out perhaps most thoroughly in the work Kitanidis and Bras /1980/ in which the authors apply the filtering approach to such a detailed conceptual model as the Sacramento model. Such tendencies, however, encounter considerable mathematical difficulties in view of the strongly nonlinear character of the dependence and in operative practice their significance is limited so far.

In spite of this, however, there exist application areas in which the use of filtering procedures is advantageous or even the sole possible one from practical aspects. This concerns mainly real-time forecasting under conditions of inaccurate and incomplete information in which the use of conceptual hydrological models is impossible. This is the case also with the problem of real-time forecasting the tributaries of such interbasins in which the tributaries are - /i/ not measured at all, /ii/ measured partially so that the measured values do not represent the considered subcatchments sufficiently, /iii/ measured sufficiently, but the values are not available for prognosis in real time.

In this contribution the problem of forecasting with incomplete and inaccurate measurements is solved with the help of a two-level procedure by a differentiated use of accessible information.

2. STATEMENT OF THE PROBLEM

From the viewpoint of identification theory the problem of short term runoff forecasting consists of two parts - the rainfall-runoff modelling and flow routing in the riverbed. Assume that the considered catchment area may be divided into interbasins limited by adequate specific profiles that may be indenpendently modelled.

Ungaged lateral inflow can be incorporated into the model in four main ways. The most common method is to assume it to be a fixed proportion of the gauged tributary flow, as used by Cameron /1980/. The second method is for lateral inflow to be implicit in the main routing model as in Wood /1978/ where autoregressive parameters are assumed to take into account all ungauged inputs. In the third method a relationship between rainfall on the ungauged section and lateral inflow may be sought /Harpin and Cluckie,1981/. Finally, water balance on the main stream may serve for deriving the magnitude of the lateral inflow and its stochastic forecasting /Minárik, 1983/.

Assume the following combination of the presented procedures. To forecast the magnitude of intermediate inflow use made both of estimations derived from the balance on the main stream, and of all measurements obtained in the interbasin area, namely within the frame of an ARMAX type stochastic model.

Assume the following time-varying model

$$z(k) = H(k)\ x(k) + v(k) \qquad /1/$$

where $z(k)$ is the estimated tributary inflow, $v(k)$ the disturbance of observations with zero mean and covariance R, $x(k)$ - unknown time-varying parameters, and $H(k)$ is

$$H(k) = [\, q_A(k-1), \ldots, q_A(k-n_1), q_B(k-1), \ldots, q_B(k-n_2),$$

$$\overline{w}(k-1), \ldots, \overline{w}(k-n_3)\,]$$

where q_A - the contribution of the interbasin on the base of water balance on the main stream, q_B - the sum of discharges measured on the tributaries, $\overline{w}$ - the errors of past estimates and n_1, n_2, n_3 corresponding orders. Parameters $x(k)$ are assumed to be stochastic processes of the form

$$x(k+1) = A(k)\ x(k) + w(k) \qquad /2/$$

where $w(k)$ is white gaussian noise sequence with zero mean and covariance Q and $A(k)$ is the transition matrix. The unknown parameters $x(.)$ can be estimated using a Kalman filter.

In the application of the Kalman filter to the identification problem of the form /1/ and /2/ the filter becomes

$$x(k/k) = x(k/k-1) + [\, K(k)\ z(k) - H(k)\ x(k/k-1)\,] \qquad /3/$$

$$K(k) = P(k/k-1)\ H^T(k)[H(k)\ P(k/k-1)\ H^T(k) + R(k)\,]^{-1} \qquad /4/$$

Eq. /3/ is initialized with $x(0/-1)$ set equal to the vector of a priori estimates of the coefficients, and $K(k)$ in the Eq. /4/ is initialized with $P(0/-1)$ set equal to the a priori covariance matrix of the coefficients. Corresponding a posteriori estimates $P(k/k)$ can be calculated by

$$P(k/k) = P(k/k-1) - K(k)\ H(k)\ P(k/k-1) \qquad /5/$$

Time propagation equations have the form

$$x(k+1/k) = A(k)\ x(k/k) \qquad /6/$$

$$P(k+1/k) = A(k)\ P(k/k)\ A^T(k) + Q(k) \qquad /7/$$

The desired vector $z(k+1/k)$ i.e. the runoff forecast can be calculated by the linear projection

$$z(k+1/k) = H(k+1)\ x(k+1/k) \qquad /8/$$

where $H(k+1)$ is in time k known.

Based on the information given by equations /1/ and /2/ it is not possible to obtain an estimate with better mean square error than the Kalman filter estimator. However, the optimal Kalman filter formulated for discrete-time linear systems requires complete knowledge of noise covariances, the initial state estimate $x(1/0)$ and the initial state error covariance $P(1/0)$. Because of the complicated nonstationary nature of the phenomenon and incomplete information about the process these quantities are either unknown or known only approximately.

The errors ensuing from the inaccurate covariance estimates and initial values together with errors in modelling and computational errors produce a reduction in the quality of forecasts. Inaccurate estimates may cause divergence of the Kalman filter in many practical situations. Two changes appear to be given for improving the situation in linear modelling. One practicability is the real-time estimation of unknown characteristics of the state space representation of the process /adaptive filtering/, the second practicability is the inclusion of additional measured quantities into the model /3/, /4/, if these measurements are available.

In case of modelling the contribution of the subcatchment there may often be available precipitation measurements from ground stations or recently also from remote sensing media.

By the underlying knowledge of the quality of rainfall data and their representativeness, it is necessary to consider in practical situations whether it is adequate to include these data into the basic model of the process /for example in form of ARMA model or within the frame of some more conceptual model/, or to exclude them from direct modelling. Only one criterion may be the pragmatic standpoint of the quality of the resulting forecasts. In case of low representativeness and bad quality of rainfall data, or if it is impossible to handle the problem of determinig effective precipitation, the problem of separating ground and underground runoff, or other problems within the frame of some conceptual hydrological model, in a satisfying manner, then it is better to utilize precipitation and other meteorological data only to influence the operation of some adaptive algorithm with a simpler basic model, e.g. of type /1/, /2/. Only the most fundamental hydrological and meteorological quantities are most often available within contemporary water resource systems in the real-time operation of forecasting algorithms.

Even in the case of large-scale information systems which are sporadically built or created within the frame of certain short-term or territorially limited experiments, the accessible conceptual procedures are not as much advanced as to allow automated operation in the control centres with the usual staff. In real-time operation there is no time for a large-scale after calibration of model dependences with frequently counterposed and physically hardly discernable influences.

3. ADAPTIVE FILTERING APPROACH

The advance of adaptive filtering is a natural evolution of the fundamental Kalman filter, which requires both the linear dynamic system and linear measurements of the state vector to be known and disturbed only by white gaussian noise with known first- and second-order statistics. Such requirements are hardly ever fulfiled in practical applications. In order to conpensate model, noise and other inadequacies, adaptive techniques can be used.

In the case of runoff forecasting noise statistics play a relevant role in the Kalman filter, as is clear from Eqs. /3/ - /8/. In particular, the noise covariance Q affects - Eq. /7/ - the determination of $P(k/k-1)$ and subsequently - Eq. /4/ - of the gain $K(k)$. In particular from Eqs. /3/, /4/, underestimation of Q may reduce the Kalman gain and hence the filtering of the previous forecast $x(k/k-1)$ by means of the new observation. The quantity

$$\nu(k) = z(k) - H(k)\ x(k/k-1) \qquad /9/$$

in the Eq. /3/ is termed innovation. It can be shown that a necessary condition for optimality of the Kalman filter is that the innovation sequence $\nu(k)$, k 1, 2, 3, ... be white. Whiteness of the innovation sequence can be used for the estimation of covariances Q and R. Several investigators /see O'Connell, 1980 for a review/ treats the problem of estimating the uncertain characteristics simultaneously with the system state. However, most of these developments are restrictive, for example to time invariant systems, computational burden, while convergence of estimates does not guaranted even theoretically.

Although parameter and noise uncertainties may effect all statistics of the estimation error, the primary measure of the increase in estimation error is the actual error covariance matrix, defined by

$$P_a(k) = E\left[(x(k) - \overline{x(k/k)})^T D (x(k) - \overline{x(k/k)}) \right] \qquad /10/$$

where D is a symmetrix positive-definite matrix, $x(k)$ is the true value of the state vector at the k step, and $\overline{x(k/k)}$ is the estimate of the state vector at the step k using invalid information about the noise covariance matrices and the system dynamics. If $P_a(k)$

reaches a finite steady-state value as $k \to \infty$, but is too large to allow the estimate to be of use, "apparent divergence" occurs. "True divergence" occurs when $P_a(k)$ becomes unbounded as $k \to \infty$.

Fig. 1a exemplifies the "true divergence" in case of using the state version of the model of the cascade of linear reservoirs to forecast the runoff with inadequately estimated values of covariance Q. Fig. 1b shows the example of divergence in the rise phase of the runoff process, of the later stabilization of estimates and of the subsequent detunig of estimates. Fig. 1b illustrates the case in which the use of a certain constant matrix Q may comply only in certain phases of the runoff process.

The selection of the appropriate model structure is the first-rate task in the modelling of real hydrological processes. The selection of the model structure can be performed on the frame of preparatory programs. In the preparatory programs subsequent verification of each eventually possible model variant for each forecast profile or profile group can be individually performed. However, the modelling problem is frequently, and considerably uncertain even for the best structure of the model. Note that the forecasting is usually performed, for each time step, Δt /e.g. 6 h./ for the time advance $T = n \Delta t$, where $n > 1$, in practice.

To limit the divergence of estimates and to improve the accuracy of estimates it appears to be necessary to solve the following two tasks:

/i/ to adequately modify or extend the filtering algorithm for the estimation of the required quantities, i.e. the estimation of vector $x(k/k)$, $x(k+1/k)$, and of the error covariance matrix $P(k/k)$, $P(k+1/k)$, for conditions of uncertaintly, with the aim to improve the numerical properties of the Kalman filter,

/ii/ to suggest a mode of the real-time estimation of unknown covariances and parameters whose knowledge is indispensible for the work of the algorithm /i/.

4. A TWO-LEVEL ESTIMATOR

The utilization of the filtering approach with fixly set characteristics does not render good results especially of periods of violent changes in the runoff process. The continuous observation of the whiteness of the innovation or residual sequence may serve as a criterion for derivation of the adaptive filtration algorithm, in which the initial estimates are corrected of the covariance disturbance matrices /O'Connel, 1980/. A great flaw of this procedure, however, lies in the typical delay of indication that the uncertainty of hydrological processes has changed, inclusive of the estimation errors following therefrom. Testing the whiteness of innovation or residual sequences calls, as a matter of fact, for long data records. In the case of hydrological processes, however, violent changes of the runoff phenomenon often occur of a few time steps.

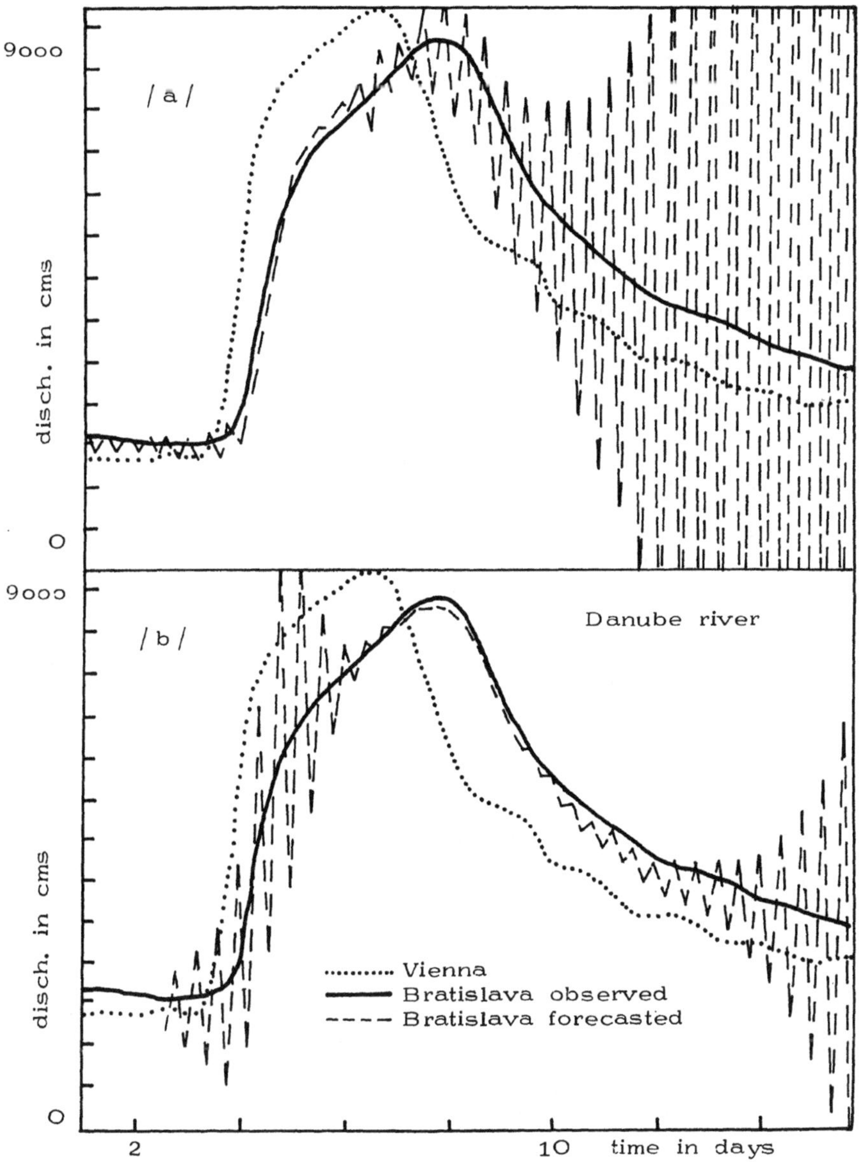

Figure 1 Example of /a/ "thrue", /b/ local divergence of the state space model of cascade of linear reservoirs - incorrect covariance Q.

In this contribution the problem of adaptive filtering is handled by using a two-level procedure /Minárik, 1983/ in which besides whiteness testing use is made for estimation also of information that cannot be utilized for quantitative modelling owing to their binary character or insufficient representativeness. The suggested method allows to utilize observations of remote sensing, radar reflections, precipitation and water-stage measurements and other hydrometeorological quantities and the continuous adaptation of their effects upon runoff.

It appears to be necessary to adequately differentiate the significance of the individual accessible data on the process. The applied two-level estimator issues from the empirical knowledge that in the case of the majority of real processes the final number of indefinite situations may be recognized. The significant conceptual or experimental information about that particular situation may be available. By analysing the uncertainties of models of long hydrological series it may be ascertained that for various phases of the runoff process certain values of the system uncertainty are optimal, that is certain values A, Q, R. In addition, it may be empirically proved that for various hydrological situations other basic system models are optimal, that is Eq. /1/ with an other structure of the observation matrix H.

The proposed estimator is working on two levels. On the first level there is a "fine" estimator which makes the final tuning of unknown parameters on the base of a "coarse" estimate. On the second level there is an estimator giving a coarse estimate of unknown parameters. The coarse estimator is based on a finite number of a priori models and its task is find out which model best fits the data.

Assume that we have a fine estimator defined by Eqs. /3/ - /7/, and a coarse estimator that is based on L a priori models of uncertain characteristics of the state space representation. The estimations $x(k/k)$ and $x(k+1/k)$ of the fine estimator serve as results of the entire algorithm for the determination of the required estimate of the instantaneous contribution of the subcatchment $z(k/k)$ and its prediction $z(k+1/k)$. Unknown quantities can be concentrated to the vector

$$a(k) = [\,Q(k),\; R(k),\; A(k)\,] \qquad /11/$$

Assume that we have L a priori models of uncertain situations defined by the parameter vectors a^i, $i = 1, \ldots, L$. The criterion of the whitness of the residual sequence for the selection of the best model can be used. The problem is to find a quantity expressing a measure of the deviation of the residual sequence

$$y^i(k) = z(k) - H(k)\; x^i(k/k) \qquad /12/$$

from the white gaussian sequence for all a^i. As it is shown in the previous section, the number of points that lie outside the confidence limits or the sum of that values can be a suitable criterion for the choise of the model a^i. The model a^i, for which this value is minimal, can be used as a coarse estimate of the unknown characteristics in the coarse estimator.

One way to discriminate the best model is to use a Bayesian approach. The application of Bayes rule yields the relationships

$$p(a^i/Z_k) = \frac{p(Z_k, a^i)}{p(Z_k)}$$

$$= \frac{p(Z_k/a^i)\, p(a^i)}{\sum_{i=1}^{N} p(Z_k/a^i)\, p(a^i)} \qquad /13/$$

where Z_k denotes in this section the sequence of all base measurements $z(0), z(1), \ldots, z(k)$, while $p(a^i/Z_k)$ is shorthand for $p(a{=}a^i/Z_k)$. In this notation $z(i)$ means $z(i) = [q_A, q_B, w]$ for the case of tributary contribution modelling. The $p(a^i/Z_k)$ are a posteriori probabilities and $p(Z_k/a^i)$ likelihood functions. It is for the recursive calculation of these quantities that the Kalman filter, or, more precisely, a bank of conditional Kalman filters, comes into its own. We have

$$p(a^i/Z_k) = \frac{p(z(k), Z_{k-1}, a^i)}{p(z(k), Z_{k-1})}$$

$$= \frac{p(z(k), a^i/Z_{k-1})\; p(Z_{k-1})}{p(z(k)/Z_{k-1})\, p(Z_{k-1})}$$

$$= \frac{p(z(k)/Z_{k-1}, a^i)\, p(a^i/Z_{k-1})}{\sum_{i=1}^{N} p(z(k)/Z_{k-1}, a^i)\, p(a^i/Z_{k-1})} \qquad /14/$$

The calculation of $p(z(k)/Z_{k-1}, a^i)$ is crucial, but is readily implemented for gaussian signal models. In this case, $p(z(k)/Z_{k-1}, a^i)$ is gaussian with mean $z(k/a^i)$ and covariance $E[\tilde{z}(k/a^i)\ \tilde{z}^T(k/a^i)]$, which we denote by $\Omega(k/a^i)$. The notation $\tilde{z}(.)$ means the innovation sequence of the i-th filter. That is /Anderson and Moore, 1979/, for a m-vector $z(k)$

$$p(z(k)/Z_{k-1}, a^i) = (2\pi)^{-m/2} |\Omega^{-1}(k/a^i)|^{1/2} \exp\{-\frac{1}{2}\tilde{z}(k/a^i)\Omega^{-1}(k/a^i)\ \tilde{z}(k/a^i)\} \qquad /15/$$

and $p(a^i/Z_k)$ can be calculated recursively from

$$p(a^i/Z_k) = c|\Omega^{-1}(k/a^i)|^{1/2} \exp\{-\frac{1}{2}\ \tilde{z}(k/a^i)\ \Omega^{-1}(k/a^i)\ \tilde{z}(k/a^i)\}\, p(a^i/Z_k) \qquad /16/$$

where c is a normalizing constant independent of a^i, chosen to ensure that

$$\sum_{i\ 1}^{L} p(a^i/Z_k) = 1$$

The quantities $\Omega(k/a^i)$ and $\tilde{z}(k/a^i)$ are available from the conditional Kalman filter covariance equations and the filter equations, resp..

By concentrating a posteriori probabilities using information from innovation processes into vector

$$s_1(k) = [p(a^1/Z_k), \ldots, p(a^L/Z_k)] \qquad /17/$$

and all probabilities of particular models based on the additional information Z_k about the process /i.e. that information not used in the base model/ into vector

$$s_2(k) = [p(a^1/Z_k), \ldots, p(a^L/Z_k)] \qquad /18/$$

we can define

$$s(k) = [s_1(k), s_2(k)]$$

and

$$S_k = [s(k), s(k-1), \ldots, s(k-M)] \qquad /19/$$

which can be used for final discrimination of the best model in time k. However, besides decision-making the coarse estimator produces coarse estimates in each k, that can be used to transfer into the fine estimator to influence its work.

In the preparatory stage it is possible to obtain the values of unknown characteristics Q, R, and A for certain typical situations in which hydrological processes often find themselves. These values may subsequently serve as basic estimates in the coarse estimator in the real time mode of identification. Thus it is possible obtain probabilities $s_1(k)$. The task of the decision-making mechanism of the coarse estimator is to find out which estimate is optimal in time k. To distinguish the type of hydrological situation in which the process is at moment, use can be made of additional information that influence is formalised by probabilities $s_2(k)$. In view of the fast changes in the behaviour of hydrological processes such additional information can be used for fast changes of coarse estimates. The advantage with the coarse estimator is that it is constantly monitoring all a priori models.

The two-level estimator includes an information link between the two estimators. There are several ways to transfer information from the coarse estimator to the fine estimator. Three possibilities are:

/i/ change the state estimate $x(k/k)$ and the error covariance $P(k/k)$ in the fine estimator by appropriate coarse estimates, as initial values for the next iteration of the fine estimator,

/ii/ change characteristics Q, R, A of the fine estimator that were determined in the last change of the a priori model by new values from the finite space of the coarse estimator,

/iii/ change /i/ and /ii/.

The task of the decision mechanizm of the coarse estimator is to indicate the necessity of the change of the work of the fine estimator, the kind of the change, and to minimize the computations.

5. APPLICATION

The suggested method was applied for the estimation and the short-term forecasting of tributary inflows in several river sections in Czechoslovakia mainly in river Váh, Rimava, and Czechoslovak part of Danube.

The model was applied to the River Váh Power System for purposes of the internal optimization of cascade production. After the Danube, the river Váh is Slovakia's largest river, with a lenght of 436 km, a catchment area of 11.601 km^2, a mean long-term runoff of 155 cms and mean utilizable potential of 1733 GWh/year. Hitherto 20 hydropower plants /mostly of day-to-day run-of-river type/, 2 barrages and 4 derivation channels have been built. The total installed power output of the cascade amounts 801 MW. The cascade is centrally controlled from the cascade dispatching. The dispatching computer system makes use of data of the telemetric system. From the hydrological point of view, there is seven interbasins limited by important specific profiles. In the catchment area there occur systematic areal changes of hydrological properties. Human interference occurs in the densely populated valley of the river with intesive economic activities.

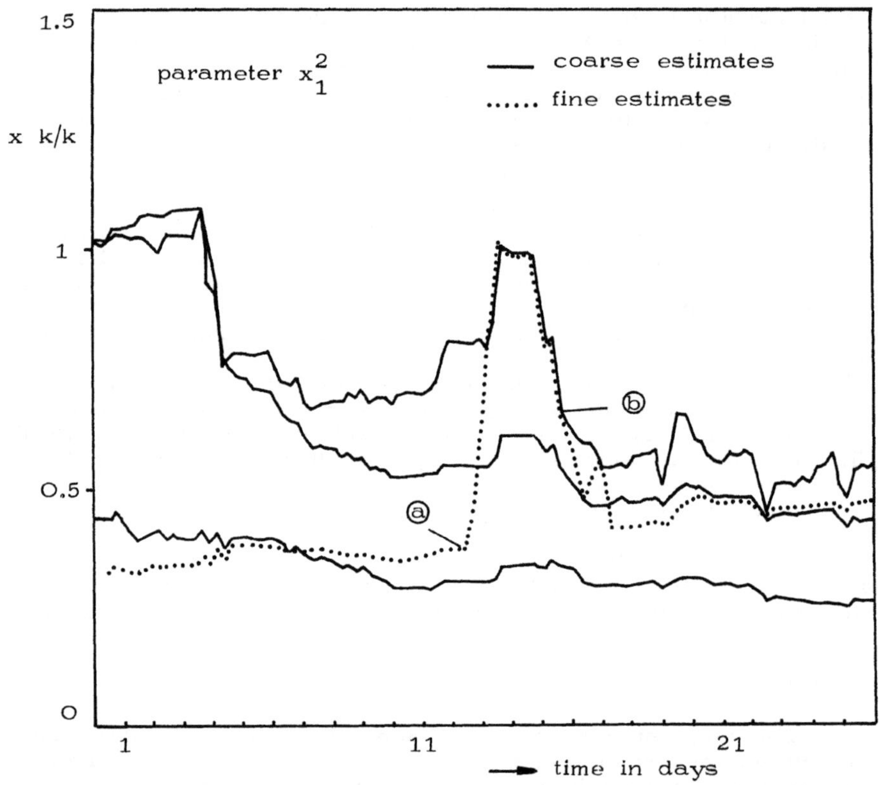

Figure 2 Coarse and fine estimates of the two-level estimator. /Váh river, section Krpeľany - Hričov/

The filtration algorithm was used for the estimation and short-term forecasting of inflows from individual interbasins. The algorithm was designed to render sufficient result during the useful life of the computer system /approximately 8 years/ with regard to the needs of power engineering and other users of the basin /Minárik, 1984/.

To derive an optimal model structure - Eq. /1/, several variants were tested of the values n_1, n_2, n_3 of matrix H. By the comparison of the mean error and mean square one the optimum type of the model /1/ was derived. In order to faciliate the choise of a priori models of the covariances in the coarse estimator, the method Leondes and Siu, (1980) for various hydrologic situations of the catchment area was used. From the time varying curve of the parameter values the Q^i, R^i, A^i models were derived keeping optimal track of the data in the various hydrological situations.

The course of coarse estimates of the parameter x_1^2 of vectors x^i, $i = 1, 2, 3$ in Figure 2 is indicated by full line. Dotted line depicts the course of the fine estimate of this parameter. In ⓐ point, after obtaining the change indication of runoff process situation /from steady state to rise/, fine estimates $x(k/k-1)$, $P(k/k-1)$ were replaced by corresponding coarse estimates $x^i(k/k-1)$, $P^i(k/k-1)$, as well as Q, R, A characteristics determined within the preceding change were replaced by corresponding Q^i, R^i, A^i values. In ⓑ point, within a new change of the situation, only Q, R, A characteristics were changed.

Type ⓐ change is suitable for sudden and single-valued changes of the runoff process, type ⓑ change for continual changes and for not completely single-valued ones.

To evaluate the quality of estimates is best the measure of coincidence of the measured level of a significant reservoir on the main stream with the calculated one on the basis of the balance of water and estimates of interbasin contribution. It follows from the analysis of the course of levels for example in the section Krpelany - Hričov /Minárik, 1984/ that the mean deviation of calculated levels after a 24 hours operation on the basis of the suggested estimater of the subcatchment contribution amounts to 0.51 m in the Hričov reservoir and is thus smaller than the deviation of 0.97 m in using contribution estimates by classical regression method.

CONCLUSION

Within the suggested method, the problem of the effective utilization of inaccurate and incomplete information on the behaviour of hydrological processes of water resource systems is handled in two parts. To immediately estimate the required hydrological forecasts use is made only of suitable quantitative measurements, namely within the ARMAX type model. Other quantitative measurements /mainly meteorological/ that are unsuitable for quantitative modelling owing to insufficient accuracy and representative features, are used together with all qualitative data, for discerning the type of hydrological situation. The suggested method has been utilized in hydrological forecasting within the water resource systems in Czechoslovakia.

REFERENCES

Anderson, B. D. O. and Moore, J. B. 1979. Optimal Filtering. Prentice-Hall, Englewood Cliffs

Cameron, R. J. 1980. An updating version of the Muskingum Cunge flow routing technique. In Proc. of Oxford Symposium, IAHS-AISH Publication No. 129

Harpin, R. and Cluckie, I. D. 1981. Real-time flow routing in the presence of high magnitude ungauged lateral inflows. Proc. International Symposium on rainfall-runoff modeling.

Kitanidis, P. K. and Bras, R. L. 1980. Real-time forecasting with a conceptual hydrologic model. Water Resource Research, Vol. 16, No. 6, pp. 1025-1044

Leondes, C. T. and Siu, T. K. 1980. Identification both of the unknown plant and noise parameters of the Kalman filter, Int. Journal Systems Science, Vol. 11, No. 6, pp. 711-720

Minárik, B. 1983. Two-level suboptimal filtering with application to water resources. In: G. A. Bekey and G. N. Saridis /Editors/, Identification and System Parameter Estimation, pp. 1409-1414, Pergamon Press, Oxford

Minárik, B. 1984. Real-time identification and control of the cascade of hydroelectric power plants, Proc. Int. Conference on Comput. Systems and Signal Processing, Bangalore, India

O'Connel, P. E., /Editor/ 1980. Real-time hydrological forecasting and Control, Institute of Hydrology, Wallingford, UK

Todini, E. and Bouillot, D. 1975. A rainfall-runoff Kalman filter model, In: G. C. Vansteenkiste /Editor/, System Simulation in Water Resources, North-Holland, Amsterdam

Wood, E. F. 1978. An application of Kalman filtration to river flow forecasting. In: Chiu, Choa-lin /Editor/, Proc. AGU Chapman Conference, Pittsburgh University Press, Pittsburgh

Wood, E. F. and Szöllösi-Nagy, A. 1978. An adaptive algorithm for analyzing short-term structural and parameter changes in hydrologic prediction models, Water Resources Researche Vol. 14, No. 4, pp. 577-581

FLOOD FORECASTING MODEL FOR CITANDUY RIVER BASIN

Kedar N. Mutreja, Superintending Engineer, J.P. Irrigation Department, 19/2 Tehri Colony, Roorekee 247 667, India

Yin Au-Yeung, Principal Engineer, PRC Engineering, Inc. 7935 E. Prentice Avenue, Bldg.40 W, Englewood, Colorado 80111

Ir. Martono, Project Manager, Project Citanduy, Pataruman Banjar, West Java, Indonesia

ABSTRACT. Three flood forecasting models were analyzed in determining the optimal model for short-term flood forecasting for the Citanduy River Basin: (1) deterministic models, (2) mathematical models, and (3) statistical models. The statistical model was selected for its simplicity and accuracy. Concurrent hourly water level data of different base stations on different tributaries were used to develop a multiple correlation model for predicting stages at the forecasting station. To prepare a forecast, the input data needed are the hourly water levels at upstream locations called the base stations. Nine such locations, each equipped with an automatic water level recorder, were identified. The accuracy of the developed model was tested at each forecasting station with independent data set, which were not used in developing the model. The forecasting accuracy was computed through the standard error of the estimate which was within 0.2 for all the forecasting stations, except at Cikawung which was 0.34.

INTRODUCTION

The Citanduy Project is located on the southern coast of the Island of Java, about 300 kilometers southeast of the capital city of Jakarta. The Lower Citanduy River Basin is comprised of three major tributaries: Cijalong, Cikawung, and Ciseel. The basin drains an area of 4,460 square kilometers. The slope of the Citanduy River ranges from about 100 meters per kilometer (m/km) in the upper reaches to about 0.2 (m/km) in the lower 40 kilometers.

There is a constant problem of flooding in the Lower Citanduy River Basin. Consequently, structural measures in the form of levees, etc. have been constructed to safeguard against a flood magnitude of once in 25 years frequency. As an integral part of the existing structural measures, a practical non-structural measure in the form of flood warning system was considered necessary and beneficial.

The drainage of the Citanduy River Basin is contained by the extinct Volcano Sawal and the hills which form the northern, western, and southern boundaries of the basin. The volume of runoff from the vol-

V. P. Singh (ed.), Flood Hydrology, 211–220.
© 1987 by D. Reidel Publishing Company.

cano and hill regions is very large for two principal reasons. First, the rainfall in the highlands is heavy, varying from an average of about 3,000 mm/year to a maximum of over 4,210 mm/year. Second, the percent of rainfall which runs off as surface runoff is large. On Volcano Sawal, where the annual precipitation is as high as 4,200 mm, two-thirds of the rainfall appears as surface runoff. In the lower hills where annual rainfall is less, the runoff is about 60 percent of the rainfall.

The drainage basin morphology of the Citanduy River consists of three main features, the most prominent being Volcano Sawal, which occupies approximately one-tenth of the drainage basin. The second is the upland region, which forms both the north and south, covering about 15 percent of the drainage area. A map of the Citanduy River System is shown on Figure 1.

FORECASTING MODEL

The models for flood forecasting being used at present can be divided into three categories: (1) deterministic models, (2) mathematical models, and (3) statistical models.

Deterministic Models

These models are based on an analysis of the hydrological and meteorological processes taking place in the river basins. These include comprehensive and sophisticated real-time rainfall-runoff models. These models have not been used for the following reasons:

1. Requires an extensive network of rain gages linked either by telephone lines or by reliable telemetry system.
2. Difficult to guarantee that the rainfall data would be available because of the problems associated with the maintenance of sophisticated and expensive telemetry system.
3. Computations will have to be made on a daily basis to update the soil moisture storage in the model. Highly trained staff will have to be on call 24 hours a day for the operation of the computer and preparation of the forecasts.
4. End result in these models is a discharge while the desired end product in flood forecasting must be at river level. Conversion of discharge to river level would involve the use of rating curves which are another source of potential inaccuracy.
5. The degree of complexity of the techniques used should be compatible with the general level of accuracy of the available data. An intricate technique will not remedy deficiencies in the data. Thus, for short-term forecasting of floods, their use is not called for.
6. Necessary to have a large computer continuously available for use at any time when forecasts may be required.
7. Rainfall-runoff models are required for river forecasting where long-term forecasts are required for the operation of the control structures such as reservoirs, etc. For flood forecasting, other simple approaches could be used.

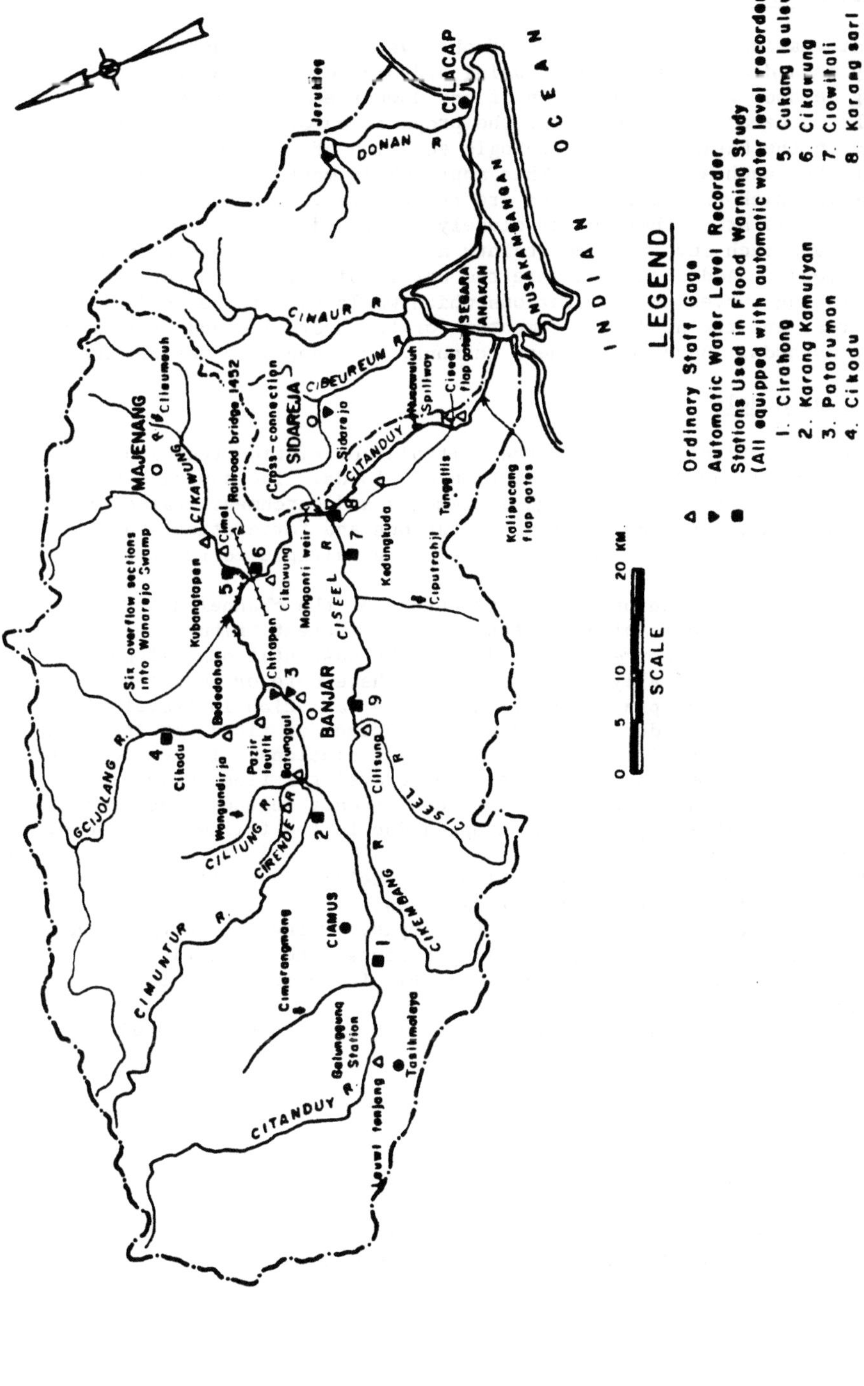

FIGURE NO. 1, CITANDUY RIVER SYSTEM

Mathematical Models

Mathematical models make use of observed discharges and their routing to a downstream location using hydrologic or hydraulic flood routing methods with the addition of any downstream tributary inflows.

The mathematical models, for the propagation of floods, are formulated on the basis of the hydrologic equation of continuity applicable for the reach as a whole. This approach is most suitable for streams in which the travel times are relatively long and the downstream inflows are either low or closely monitored.

The approach involves uncertainty about the river cross sections during floods. Added reasons were the unreliability of many of the discharge rating curves at the stations which would have to be used in the procedures, and the need of a computer for the hydraulic and hydrologic routing methods. Because of these reasons, this approach was not used.

Statistical Models

The statistical model uses the multiple correlation technique. This is commonly used in the prediction of floods. The approach varies according to the behavior of regime elements and catchment characteristics. Basically, only gages of base stations and forecasting stations are utilized in different forms in developing multiple correlation models.

Whenever more than one stream/tributary of the channel affects the event forecasts, stages of base stations of different tributaries are used in the multiple correlation models. The base stations, along the tributaries, are generally located so as to have more or less equal travel time as far as practicable, but due to variation in travel times of the same tributary during the different stages of flood, it is difficult to ensure an equal lag between different stations. To develop the multiple correlation, the average time of travel for flood waves in each tributary is determined and different sets of data regrouped as needed for the model. This approach was adopted due to the following reasons:

1. Approach is simple.
2. Does not necessarily need an elaborate telemetry system.
3. Does not need large computer facilities.
4. Data base needed for the model is simple. Owing to the existing network of automatic water level recorders, the short time interval data of river stages were available for a number of past floods.
5. Capability of utilizing direct river levels without the need to produce discharge rating curves.

DEVELOPMENT OF FLOOD FORECASTING MODEL

Outline of Adopted Model

Concurrently, hourly water level data of different base stations X_1, X_2, and X_3 on different tributaries have been used to develop a multiple correlation model for predicting water level, Y, at the forecasting station. The model used is:

$$(1) \qquad Y_t = a + bX_{1,\ t-T_1} + cX_{2,\ t-T_2} + dX_{3,\ t-T_3}$$

where:

Y_t	=	Dependent variable and is the water level to be forecasted at the forecasting station at time, t (MSL)
$X_{1,t-T_1}$	=	Independent variable and is the observed water level at the base station of first tributary at time $(t-T_1)$ with T_2 as the travel time between this station and the forecasting station (MSL).
$X_{2,\ t-T_2}$	=	Independent variable and is the observed water level at the base station of second tributary at time $(t-T_2)$ with T_2 as the travel time between this station and the forecasting station (MSL).
$X_{3,\ t-T_3}$	=	Independent variable and is the observed water level at the base station of third tributary at time $(t-T_3)$ with T_3 as the travel time between this station and the forecasting station.
a,b,c,d	=	Multiple correlation coefficients.

It should be noted from Equation (1) that the advance time for the forecast at the forecasting station is the least of the travel times, T_1, T_2, T_3. Therefore, to get a more advanced time, it is desirable that T_1, T_2, and T_3 are approximately equal. This could be achieved by installing, at some future date, automatic water level recorders at different locations with almost the same travel times to the forecasting station. However, for the present time, the existing data were utilized in predicting flood levels under different steps. The procedures involved in the development of the model are: (1) identifying flood forecasting stations; (2) identifying potential base stations; (3) preparation of data base; (4) estimation of travel times; (5) development of flood forecasting model; and (6) accuracy of forecasting.

Identifying Flood Forecasting Stations

The requirements for an ideal forecasting station are: (1) it must be strategically located; (2) it must be based on data from a gaging station which is equipped with an automatic water level recorder; and (3) it must have sufficient long-term records with continuous water level recording charts.

Considering these requirements, the four forecasting stations selected were Pataruman, Cikawung, Ciawitali, and Karangsari. However, for the purpose of illustration, the sample computations are being given only for the Cikawung forecasting station.

Identifying Base Stations

For forecasting the flood levels at the forecasting station, the base stations in the upper reaches of the watershed were selected. The basic requirements for an ideal base station are: (1) it should be strategically located; (2) it should be equipped with an automatic water level recorder; (3) the gage should be easily accessible; (4) the gage should be located in a straight reach of the river; (5) the record should be long and concurrent with the forecasting stations; and (6) travel time to the forecasting station should be about four to five hours. Too long travel time introduces more stochasticity into the model while too short travel time reduces the advance of forecast time.

Based on the above criteria, the base stations selected for Cikawung forecasting station are given in Table I.

Table I
Base Stations for Cikawung Forecasting Station

S#	Name of Gage	Type of Gage	Name of River	Catchment Area (km^2)	Gage When Established Automatic	Zero of the Gage (MSL)
1	Pataruman	Automatic	Citanduy	1,406	May 20, '83	20.00
2	Cikadu	Automatic	Cijalang	382	May 1972	96.895
3	Cukangleuleus	Automatic	Cikawung	670	Oct.14, '80	9.00

Preparation of Data Base

The points considered for preparing the data base were: (1) data collected only from flood records, (2) concurrent data from base stations as well as the forecasting station, (3) the data after the year 1980 has been used in the analysis as the data thereafter represent the present situation of flood control measures with levees constructed downstream from Banjar, and (4) water stages have been computed into mean sea levels for use in the model to avoid confusion due to change in zero elevations of different gages at some future date.

In the forecasting model of Equation (1), different independent variables X_1, X_2, and X_3 are the water levels at different base stations observed at different times. For example, X_1 is observed T_1 hours before the time of forecast, t at the forecasting station where T_1 is the travel time between this base station and the forecasting station.

Thus, the water levels used in the model of Equation (1) are observed at different times depending on the travel times between the base stations and the respective forecasting stations. To estimate the travel time, the approaches commonly used are: (1) streamflow routing, (2) Kirpich formula, and (3) study of historical record.

The technique of streamflow routing was discarded because it required use of: (1) a rating curve which introduces an element of uncertainty; (2) cross sections which change from flood to flood and thereby introduce another source of error; (3) time-consuming analysis at the time of flood forecasting; and (4) a computer which is an expensive proposal.

Kirpich gave a formula for time of concentration as below:

(2) $$t_c = 0.0078\ L^{0.77}\ S^{-0.385}$$

where:

t_c = Time of concentration (min).
L = Maximum length of travel of water (ft).
S = Slope, equal to H/L where H is the difference in elevation between the most remote point on the basin and the outlet (ft).

Basically, the formula is for computing the time of concentration. Nevertheless, it was used for estimating the travel time between the stations by taking L and S of this reach. However, the results obtained are generally very approximate. Travel times of past historical events were studied between different reaches. In fact, travel time changes from flood to flood. The average travel time of several observed floods has been computed. However, this alone does not give very convincing results due to contributions from different tributaries at forecasting stations.

The authors tried a statistical approach to estimate the travel time as given by (Mutreja 1985). Let there be two sections, 1-1 and 2-2, in a river with a slope, S, as shown in Illustration 1. The distance between the two sections is L. If the travel time between Section 1-1 and Section 2-2 is T hours, the cross correlation between the gates at these two sections at a lag of T hours would be the maximum. This property has been utilzied to estimate the travel time of the reach by processing the historical water level data.

Illustration 1: Travel Time

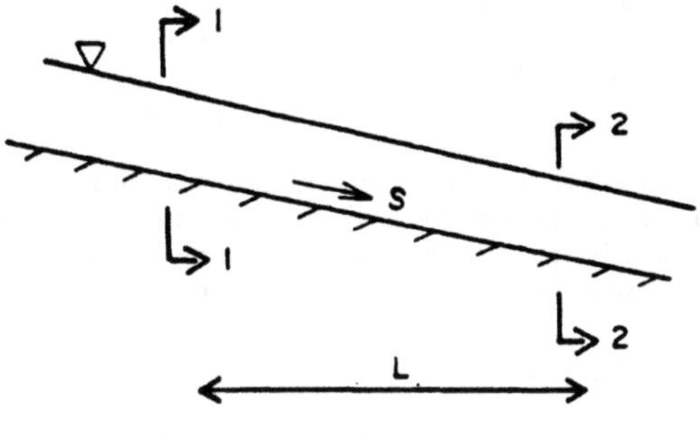

The procedure consists of collecting the water level data of floods at Section 1-1 for the N-th hour and at Section 2-2 for the (N+T)-th hour in a tabular form. By taking different values of T, prepare different data tables such that each data table corresponds to one value of assumed T. To compute the cross correlation of the water level data of these two sections at different lags, the cross correlation of the data on each table is computed. The value of T corresponding to the data table giving the maximum cross correlation is the travel time of the reach. The travel times estimated for different base stations by this approach are given in Table II. Based on the judgment of the hydrologist, the travel times are then adopted as shown in Table II.

Table II
Travel Times for Cikawung Forecasting Station

Base Station	Statistical Approach (h)	Historical Record (h)	Kirpich Formula (h)	Adopted Travel Time (h)
Pataruman	5.0	5.7	18.57	5.0
Cikadu	6.0	6.4	12.66	6.0
Cukangleuleus	1.0	0.66	5.55	1.0

CIKAWUNG FLOOD FORECASTING MODEL

Considering the travel times adopted as per Table II, the flood forecasting model of Equation (1) modifies for Cikawung station as:

$$Cik_t = a + b\ Pat_{t-5} + c\ Ckd_{t-6} + dCkl_{t-1} \tag{3}$$

where:

- Cik_t = the water level to be forecasted at Cikawung at time, t (MSL)
- Pat_{t-5} = the observed water level at Pataruman five hours before time, t (MSL)
- Ckd_{t-6} = the observed water level at Cikadu six hours before time, t (MSL)
- Ckl_{t-1} = the observed water level at Cukangleuleus one hour before time, t (MSL)
- a,b,c,d = multiple correlation coefficients

To estimate the multiple correlation coefficients a,b,c, and d, the observed historical water level data were analyzed. It is clear from Equation (3) that the final forecast at Cikawung could be made one hour in advance. To gain more time, forecasts would be issued in two steps.

Model for First Step Forecast

This will be issued five hours in advance and shall be based only on the observed water levels at Pataruman and Cikadu base stations. The model used is:

$$Cik_t = -118.06 + 0.47\ Pat_{t-5} + 1.2\ Ckd_{t-6} \tag{4}$$

Model for Second Step Forecast

This will be issued one hour in advance and shall be based on the observed gages at all the base stations of this forecasting station.

TABLE III

Accuracy Test on Independent Data for Cikawung Forecasting Station

Date	Time of Warning t (Hours)	Water Level at Pataruman at Time (t-5) Hours (MSL)	Water Level at Cikadu at Time (t-6) Hours (MSL)	Water Level at Cukangleuleus at Time (t-1) Hours (MSL)	Observed Water Level at Cikawung (MSL)	Forecasted Water Levels in Different Steps (MSL) 1st Step (10 Hours Advance)	2nd Step (5 Hours Advance)	3rd Step (1 Hour Advance)	Percentage Error of Final Forecast With Respect to Water Depth (%)
19 Feb. 84	21.0	21.20	98.56	13.56	11.74	N.P.	10.188	11.773	0.39
	22.0	21.31	98.75	13.57	11.81	N.P.	10.456	11.925	1.36
	23.0	21.35	99.25	13.61	11.89	N.P.	10.844	12.137	2.88
	24.0	21.32	99.07	13.63	11.94	N.P.	10.844	12.137	2.29
20 Feb. 84	1.0	21.31	99.07	13.64	11.97	N.P.	10.840	12.141	1.98
	2.0	21.34	99.40	13.66	12.02	N.P.	11.250	12.333	5.60
	3.0	21.36	100.07	13.68	12.02	N.P.	12.063	12.684	7.64
	4.0	21.37	99.75	13.68	11.94	N.P.	11.684	12.534	6.90
	5.0	21.15	99.40	13.68	11.89	N.P.	11.161	12.252	4.23
	6.0	21.07	99.23	13.67	11.80	N.P.	10.919	12.120	3.78
	7.0	21.00	99.13	13.64	11.74	N.P.	10.766	12.009	3.20
	8.0	20.94	99.02	13.63	11.65	N.P.	10.606	11.916	3.20

Standard Error of Estimate = 0.34

N.P. = Not possible to predict because Pataruman water level cannot be predicted as the Cirahong gage was not available because of its being out of order during this period.

The model used is:

$$(5) \qquad Cik_t = -58.571 + 0.518\ Pat_{t-5} + 0.482\ Ckd_{t-6} + 0.974\ Ckl_{t-1}$$

ACCURACY OF FORECASTING

The accuracy of the developed model has been checked with independent data. For this, a number of data sets not used in the development of the models, were reserved for making independent checks on the derived relationships of Equations (4) and (5). Table III shows the results obtained from an independent check of one flood of February 1984. The error of forecast, with respect to the water depth, is less than eight percent. The accuracy expressed as the standard error of the estimate has been computed as 0.34 which is satisfactory.

In order to simplify the application of these procedures, separate nomographs for each forecasting step have been prepared. Figure 2 shows a typical nomograph of second step forecast at Cikawung. This can be used at the Flood Operations Center by non-skilled technicians, if necessary.

To read from nomograph, take Point A on abscissa of first quadrant corresponding to the observed water level at Pataruman. Locate Point B on curve (1) and Points C and D in the direction of arrows. Draw line DF parallel to curve (3). Locate Point E on the ordinate of first quadrant corresponding to the observed water level at Cikadu and draw a horizontal line from E to location Point F. From F, draw a vertical line to locate Point G on curve (4). From G, draw a horizontal line to location Point H which is the forecast of water level at Cikawung.

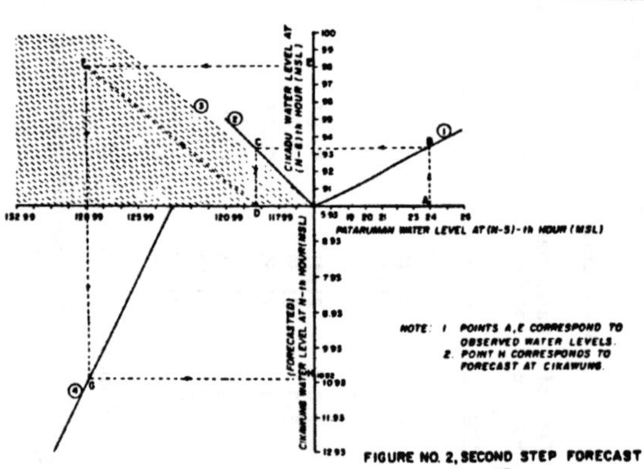

FIGURE NO. 2, SECOND STEP FORECAST AT CIKAWUNG FORECASTING STATION

REFERENCE

Mutreja, K.N., (1985). "Applied Hydrology", Tata McGraw-Hill Publishing Company, 12/4 ASAF Ali Road, New Delhi, India 110002.

UNCERTAINTY AND CONFIDENCE IN ESTIMATING FLOOD FREQUENCIES FROM HURRICANES AND NORTHEASTERS AT UNGAGED COAST LOCATIONS – A NEW METHODOLOGY

Stergios Dendrou
ZEI Inc., Suite 703
5111 Leesburg Pike
Falls Church, VA 22041

Basil Dendrou
ZEI Inc., Suite 703
5111 Leesburg Pike
Falls Church, VA 22041

ABSTRACT. Coastal flooding along the Atlantic coast north of Cape Hatteras is caused by hurricanes and by northeasters. The best available tidal gage records are about 100 years long, so that estimation of flood levels of a 50 year return period and above is fraught with statistical error. This, together with the fact that hydrodynamically consistent flood profiles are usually required everywhere along the coastline of communities (not just at the gage), led to the development and use of the Joint Probability Method (JPM). This is a simulation based method, whereby surges are simulated via a numerical hydrodynamic model. This model, in turn, is used to derive flood level statistics (frequencies) from statistics of the meteorological cause. Two elements necessary in applying the JPM methodare a parameterization of the storm, and the extraction of statistics for these parameters from the meteorologic record. A synthetic sample of storms is then used to simulate representative surges covering the entire range of flood levels. In the 15 years of use of the JPM method the sample size of synthetic storms has been determined heuristically to vary from 300 to 800. The central question addressed in this paper is: given the required accuracy of estimation of flood levels, what is the optimal sample size of synthetic storms which will achieve the required accuracy without producing redundant information? The question is approached from the viewpoint of confidence intervals on the estimated frequency of exceedence of a given flood level. In contrast, gage analyses provide estimates of flood levels for given exceedence frequencies and the corresponding confidence intervals are given on the flood levels rather than on the frequencies. The flood frequency estimation error expressed as a confidence interval is complemented by determination of the accuracy of the estimate. This is achieved by use of a second-moment first order technique in conjunction with multiple simulations according to a stratified sampling scheme. Thus, the level of acceptable error controls the number of synthetic storms that need to be simulated (the sample size), and the required accuracy dictates which synthetic storms to simulate.

1. INTRODUCTION

Coastal flooding originates from meteorologic phenomena such as hurricanes and northeasters. These storms may be quantitatively described by a barometric pressure field and the attendant wind field,

V. P. Singh (ed.), Flood Hydrology, 221–230.
© 1987 by D. Reidel Publishing Company.

both centered around the eye of the storm and following the track of the storm. All other characteristics, size, intensity, duration, etc., vary with the type of storm. These meteorologic conditions, in turn, produce a rise in water levels which results in a surge that is driven by the storm toward the coast. Flood studies usually consist of determining the frequency of the flood levels based on the frequency of occurrence of such storms and their related meteorologic parameters. Two types of approaches exist: those based on historical records at tidal gages and those based on a cause-effect simulation of surges from the meteorologic causes. The second category involves numerous numerical simulations. The Joint Probability Method (JPM), Myers (1970), is a simulation-based method that has been used over the past 15. The number of synthetic storms (sample size) used in the numerical experiments and their selection are derived heuristically from experience. The purpose of this paper is to provide a method for systematic selection of the synthetic sample of storms. This method uses the concepts of error in the frequency estimate and accuracy of the flood level estimate, and it is based on the theory of stratified sampling. The statistical formulation of the problem is presented first, followed by a description of the simulation based methods. Then, error evaluation on the frequency estimate and accuracy of the corresponding flood level are introduced separately. They are both used jointly in a coherent methodology for the selection of the synthetic sample of storms. An example is provided from a case study for New York City, for both hurricanes and northeasters.

2. PROBLEM FORMULATION

The frequency of flooding, whether it be coastal or riverine, is conventionally defined in terms of the return period. Return period is, in standard statistical terms, the expected value (or mean) of the waiting time between exceedences of a flood level H. For example, the 100-yr level, on the average, is equaled or exceeded once every one hundred years. The probability that there may be more than one exceedence is finite (meaning non zero). The return period also has an alternate, equivalent interpretation, which derives from the basic statistical model used in describing extreme events. This model is as follows: assuming that there is a one-to-one correspondence between the meteorological causes (storms) and their associated high water elevations, their joint occurrence may be viewed as a Bernoulli point stochastic process. The probability of exceeding a flood level H a given number of times in a given period of years is then binomially distributed, and conversely, the trial number N (number of years) at which the first success occurs has a geometric distribution with mean value 1/P(H), where P(H) is the probability of exceeding level H in any given year. The trial number of years, N, is the return period and it is seen to be equal to the inverse of the average probability of exceeding flood level H in any year. Thus, the 100-year level is that level whose probability of being exceeded in any year is one percent (0.01). Flood studies by convention deal with the determination of the relationship between flood levels H and the corresponding yearly probability of exceedence P(H), or its inverse, the return period. In gage analyses, the N yearly peak levels from the historical record of a gage spanning N years are ranked from high to low, and the corresponding yearly probability of exceedence P(H) is estimated as the rank over the length of record N or some other "plotting formula," Chow (1964) (e.g., the Hazen formula which provides unbiased estimators). In gage analyses, confidence intervals are constructed around the flood levels of the historical record as illustrated in Figure 1.

In contrast with gage analyses, simulation based methods consist of the following three basic steps: (1) extract statistics of the meteorologic properties of hurricanes from the historical record; (2) simulate surges from the storm forcing; and (3) infer statistics on surge levels from meteorologic statistics. The radius to maximum winds, the central pressure depression and the forward storm velocity define the surge producing potential of the storm as illustrated in Figure 2. The frequency of the storm is computed as the average frequency of storms in events per year, multiplied by the probability that the storm jointly possesses a specific set of characteristics. Where statistical independence is not a valid assumption, conditional probabilities may be used. This procedure is summarized for hurricanes in Table I. The expected frequency of a synthetic storm, and therefore, the frequency of the peak surge that it produces at any location, are computed as the product of the average frequency of storms in events per year times the joint probability of the storm parameters. This is shown in Table I under "Frequency of Synthetic Storm i." The probability distribution for the central pressure depression, the radius of maximum winds, and the forward velocity are derived from a statistical analysis of historical hurricanes that have affected the area. Probability distributions for the direction of storm movement and the latitude at which the storm crosses the study area are also developed from this sample of historical hurricanes. Thus, for a range of storm tide elevations, the frequency density is simply the summation of the frequencies of the individual synthetic storm tides that fall into that particular range, shown as "Flood Event Frequency, Level K" in Table I. Conversely, the cumulative flood frequency for a given elevation is computed as the summation of the individual elevation frequency densities for all elevations greater than, or equal to, the given elevation as shown in Table I. A similar method has been developed for northeasters or winter storms, Dendrou et al. (1983).

3. ERROR IN FREQUENCY ESTIMATES

The error in estimating frequencies by the simulation based method is evaluated in terms of confidence intervals on the frequency estimates. Estimation of the confidence intervals associated with the frequency-elevation curves evaluated by the JPM requires invoking the basic "Bernoulli Trial" binomial distribution model for the probability of realizing a flood event exactly X times in N independent trials (years), Benjamin & Cornell (1970). Conversely, the trial number N (number of years), at which the first success occurs has a geometric distribution, with mean value $1/P(H)$ and variance $(1-P)/P^2$. The mean (expected) value is called <u>return period</u> and is the inverse of the annual exceedence probability P(H). Thus, the JPM provides estimates of the probabilities P(H) for all levels H by carrying experiments (Bernoulli trials) in parallel for all levels H. Since P(H) is binomially distributed, the standard error of the estimate is provided by the standard deviation of the binomial distribution, Hammersley et al. (1967): $\sigma_p = (P(1-P))^{1/2}/N$, where P is the annual exceedence probability evaluated by the last equation in Table I, separately for northeasters and hurricanes; and N is the sample size in equivalent number of years calculated as the product of the number of synthetic storms times the average return period (years). The standard error on the total probability exceedence of all events, hurricanes, and northeasters is provided by the law of the summation of two independent random variables, which states that the variance of the sum is equal to the sum of variances.

Application of the standard error estimation for the exceedence probabilities at the Battery, New York is shown in Table II. Figure 1 compares the exceedence probability estimates and the associated confidence intervals respectively from the gage analysis and from the Joint Probability Method. The JPM method is clearly superior within the range of the historical record and provides a systematic way for estimating return periods beyond the historical record. It is important to notice that the size of the synthetic sample also depends on the average frequency of storms. For example, for infrequent storms such as hurricanes in the latitudes of New York, a small sample (250 storms) may be representative of over one thousand years of record as shown in the second footnote of Table II. More frequent storms require larger synthetic samples. Therefore, we established that the error on the frequency provides a means to determine the required size of the sample of synthetic storms.

4. ACCURACY OF FLOOD LEVEL ESTIMATES

The accuracy of flood level estimates as obtained from the simulation based method depends on the simulation model used as well. In coastal flood studies, the hydrodynamic simulation model provides a numerical solution to the 2-D, vertically integrated, Navier-Stokes equations. In the x-direction they take the form:

$$\frac{\partial u}{\partial t} + \frac{\partial u}{\partial x} + v\frac{\partial u}{\partial y} = -g\frac{\partial H}{\partial x} + fv - \frac{\tau_{bx}}{\rho H} + \frac{\tau_{wx}}{\rho H} - \frac{1}{\rho}\frac{\partial p}{\partial x} \qquad (1)$$

$$\frac{\partial H}{\partial t} + \frac{\partial uH}{\partial x} + \frac{\partial vH}{\partial y} = 0 \qquad (2)$$

where u, v are the horizontal velocity components, f is the Coriolis parameter, p is the atmospheric pressure and bx and wx, respectively, the bottom friction and the wind stress. The wind stress and the pressure gradient are the forcing terms in Eq. (1) and may be represented as a function of the storm parameters, namely the wind velocity and central pressure depression. The numerical solution of Equations (1)-(2), by a numerical technique such as finite elements or finite differences, takes the form:

$$A \begin{bmatrix} u,v \\ H \end{bmatrix} = F \quad \text{or} \quad \begin{bmatrix} u,v \\ H \end{bmatrix} = A^{-1} F \tag{3}$$

where A is the matrix of hydrographic parameters and F is the forcing term, a function of the meteorological parameters x_i. Focusing attention on the peak water elevations, the first-order uncertainty analysis, Papoulis (1965), takes the first moment of H as the deterministic solution of Equation (3), and approximates the second moment as follows:

$$\mathrm{Var}(H) = \sum_i \sigma_{x_i}^2 \left\{\frac{\partial H}{\partial x_i}\right\}^2 \tag{4}$$

where the partial derivative on the r.h.s. may also be obtained as the solution of the following linear system:

$$A \cdot \frac{\partial H}{\partial x_i} = \frac{\partial F}{\partial x_i} - \frac{\partial A}{\partial x_i} \cdot \begin{bmatrix} u,v \\ H \end{bmatrix} \tag{5}$$

All terms on the r.h.s. are known, and the solution of Equation (5) involves the inversion of the same matrix A, as in Equation (3).

The variance of H evaluated by Equation (4) or its square root, the standard deviation, provides an estimate of the accuracy of the flood levels in relation to the standard deviation σx_i of the meteorologic parameters. This variance depends both on the variance of the meteorologic parameters (first term on the r.h.s. of Equation (4)) and on the sensitivity of the hydrodynamic model (second term on the r.h.s. of Equation (4)). Often, the accuracy of the flood level estimates is dictated by the overall study objectives. For example, for flood insurance studies, the required accuracy is $\pm$ 1/2 ft. Thus, Equation (4) provides a means to ensure that the synthetic sample of storms achieves the required accuracy. Equation (4) can also be used to optimize the selection of the synthetic storms. As illustrated in Figure 3, the error of frequency estimates and the accuracy of flood level estimates define a two-dimensional error band, where the horizontal axis defines the frequency error, and the vertical axis the flood level accuracy. Since the required accuracy is usually constant over the entire range of flood levels, a stratified sampling technique, Rubinstein (1981), can be used in conjunction with Equation 4 for an optimal selection of synthetic storms. Each meteorologic parameter is partitioned into ranges for which the median value and the standard deviation are determined. These values define the r.h.s. of Equation 4 and thus allow estimation of the accuracy of resultant flood level. Unlike the traditional Monte Carlo Method where numerous random simulations are performed, only one simulation per range is performed, allowing for complete control over the required computational effort and the resultant accuracy.

5. SUMMARY AND CONCLUSIONS

Simulation based methods are almost always necessary for coastal flooding studies because tidal gages do not adequately cover all points of interest along the coastline, and are seldom long enough for the return period of interest. For example, the regulatory level of the Flood Insurance Administration (FIA) of the Federal Emergency Management Agency (FEMA) on which land zoning ordinances and flood insurance rates are based is the 100-year level. A 100-year record (among the largest existing records) would provide a 100-year flood level estimate with an accuracy of + 1 ft (0.3 m). The required accuracy for FIA is $\pm$ 0.5 ft. Simulation based methods require many expensive hydrodynamic simulations. A stratified sampling technique is proposed here that streamlines the effort involved in simulation based methods and scales it according to the required level of accuracy as follows: (1) Determine the approximate sample size by evaluating the error on the frequency estimates; (2) Segregate the meteorologic parameters into ranges such that all combinations approximately form a sample of the previously determined size; (3) For each range of the meteorologic parameters, determine the median and standard deviation; (4) Use these statistics in conjunction with the hydrodynamic simulation model to evaluate the corresponding flood level accuracy, according to Equation 4: (5) Adjust ranges until desired accuracy is achieved. This procedure can result in substantial savings as compared with the Monte Carlo technique. More importantly, it provides objective statistical criteria to use in achieving the required level of accuracy.

6. REFERENCES

Benjamin, J.R., and C.A. Cornell, 1970. Probability, Statistics, and Decision for Civil Engineers, McGraw-Hill, New York.

Chow, V.T., 1964. Statistical and Probability Analysis of Hydrologic Data, Section 8-I, Figures 8-I-9, 8-I-10, in Handbook of Applied Hydrology, Ven Te Chow Editor, McGraw-Hill.

Dendrou, S.A., Taylor, R.S., and Dendrou, B.A., 1980. Second Moment-Inference Analysis Approach for Probability Evaluation of Storm Surges. in: C.Y. Kuo (Editor), Urban Stormwater Management in Coastal Areas, pp 23-32, ASCE, New York.

Dendrou, S.A., Moore, C.I., and Myers, V.A., 1983. Frequency of Northeaster Tides in Coastal New York. in: ASCE Hydrauilcs Division Specialty Conference, Cambridge, Massachusetts, 9-12 August.

Gumbel, E.J., 1954. Statistical Theory of Extreme Values and Some Practical Applications, U.S. Department of Commerce, National Bureau of Standards, Applied Math, Series No. 33.

Hammersley, J.M. and D.C. Handscomb, 1967. Monte Carlo Methods, Metheum and Co. Ltd.

Myers, V.A., 1970. Joint Probability Method of Tide Frequency Anaysis. ESIA Technical Memorandum WBTM, Hydro II, National Weather Service, NOAA, Silver Spring, Maryland.

Papoulis, A., 1965. Probability, Random Variables and Stochastic Processes, McGraw-Hill, New York.

Rubinstein, R.Y., 1981. Simulation and the Monte Carlo Method, Wiley-Interscience, pp 118-119, 131-135, 187-193.

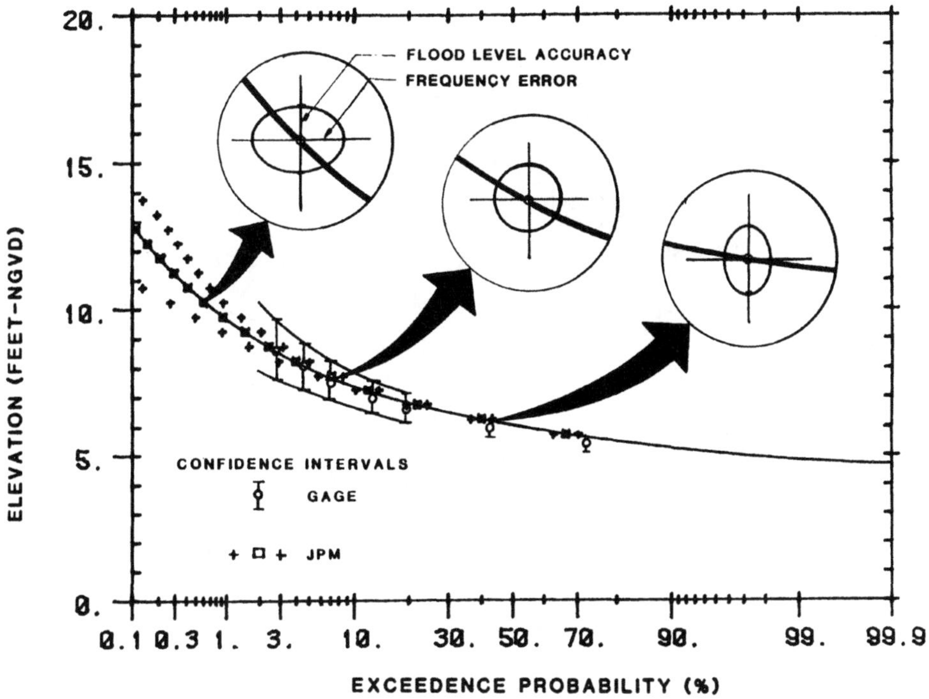

Figure 1 Illustration of Confidence Intervals

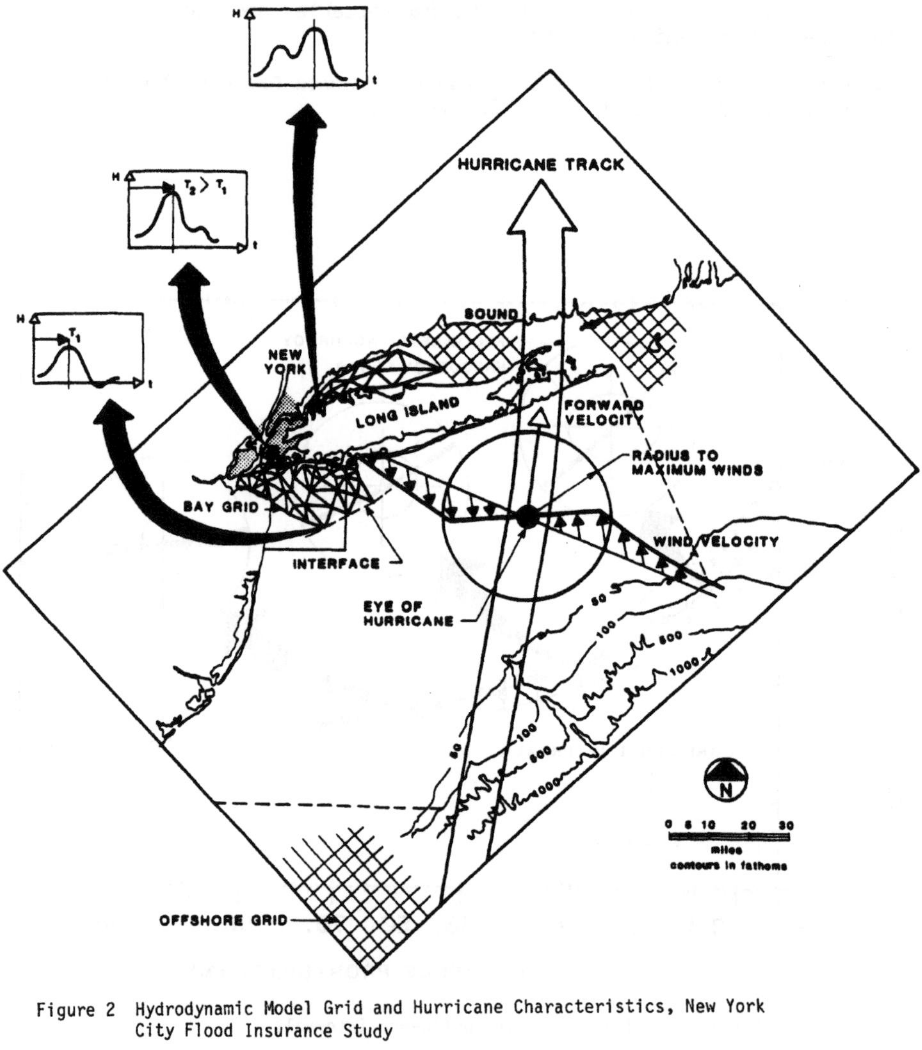

Figure 2 Hydrodynamic Model Grid and Hurricane Characteristics, New York City Flood Insurance Study

TABLE I

OVERVIEW OF JOINT PROBABILITY METHOD FOR HURRICANES

Storm Properties	Statistical Parameters	Probabilities
Storm Frequency	Track frequency	f - Frequency (events/year)
Storm Intensity	Central Pressure Depression, P_o	$P\ (P_i)$
Forward Speed	V_f	$P\ (V_{fi})$
Storm Size	Radius to maximum winds R	$P\ (R_i)$
Storm Track	Storm Angle, θ	$P\ (\theta_i)$
	Crossing Latitude, X	$P\ (X_i)$
Astronomic Tide	Tide Amplitude, A	$P\ (A_i)$
	Phase, Ph	$P\ (Ph_i)$

- Frequency of Synthetic Storm "i" (density)

$$F_{si} = f \cdot P(P_i) \cdot P(V_{fi}) \cdot P(R_i) \cdot P(\theta_i) \cdot P(X_i)$$

- Flood Event Frequency, Level K (density)

$$F_{hk} = \Sigma F_{si} \cdot P(A_j) \cdot P(Ph_j) \quad \text{(all contributing i's and j's)}$$

- Cumulative Flood Event Frequency, Level K

$$P(H_K) = \sum_{k=1}^{K} F_{hk}$$

TABLE III

CONFIDENCE INTERVALS ON THE JOINT PROBABILITY METHOD

	Northeasters*			Hurricanes**			Total	
H (ftMSL)+	P(H)	σ_n^2 X10^{-4}	σ_n	P(H)	σ_h^2 X10^{-4}	σ_h	P(H)	σ
14.75	$1.1x10^{-5}$	0.00055	0.00023	0.00019	0.0012	0.00035	0.0002	0.0004
14.25	$4.2x10^{-5}$	0.0021	0.00046	0.00025	0.0017	0.00041	0.0003	0.0006
13.75	$9.5x10^{-5}$	0.0047	0.00069	0.00037	0.0024	0.00049	0.0005	0.0008
13.25	$2.05x10^{-4}$	0.010	0.0010	0.00052	0.0034	0.00059	0.0007	0.0012
12.75	$3.28x10^{-4}$	0.016	0.0013	0.0074	0.0049	0.00070	0.0011	0.0015
12.25	$4.58x10^{-4}$	0.023	0.0015	0.0010	0.0068	0.00083	0.0015	0.0017
11.75	$6.34x10^{-4}$	0.032	0.0018	0.0014	0.0095	0.00097	0.0021	0.0020
11.25	$8.85x10^{-4}$	0.042	0.0021	0.0026	0.0135	0.00116	0.0029	0.0024
10.75	0.00120	0.059	0.0024	0.0029	0.0193	0.00139	0.0041	0.0028
10.25	0.00167	0.083	0.0029	0.0044	0.0287	0.00169	0.0060	0.0033
9.75	0.00296	0.148	0.0038	0.0063	0.0417	0.00204	0.0093	0.0044
9.25	0.00537	0.267	0.0052	0.0094	0.0619	0.00248	0.0148	0.0057
8.75	0.0101	0.499	0.0071	0.0137	0.0894	0.00299	0.024	0.0077
8.25	0.0191	0.936	0.0096	0.020	0.131	0.00362	0.039	0.010
7.75	0.0351	1.69	0.013	0.036	0.228	0.00478	0.071	0.014
7.25	0.0650	3.04	0.017	0.055	0.345	0.00587	0.120	0.018
6.75	0.132	5.73	0.024	0.085	0.519	0.00720	0.218	0.025
6.25	0.265	9.75	0.031	0.133	0.764	0.0874	0.399	0.032
5.75	0.453	12.39	0.035	0.210	1.09	0.0105	0.664	0.037
5.25	0.681	10.87	0.033	0.339	1.48	0.0121	1.019	0.035
4.75	0.858	6.1	0.025	0.565	1.62	0.0127	-	-
4.25	0.966	1.66	0.013	0.99	0.0655	0.00256	-	-

* N_n = 42 simulations/ 0.21 events/year = 200 years

** N_h = 252 simulations/0.167 events/year = 1512 years

+ 1 ft = 0.3048 m

STAGE-FREQUENCY CURVES FOR FLOODING DUE TO WAVE OVERTOPPING OF SEAWALLS

Thomas A. Hardy
Coastal Engineering Research Center
US Army Engineer Waterways Experiment Station
Vicksburg, MS 39180

Currently
Department of Civil and Systems Engineering
James Cook University
Townsville, Qld 4811 Australia

ABSTRACT. Curves relating flood levels with their frequency of occurrence are established for a coastal area just north of Boston. The principal cause of flooding is the overtopping of seawalls by storm waves. The project procedure involves the use of five modeling systems, including numerical storm surge, numerical spectral wave, physical wave overtopping, flood routing, and probability models. The main focus is on the probability model with the other four models only briefly described.

1. INTRODUCTION

The Coastal Engineering Research Center (CERC) of the US Army Waterways Experiment Station (WES) has been actively involved in determining the frequency of coastal flooding. The final product of such a study is a stage-frequency curve which relates the elevation of flood waters to the average waiting time between floods of equal or greater severity. The ordinate of these curves is stage (water level elevation) and the abscissa is return period. A typical study would involve flooding caused by the inundation of low lying coastal areas by the combined effects of storm surge (the component of total water level caused by wind and pressure effects) and astronomical tide. A recently completed study took this process one additional step, by developing the frequency of flood levels caused when seawalls were overtopped by storm waves.

The US Army Engineer Division, New England (NED) requested that CERC conduct numerical and physical model studies to determine the frequency of flood levels at various locations in Revere, Saugus, and Lynn, Massachusetts. This paper discusses one segment of these studies, the establishment of stage-frequency curves at Roughans Point.

The study area is located in the city of Revere, Massachusetts which is located immediately north of Boston. Roughans Point (Figure 1) is a 55 acre residental area located on a point of land which is partially protected by seawalls from coastal flooding along both the northern and eastern boundaries. The concave shaped surface of the

V. P. Singh (ed.), Flood Hydrology, 231–240.
© 1987 by D. Reidel Publishing Company.

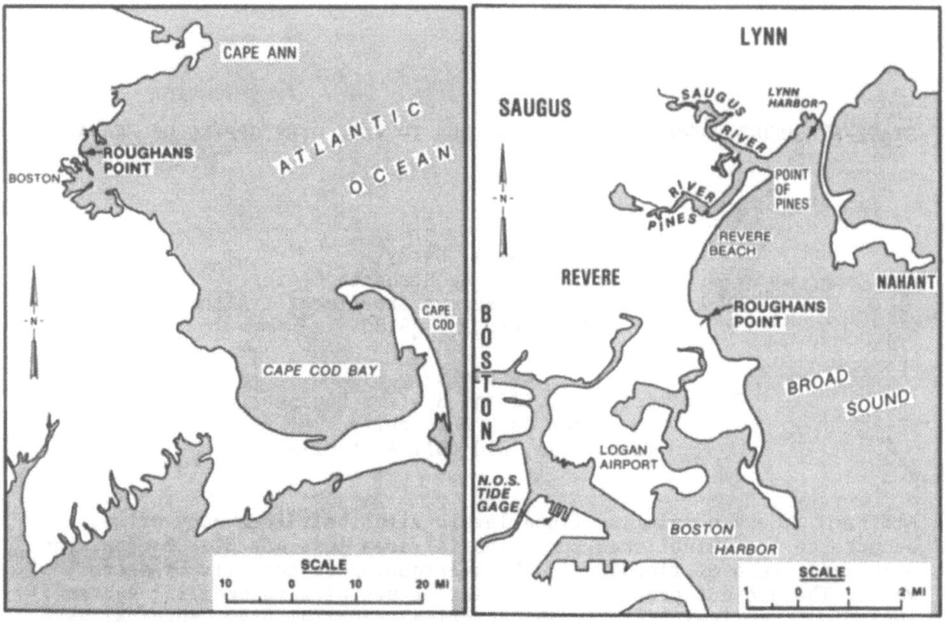

Figure 1. Project location

interior of Roughans Point is below the elevation of a spring tide at many locations. Although protected from inundation from still-water level resulting from the combination of storm surge plus astronomical tide, damage resulting from flooding caused when waves overtop the seawalls is a frequent occurrence.

Wave overtopping is a complex phenomenon that is governed by a number of sometimes interdependent variables such as tide, storm surge, wave height and period, and structure geometry. To study this problem five models were used in conjunction. A numerical storm surge model calculated the still-water level. A numerical wave model estimated the characteristics of the waves that would attack the structures at the water levels predicted by the storm surge model. The results from a physical model study produced an analytical relationship for determining overtopping rates given water level, wave height and period, and structure geometry. A flooding routing model calculated the maximum stage in the interior of Roughans Point caused by each simulated event. A probability model chose events to model, assigned probabilities to those events, and constructed stage-frequency curves.

The main focus of this paper is on the probability model. The other models are not explained in detail. For a more complete description of the storm surge, wave, and flood routing modeling see Hardy and Crawford (1986). For details of the physical modeling see Ahrens and Heimbaugh (1986).

2. PROBABILITY MODEL – EVENT SELECTION AND PROBABILITY ASSIGNMENT

The probability model was essentially an assemblage with four specific tasks: select events for modeling by the other models, assign probabilties to these events, create stage-frequency curves, and determine a measure of the accuracy of the final result.

A flood event is multi-dimensional. The severity of the flood is determined by several factors, among these are the magnitude and duration of winds, waves, and water levels. Because of the difficulty of ranking multi-dimensional entities, as well as the lack of available data for doing so, it is necessary to reduce the dimensionality. Therefore only one dimension, the maximum still-water level occurring during an event was used to measure the severity of an event. This measurement was chosen for two main reasons. First it was deemed the most important, and second, there was a large volume of data available. NED has established a stage-frequency curve relating the maximum still-water level with its frequency of occurrence at the Boston tide gage (NED 1983). This stage-frequency curve was used as the basis for both event selection and the assignment of probability to simulated events.

Since the tidal range at Boston (mean range = 2.90 m and maximum range = 4.45 m) is much larger than the largest recorded surge (1.5 m), the tide is a very important component of the still-water level. If a severe storm were to occur during a neap tide, the flooding effects at Roughans Point would be considerably reduced over those caused by a similar storm occurring at a spring tide. Both the lower wave heights caused by the shallower water of the neap tide as well as the increased freeboard of the structure would combine to greatly reduce the rate of wave overtopping. Therefore, in developing an ensemble of events, the possible tides are as important as the possible storm characteristics of surge, wind, and waves.

There are three steps in the development of the ensemble of possible events. First, an ensemble of storms representing all possible storms is developed. Next, an ensemble of tides representing all possible tides is developed. Finally, the two ensembles are combined to produce an ensemble of events representing all possible events. This process is based on the assumption of the statistical independence of tide and storm, i.e. a storm has an equal chance of occurring with every tide possible during storm season.

2.1 Choosing Representative Storm Surge Time Series

Data in the vicinity of the study area are essential for identifying input to the numerical modeling and for assigning probabilities. This project was fortunate in having convenient sources for the necessary data. The National Ocean Service's (NOS) Boston tide gage has been in continuous operation since 1922. This gage is located at Commonwealth Pier in Boston Harbor, which is less than five miles from the study area. Wave hindcast information was available for deepwater adjacent to the study area from the WES Wave Information Study (WIS). Hourly wind data were available from Logan International Airport which is also less than five miles from the study area. A twenty year period from 1956-1975 was chosen from which to gather data for use in the numerical modeling.

This period was selected because information was available from the above mentioned sources in all the necessary data categories of water level, wind, and wave.

By defining storm surge as the difference between measured water level and predicted tide, a partial duration series of storm surges (26 storms) was extracted from the Boston tide gage record. A minimum value of storm surge equal to 0.76 m (2.5 ft) was chosen, so that only those storms which had a reasonable probability of causing significant flooding were represented. Two additional storms from outside the 1956-1975 period were included in the storm ensemble, November 29, 1945 and February 6, 1978. These storms caused the first and second highest surges recorded at Boston and all the necessary data were obtainable. Futhermore the February 1978 event caused the highest still-water (3.14 National Geodetic Vertical Datum (NGVD)) on record at Boston. Adding these two storms helped insure that the top end of the storm ensemble was representative of what could occur at Boston. Therefore, a total of 28 storms were chosen to represent the surge time histories.

2.2 Creating Synthetic Surge Plus Tide Events

Using tidal constituents from NOS analyses of the Boston tide gage, 19 years of hourly tide heights for the winter season (October 15 to April 30) were predicted. Each of the 28 chosen storm surge time series was assumed to have an equal chance of starting at any of the more than 85,000 hours in the 19 year tidal cycle. This results in more than two million possible surge plus tide time series that could be created. Obviously, it would be economically impossible to simulate any but a very small portion of these possible events. Therefore, in order to form a representative sample of the total possible ensemble, a random selection process was devised.

Each of the 28 surge time series was combined with a large number of tide time series by randomly choosing a starting point in the tide time series. Each of the 26 storms in the original partial duration series was combined with 500 tides. Since the two additional storms were not part of the partial duration series they were combined with fewer tides. From an annual maximum series it was determined to combine the 1978 storm with 340 tides and the 1945 storm with 170 tides. Approximately 13,500 surge plus tide event time series were created by this process. Events to be modeled were selected from this file of possible surge plus tide events.

Based upon previous experience (Prater, Hardy, and Butler, in preparation) it was estimated that by simulating 50 events the frequency of still-water level would be accurately represented throughout the study area. However, the extra variables involved in simulating wave overtopping would cause added uncertainty in the final frequency curves. Therefore, it was decided to simulate 150 events in order to increase the confidence that the frequency curves for flooding from wave overtopping were accurate.

The selection procedure involved three steps. First, the height increments where simulations were necessary were chosen. As previously mentioned, the highest still-water level on record for the Boston area is 3.14 m NGVD. As predicted by the NED frequency analysis the 500-year level is 3.41 m NGVD and the annual level is 2.41 m NGVD. Events were selected to duplicate the NED curve (at Boston) below the 500 year level. Therefore, given the small range in elevation, selections were made every 0.03048 m (0.1 ft) at the lower elevations and every 0.06096 m (0.2 ft) at the higher elevations. The second part of the selection process was the actual select of events. The 13,500 events created by the combination of surge and tide time series were ranked by the maximum water level that would occur during the event (at the Boston gage). At each of the height increments chosen above, events were randomly selected for simulation.

The final part of the selection process was the assignment of probability to the selected events. The probability represented by each increment of elevation of the NED curve was calculated by taking the difference of the exceedance probabilities of the end points of the increment. If more than one event was selected to represent a given height increment, then the probability of that increment was divided equally among events chosen from it.

In conclusion, the essence of the selection process is to choose events for simulation so that the stage-frequency curve will be transfered from a location where it is known (Boston) to locations where it is unknown (the study area).

3. SIMULATION OF EVENTS

The selected events were simulated by three numerical models and one physical model in order to determine flood levels for the interior of Roughans Point. These models will be briefly described in the following section.

3.1 Brief Descriptions of the Component Models

The WES Implicit Flooding Model (WIFM) was used as the hydrodynamic storm surge model. A detailed description of WIFM is given in Butler (1978), and the application of WIFM to coastal studies is shown in numerous reports including Butler (1983). WIFM solves the vertically integrated time-dependent shallow water wave equations of fluid motion using an alternating direction implicit finite difference algorithm. Each of the 150 events was simulated by the surge model. The results of these simulations were the water level time series for each event at a location adjacent to Roughans Point. These time series provided input to the computer codes which calculated wave overtopping rates and interior flood levels at Roughans Point. The storm surge model transfered the surge plus tide levels (synthesized using the Boston gage) to Roughans Point.

A numerical, steady state, shallow water, directional spectral wave model, ESCUBED (Hubertz in preparation) was used to simulate the waves which would attack the structures at Roughans Point during each of the 150 events. Unlike the most common method of wave modeling which is monochromatic and unidirectional, ESCUBED propagates waves with a spectrum of frequencies and directions. This directional spectrum gives a more realistic storm wave climate which is important in the calculation of wave overtopping rates.

A two dimensional physical model study (Ahrens and Heimbaugh 1986) using a spectral wave generator was conducted to determine a method to calculate the wave overtopping rates at Roughans Point. Model tests could not be run for every wave condition during the 150 events, therefore representative values of water level, wave height, and wave period were tested. An analytical relationship was developed giving the overtopping rate as a function of water level, wave properties, and structure geometry. Physical model tests on existing as well as alternative proposed wall structures were conducted. Results from the configuration called "wide berm + 1 ft cap" (Figure 2) are used for illustration in this paper.

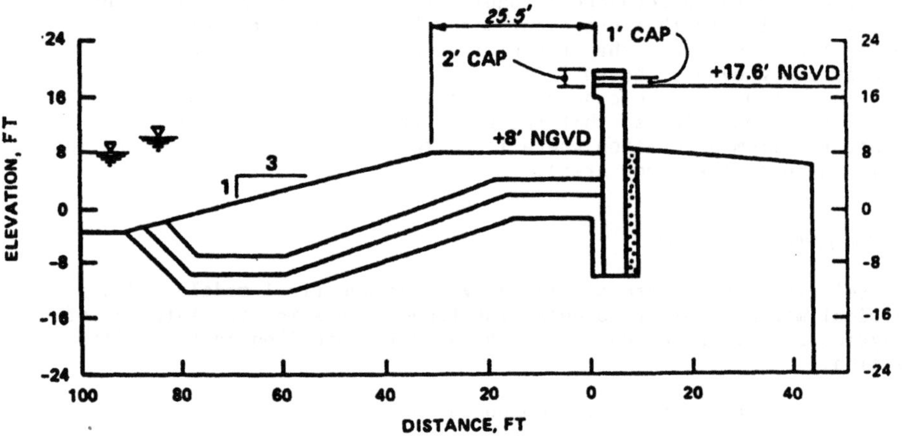

Figure 2. Structure cross section - "wide berm + 1 ft cap"

Since the maximum water level in the interior of Roughans Point during any event depends upon the timing of the overtopping and the drainage characteristics of the interior a flood routing program was developed. This computer code used the time series of overtopping rates calculated from the results of the other modeling efforts and determined the water level at 15 minute increments. Details of the flood routing model are in Hardy and Crawford (1986).

3.2 Calibration and Verification of the Combined Models

Sufficient data were available during two historic events, February 1972 and February 1978, for a calibration and verification of the combined model system. The maximum interior water levels caused by these two events were estimated from the flood evidence and eye witness accounts. To calibrate the combine models, the 1978 event was simulated. Then the overtopping rates and maximum water level were calculated. Adjustments were made to the overtopping calculations and the flood routing model to calibrate the calculated interior water levels to those measured during the 1978 event. The 1972 event was then simulated to verify the combined models. Full details of the calibration and verification are contained in Hardy and Crawford (1986). The results of these simulations are as follows:

STORM	CALCULATED (m NGVD)	OBSERVED (m NGVD)
1978	3.62	3.60-3.66
1972	2.91	2.68-2.74

The results of the calibration and verification were judged to be acceptable. The 0.18 m difference between observed and calculated water levels for the 1972 storm seems reasonable when considering the lack of precision in determining the actual flood level and the accuracy with which storage areas and drainage could be determined.

4. PROBABILITY MODEL – STAGE-FREQUENCY CURVES

Probabilty was assigned to each of the events selected for simulation as described previously. By assigning the probability of the event to the maximum flood level caused by that event, a stage-frequency curve can be constructed by the following method. First an array of possible stages was established with a discretization interval of 0.03048 m (0.1 ft). Next, for all 150 events, the probability masses assigned to each event are accumulated in in the height intervals which bracket the maximum water level for each event. Exceedance probabilities at each interval can then be calculated by adding the probabilities of an interval to the sum of the probabilities of the intervals above it.

Plotting of this raw exceedance file resulted in a jagged curve since not all of the height increments received probability due to the limited number of events modeled. Smoothing was accomplished using linear regression of the curve when plotted on an appropriate probability paper. Using Equation 1 for Weibul probability paper, a "c" value of 0.8 was chosen. The fit between results and the regression line was excellent, resulting in a regression coefficient that was usually above 0.99 and always above 0.95 depending on the structure tested.

$$x_{new} = [-\ln(x_{old})]^c \tag{1}$$

5. RESULTS

A representative stage-frequency curve for the interior flood levels at Roughans Point is presented in Figure 3.

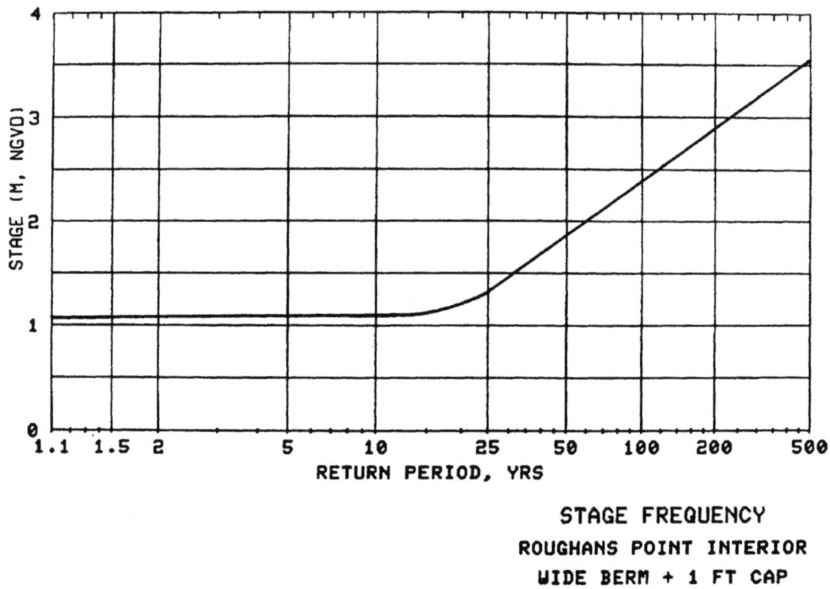

Figure 3. Roughans Point stage-frequency

5.1 Determining Error Bands for the Selection Process

The final products of this study are curves which depict stage vs return period. At any one return period, say 100 years, the curve is merely an estimate of the true flood level. The calculations which resulted in the final curves involved the use of multiple data sets and multiple modeling systems. Even if it were possible to determine the confidence interval of each of these separate processes it would not be possible to combine these into final confidence bands.

The selection process that determined which events to simulate and assigned probability to these events was designed specifically for this project. As a result of the limited experience with this technique it was desirable to estimate the probable error associated with the selection technique. For this reason, the 150 events were selected in three groups of 50. Each of these groups was designed to represent the whole range of stages and was simulated separately resulting in three stage-frequency curves.

Assuming that for any given return period the calculated stage is a normally distributed random variable, an estimate of the probable error, PE, can be calculated using the three stage-frequency curves. Equation 2 states the relationship between PE (the 50% error bound) and the standard deviation, σ. Equation 3 shows the relationship between the range of stages, RA, calculated at a given return period and σ (Beyers 1966).

$$PE = 0.6745\sigma \quad (2)$$

$$\sigma = 0.5908\ RA \quad (3)$$

A stage-frequency curve with probable error bands for the selection process was produced by the following method. First, at selected return periods, RA was calculated by substracting the lowest from the highest stage. Second, the PE was estimated using Equations 2 and 3. Third, a single curve was produced by processing all 150 events as one set. Finally, the PE bounds were placed about this combined curve. Figure 4 shows the results of this process. If the error were to to be unacceptable, then simulating more events would be necessary to narrow the error bounds. It should be emphasized that the error bands only depict the accuracy of selecting a limited number of events to represent all possible events and not the error of the total project. Note that the event selection for the final stage-frequency curve made up of 150 events is more accurate than the error bands suggest since the bands were calculated from sets with only 50 events.

6. SUMMARY

A study to determine the frequency of flood levels caused by wave overtopping of seawalls was described. The study required the conjunctive use of five modeling systems, however the main focus of this paper was on only one, the probability model.

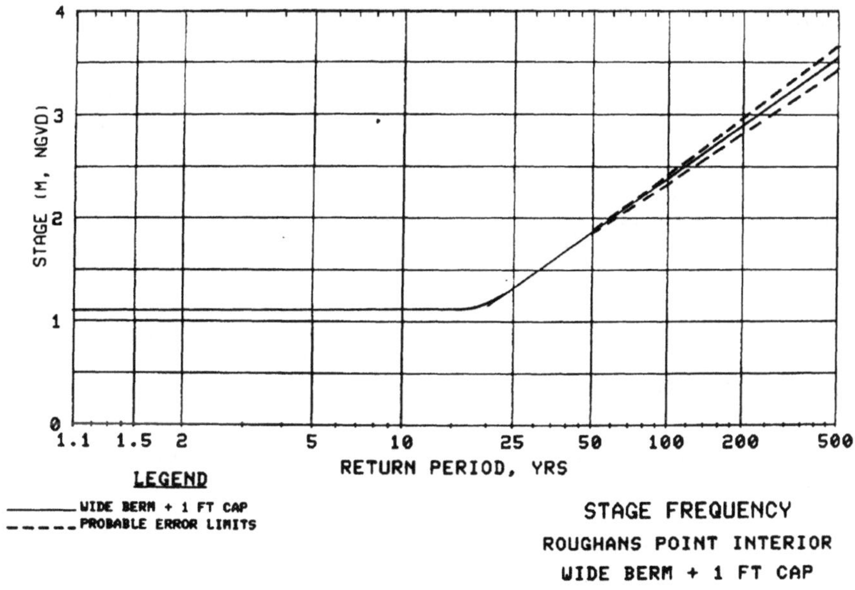

Figure 4. Probable error of the selection process

The probability model had four main tasks: selection of flood events to be modeled, assignment of probability to these events, generation of stage-frequency curves, and providing a measure of error in the selection process. The maximum still water level which occurs during an event was chosen as the key variable. Surge time series extracted from historical data were combined randomly with a starting point in a 19 year tide time series to produce a large ensemble of synthetic storm plus tide events. Events were selected from this ensemble for simulation and were assigned probabilities according to portion of a predetermined frequency curve. This frequency curve was transfered from Boston to the study area through the modeling process. Probable error bounds calculated for the selection process showed that the process could represent all possible events with a small number of simulated events.

7. REFERENCES

Ahrens, J.P., and Heimbaugh, M.S. 1986. 'Irregular Wave Overtopping of Seawall/Revetment Configurations, Roughans Point, Mass.' Technical Report, US Army Engineer Waterways Experiment Station, Vicksburg, Miss.

Beyers, W.H. (editor) 1966. Handbook of Tables for Probability and Statistics, pp 270-272, The Chemical Rubber Company, Cleveland, Ohio.

Butler, H.L. 1978. 'Coastal Flood Simulations in Stretched Coordinates,' Proceedings, 16th International Conference on Coastal Engineering, Hamburg, Germany.

Butler, H.L. 1983. 'Lake Pontchatrain and Vicinity Hurricane Protection Plan, Report 3, Numerical Model Investigation of Plan Impact on the Tidal Prism of Lake Pontchartrain,' Technical Report HL-82-2, US Army Engineer Waterways Experiment Station, Vicksburg, Mississippi.

Hardy, T.A., and Crawford, P.L. 1986. 'Frequency of Coastal Flooding at Roughans Point, Broad Sound, Lynn Harbor, and the Saugus-Pines River System,' Technical Report CERC-86-8, US Army Engineer Waterways Experiment Station, Vicksburg, Mississippi.

Hubertz, J.M. in preparation. 'A Steady State, Shallow Water, Directional Spectral Wave Model: Model Description and Users Guide,' Technical Report, US Army Engineers Waterways Experiment Station, Vicksburg, Mississippi.

Prater, M.D., Hardy, T.A., and Butler, H.L. in preparation. 'Fire Island to Montauk Point, New York, Storm Surge Study,' Technical Report, US Army Engineer Waterways Experiment Station, Vicksburg, Mississippi.

US Army Engineer Division, New England. 1983. 'Roughans Point, Revere, Massachusetts, Coastal Flood Protection Study, Water Resouces Interim Report,' Volume II.

SPATIAL AND TEMPORAL FACTORS CONTROLLING OVERTOPPING OF COASTAL RIDGES

William F. Tanner, Regents Professor, Geology Department,
Florida State University, Tallahassee, Fla. 32306 U.S.A.

ABSTRACT. On low coasts there may be one of three basic sand or gravel ridge types: swash-built ridge, dune ridge or storm surge ridge. The first of these commonly occurs in beach ridge plains, with up to 200 or more ridges, more-or-less parallel with each other.

Common processes for storm overtopping include Langmuir superelevation, wave crossing, wave energy funneling (bathymetry) and Coriolis effects (each up to 1.0 m or more), perhaps additive; all are short-term items (seconds to days). Edge waves make no energy and are unimportant.

Long-term changes have climatological and/or geological controls. The number of controls and interactions are unknown. A few are listed to show some of the complications. In the last 5,000 yrs, big rapid changes have occurred at about 1,000-yr intervals, long after ice sheet melting was over. Many major geological catastrophes have unknown periods and cannot be forecast.

1. RIDGE TYPES

1.1. Swash Ridges

Many coasts are marked by one or more sand or gravel ridges, paralleling the shore at the time of construction. If many ridges are present, much information on sea level changes is available, going back thousands of years. The controls are in good part geological, hence the topic is to be addressed geologically. Much short-term information comes from coastal ridges, so these will be treated first.

The simple (swash-built) beach ridge is made of sand (or gravel), and occurs in groups of up to 200 and more. Each member of the group was at one time the water front. The full set is a beach ridge plain. The ridge closest to land is the oldest, with rare exceptions. Some plains contain two or more sets of ridges. Those within a single set tend to be parallel to each other, to have heights and altitudes which are similar, and to look alike. One set may be distinguished from an adjacent set, on the same plain, by having different height, altitude, crest line orientation, ridge geometry, plant cover, or granulometry.

Any one ridge is linear and gently curved. Height above adjacent swales is tens of centimeters, rarely 1-2 m. Width is typically tens of meters, perhaps 30-80. Wind work, modifying the geometry of the ridge,

V. P. Singh (ed.), Flood Hydrology, 241-248.
© 1987 by D. Reidel Publishing Company.

and/or a plant cover, may make the task of finding it very difficult.

Swash ridges are commonly built of quartz sand. In warm climates they may be made of shells and shell debris, which permit C-14 dating, for good time control (Stapor & Mathews 1980). The swash ridge was built almost entirely by wave run-up. Wind contributions (eolian decoration) and overwash were minor, commonly less than 30%, in most cases much less. Cross sections through swash ridges show beach type cross bedding.

Growing ridges are hard to find because the era of beach ridge construction is almost over (Tanner and Stapor, 1972; Tanner, 1975), but there are a few examples. On the Pacific coast of Mexico, logs floated in from the sea are stranded initially on the lower part of the beach but higher waves gradually move them to the crest.

With heights, typically, of 0.2m to 1m, older ridges are no protection against flooding. The value of the swash ridge plain lies in the detailed historical information that it provides concerning sea level, storminess, ridge topping and shoreline modifications of recent time.

1.2. Dune Ridge

The dune ridge may develop from a beach ridge, by "dune (or eolian) decoration," producing a hummocky surface. At mid- and low tide the beach is exposed to the wind, perhaps the sea breeze. The longer this time of exposure, the higher the probability that the youngest beach ridge will be built up into a dune ridge, which then has a distinctive geometry. A dune ridge therefore probably indicates a pause in construction of the beach ridge set. Instead of the 20 or so years needed for the construction of the average swash ridge, more than 100 years might be required for a well-developed dune ridge. This presupposes some kind of quickly-established plant cover, which then protects the older swash ridges from eolian modification. In a few places, migrating dunes appear on the dune ridge, and then grow landward, burying a narrow strip of the beach ridge plain, but commonly leaving enough of each swash ridge crest that the investigator is not seriously hampered.

1.3. Storm Surge Ridge

Swash ridges occur (at certain places) by the scores or hundreds, and therefore provide small time slices (years or decades). The storm surge ridge, however, ordinarily stands alone, and represents much of Holocene time: some thousands of years. The geometry differs from that of the other two types. It has a whale-back shape (convex up), and may have obvious overwash fans and/or deltas. Maximum height may be as much as about 10 m. The material may be any mix of sand and gravel. Bedding is concentric and concave-down, or land-ward, and may include cut-and-fill.

Where there is ample sand for accretion (equilibrium of abundance; Tanner and Stapor, 1972), many swash ridges may be built, with deposition at various rates, up to 15 m/yr, or more. A single ridge persisting at the beach is probably of storm type. Well-studied storm surge ridges have been described from Ireland (Carter and Orford, 1981; Carter, 1983; Carter, et al, 1984).

A good example of topping could be observed after the landfall of

Hurricane Camille, in southern Mississippi, in 1969 (Tanner 1970). The upper surface is an undulatory strip seven to nine meters high. Maximum height is 31 ft (9.3 m), and width is about two Km. During Camille, wave damage to houses and buildings extended to 3-to-5 meters above ground level. Boats up to about 18 m long were swept to the top of the ridge. Therefore maximum water depth over the top was about 2-to-3 m.

Another storm surge ridge of interest is located at the town of Farol Sao Tome, near the mouth of the Rio Paraiba do Sul, Rio de Janeiro state (Brazil). The ridge is 7-8 m high, has a seaward slope of about 11°, is about a kilometer wide, and is made of sand and fine gravel. Small berms, 30-40 cm high, line the top, all parallel with the coast. In both Brazil and Mississippi small ridges on top indicate repeated high energy events. The single massive ridge in each area shows that topping has been common.

The dune ridge is distinctive because of its basic wind history. The swash ridge and the storm surge ridge are different, not because of the fluids involved, but because the former has grown seaward at rapid rates so that NO ONE ridge is affected by many storms, whereas the latter has grown upward as a result of many high-energy events in the same place. This explains why shape, height, internal structure and surface features are different.

1.4. Beach Ridge Plain

The St. Vincent I. (Fla.) beach ridge plain has yielded a detailed sea level change history (Tanner 1974; Stapor and Tanner, 1977; Demirpolat, Tanner and Clark, 1986). There are more than 180 ridges, the earliest roughly 3,500-4,000 years old (Stapor, 1973). Mean growth was about 20 yrs/ridge, but the coast is now eroding. The history of the St. Vincent I. area (Tanner, 1983) includes a high stand at about +2 m, roughly 3,500-6,000 years ago, shown by a wave-cut cliff landward of the lagoon. Next was a drop of about 4 m, probably in less than 100 years. Since then the local history includes:

Rise, 2-3m, about 2,600 B.P., probably in less than 100 yrs.

Drop, 3-4m, about 1,800 B.P., in 50 yrs or less.

Rise, 2m, about 600-700 B.P., in 50 years or less. Detailed mapping showed that certain entire ridge sets were built on distinctive levels. The island average (20 yrs/ridge) suggests that changes between levels took place very rapidly indeed. Each change (1-5m) fitted into as little as 20-30 years, surely less than a century. A four meter rise, in 30 yrs, would be socially, economically and politically disastrous; yet this is what we had in Recent time (Holocene).

2. WHY OVERTOPPING?

2.1. Not Sea Level Change

Storm surge ridge history includes repeated topping, and this cannot have been due to long-term sea level rise. Therefore we must look for other causes. If simple greenhouse models turn out to be good predictors, additional rise is to be expected. However, these simplistic models have flaws which may make them useless for forecasts.

Subsidence, unlike sea level rise, gives only regional apparent increase in water depth. Subsidence is local, cannot be extrapolated broadly, and produces no repeated surges to the same high position. We therefore seek other causes for topping.

2.2. Langmuir Cells

The Langmuir cell is a single long horizontal roll, in the upper part of ocean or lake, having the long axis parallel with the water surface and wind vector. Such cells always exist in sets, typically in hundreds or thousands. At any moment even-numbered cells have one rotation (e.g., advance of a right-handed screw), odd numbered cells the opposite. They were first given careful study by Langmuir (1938). Because adjacent rolls, or cells, spiral with opposing senses, there are vertical sheets of water, between them, having vertical motion: even-numbered sheets move upward, odd-numbered sheets down. This motion introduces deeper water (not previously subject to wave generation) to the surface, thus creating a "slick" in the wave field. Downward motion traps floating foam, sea-weed and trash in parallel strips: windrows.

Weller et al (1985) reported data from a vertical floating platform ("FLIP") in the Pacific. Measurements were taken down to about 35 m: the first direct observations of three-dimensional flow in Langmuir patterns in the open ocean. Vertical velocity had magnitudes up to 0.2 m/sec, more commonly about 0.1 m sec. This should be compared with surface waves about 1.0 m high, and wind speeds about 3-14 m/sec. Maximum downflow velocity was directed at an angle in the vertical plane of about 45°, indicating horizontal transport (in the downwind direction) about equal to downwelling.

Observations of Langmuir cells in shallow water show ellipticities up to as much as 5:1 or even 10:1 (e.g., Faller and Caponi 1978; Hunter and Hill, 1980). The writer has seen Langmuir cells, in a small lake with fetch less than 10 Km, approach the beach to within 0.3 m of the water's edge, where bottom sediment could be followed as evidence of vertical circulation (upward flow). This rise is associated with superelevation of the water surface; with converging flow, wind sweeps surface water into low ridges, where excess mass causes sinking.

A comparison of wave energy and Langmuir circulation energy, using kinetics, can be based on the FLIP data. These provide an estimate of wave energy, per square meter of surface, of 668 (S.I. units). For Langmuir circulation, the estimated value of kinetic energy, per unit area, is 250 to 562 (S.I. units). Therefore Langmuir cell energy appears to be about 40 to 80 % of wave energy. It is concluded that fair weather Langmuir superelevation is on the order of 10 cm, but in storms is a large multiple of 10 cm, probably one or more meters.

2.3. Wave Crossing

Waves in the ocean, lakes and lab, cross each other. I have seen, from the air above Lake Michigan, as many as four sets of waves, at one time. This is not surprising, for two reasons: (1) Benard cell (atmospheric convection) winds may blow in two or more directions for 10 to 100 Km, or so, simultaneously, generating two or more wave sets (Agee, 1984), having periods of seven to 10 sec, heights in the meters, and wave lengths over 100 m. (2) Hurricane waves mapped by air-borne radar have many orientations (Elachi, et al, 1977). In Hurricane Gloria-1976, periods and heights (calculated from their published map) are 9-to-12 seconds and 10, 20 or 30 m.

Wave crossing is easy to study over the transverse bar (Niedoroda, 1972; Niedoroda and Tanner, 1970). These relatively small bars (e.g., 10 m to 1000 m long) are common on sandy low-energy coasts. They occur in sets, parallel with each other; the long axis is normal to incoming wave crests. A single transverse bar causes each wave to cross itself at least once before it reaches the surf. At the point of crossing, wave energy density is added and a high local peak (rooster tail) appears. The rooster tail is breaker-like due to the fact that the combined wave is too high (trough to crest) for the available water depth. A rooster tail moves landward along the bar axis.

Very long orthogonal sand bodies should act during storms like transverse bars do in fair weather: focussing waves so that energy is drained from inter-bar reaches and concentrated on bar top. When two wave sets are combined, the result is adding of wave energy density ($E = grH^2/8$; g is acceleration of gravity, r is mass density of water, and H is wave height). To evaluate the combination, one adds E, not H. This means that two 1.0 m waves produce, at crossing, a combined height of slightly more than 1.4 m, giving a differential of about 40%. This greater height may result in local topping of a ridge. The space between storm rooster tails is on the order of a few Km.

2.4. Other Effects

Langmuir cells and storm rooster tails may combine with storm surge, astronomical tide, and coastal ridge irregularities, to produce local topping. Additional complications include wave energy funneling because of irregularities in bathymetry. On a low sandy coast, bottom features such as old river channels, drowned islands and shoals, may cause local wave refraction, hence funneling of wave energy to certain points on the coast. Sleath (1984) has given a good treatment, with examples.

Because of many different causes, funneling cannot be treated in simple fashion. However, given the bathymetry, one can track any wave set shoreward, and observe the effects. An advanced refraction program (WAVENRG; May, 1974) will show where wave orthogonals are concentrated, and will also print the energy density at each point. This is crucial: bunched orthogonals may or may not represent energy concentration.

All of these mechanisms may be added, cumulatively. The result may be one or more meters of superelevation, plus augmented wave height, plus a storm surge. Storm waves have short wave lengths, so the combined momentary rise can be very great, close to the shoreline.

Coriolis effects have been treated by Csanady (1982). For maximum results, one needs wind vectors along the coast, and a cross-wise water transport due to rotation of the planet. Modest winds (4 or 5 m/sec) have been observed to produce sea level changes of a few tens of centimeters, and winds of 10-15 m/sec cause water surface changes of 40 or more centimeters. These are not storm wind velocities, and therefore the elevation in a storm may exceed a meter (Sandstrom, 1980), added to the other effects here.

Edge waves do not produce, steer, or deliver any important package of energy. A formulation by Guza and Inman (1975) showed that waves with storm heights and periods do not make edge waves.

3. SEA LEVEL HISTORY

Global and local processes control sea level change; the result is complicated. There is no one algorithm (or set) having modest-to-good explanatory power. The actual record of sea level change data is in many places only decades long, which is far too short for determining the full set of causes.

The number of controlling factors is unknown. Of those that are known well, not all work equally well throughout both glacial and interglacial time. Among other handicaps, two are important: (1) The surf zone at a given position will obliterate evidence for an earlier similar position. The number of known positions, from Plio-Pleistocene time, is less than we might infer from other data: a good deal of information is missing. (2) There is no reason to think that the geoid has time-invariant geometry.

The most recent major icecap growth started 110,000 to 75,000 years ago; peak size was reached 25,000 to 20,000 B.P.; and melting occurred 20,000 to 10,000 B.P.(Wisconsin glaciation). The third part of this history is fairly well documented; changes of sea level, for short times, must have occurred at rates of 1-to-10 cm/yr, or 1-to-5 m/100 yrs. We do not expect such rapid changes in the next few decades, but on the other hand we might get them anyway.

Rapid irregular rise changed to slow (irregular?) rise about 10,000 B.P., and slow rise changed to fluctuation roughly 6,000 years ago. At that time, the ocean surface stood some 2-to-3 m higher than now. Since then there have been various events, depending on the investigator: from zero to 15 or 20 of them (Stapor and Tanner, 1977), and probably about five to eight.

There are other controls also, such as epic volcanic events, which in the past exceeded the largest historical eruptions by an order of magnitude; the most recent took place about 75,000 years ago (Anon, 1983). Therefore we have no written history on which to base an estimate of what to expect. Further, there is not even agreement as to what is happening now: a new ice age may lie in the near future (Duphorn, 1976), or 100,000 or more years away (Tanner, 1985).

In practice, most controls are ignored and forecasts are based on tide gauge records and storm statistics. These go back such a tiny bit into the past, with such small changes, that they can not be used to predict rare events. Sea level changes appear to be on a roller coaster, and past history may not be a good guide to the future.

4. CONCLUSION

There are two major classes of potentially disastrous sea level changes: short term (wave, tide, storm surge, Langmuir superelevation) and long term (changes in land or ocean level). The first is marked by short intervals (seconds to days), and an assurance that there will soon be a return to normal. The coastal resident may see this as something to be lived through. The second is longer (years, millenia, more), and carries no assurance of a return to normal. The coastal resident might not even note such an event. OUR main interest is temporary overtopping; here we are unable to predict very well.

Members of the second are understood in general, but it is NOT clear that we have identified the full group, and it IS clear that for many members we do not know rates or magnitudes of temporary departures from normal. In summary: (1) Long term changes, in the most recent 5,000 years, have been big and fast by human standards: meters of change in decades. We must not think things will be different in the future. (2) Short term changes may be 5-10 m, perhaps beyond.

(A detailed version may be requested from the author.)

5. REFERENCES

Agee, E.M. 1984. Observations from space and thermal convection: a historical perspective. Bull. Amer. Meteor. Soc. v.65 p.938-949.

Anonymous. 1983. Explosionen in der Grossenordnung von Atombomben. Frankfurter Rundschau v.39 no.198 p.13.

Carter,R.W.G.1983.Raised coastal landforms as products of modern process variations, and their relevance in eustatic sea-level studies: examples from eastern Ireland. Boreas, v.12 p.167-182.

Carter, R.W.G., and J.D.Orford, 1981. Overwash processes along a gravel beach in southeast Ireland. Earth Surface Processes and Landforms, v. 6 p.413-426.

Carter, R.W.G.,T. W. Johnston,and J. D. Orford, 1984. Stream outlets through mixed sand and gravel coastal barriers: examples fromsoutheast Ireland. Zeitschrift fur Geomorphologie, vol. 28, p. 427-442.

Csanady, G. T., 1982. Circulation in the coastal ocean. D. Reidel Publishing Co., Dordrecht, Holland, p.243-266.

Demirpolat, S., W. F. Tanner and D. Clark, 1986. Subtle mean sea level changes and sand grain size data. P.113-128 in: Suite Statistics and Sediment History, Proc. 7th Symp. on Coastal Sedimentology, W. F. Tanner, ed.; Fla. State Univ., Tallahassee, 228 p.

Duphorn, K., 1976. Kommt eine neue Eiszeit? Geologische Rundschau, v.65 p.845-863.

Elachi, E., T. W. Thompson and D. King, 1977. Ocean wave patterns under Hurricane Gloria: Observation with an air-borne synthetic-aperture radar. Science, v.198 p.609-610.

Faller, A. J., and E. A. Caponi, 1978. Laboratory studies of wind-driven Langmuir circulation. Jour. Geophys. Res., v.83 p.3617-3634.

Guza, R. T., and D. I. Inman, 1975. Edge waves and beach cusps. Jour. Geophysical Research, v.80 p.2997-3012.

Hunter, R. E., and G. W. Hill, 1980. Nearshore current pattern off South Texas: An interpretation from aerial photographs. Remote Sensing of Environment, v.10, p.115-134.

Langmuir, I., 1938. Surface motion of water induced by wind. Science, v.87 p.119-123.

May, J. P., 1974. WAVENRG; A computer program to determine the distribution of energy dissipation in shoaling water waves, with examples from coastal Florida. In: Sediment Transport in the Near-Shore Zone, Proc. 1st Symp. on Coastal Sedimentology, W. F. Tanner, ed.; Fla. State Univ., Tallahassee; p. 22-80.

Niedoroda, A. W., 1972. The generation of currents associated with transverse bars. Unpubl. Ph.D. dissertation, Fla. State Univ., Tallahassee; 88 p.

Niedoroda, A. W., and W. F. Tanner, 1970. Preliminary Study of transverse bars. Marine Geology, v.9 p.41-62.

Sandstrom, H., 1980. On the wind-induced sea level changes on the Scotian shelf. Jour. Geophysical Research, v.85 n.C-1 p.461-468.

Sleath, J. F. A., 1984. Sea Bed Mechanics. Wiley; New York, 335 p.

Stapor, F. W., 1973. Coastal sand budgets and Holocene beach ridge plain development, northwest Florida. Unpubl. Ph.D. dissertation, Fla. State Univ., Tallahassee, 219 p.

Stapor, F. W., and T. D. Mathews, 1980. C-14 chronology of Holocene barrier islands, Lee County, Florida: A preliminary report. Shorelines Past and Present, Proc. 5th Symp. on Coastal Sedimentology, W. F. Tanner, ed.; Fla. State Univ., Tallahassee, p.47-68.

Stapor, F. W., and W. F. Tanner, 1977. Late Holocene mean sea level data from St. Vincent Island, and the shape of the late Holocene mean sea level curve. In: Coastal Sedimentology, Proc., 3rd Symp. on Coastal Sedimentology, W. F. Tanner, ed.; Fla. State Univ., Tallahassee; p.35-68.

Tanner, W. F., 1970. Significance of Camille (1969). Southeastern Geology, v.12 p.95-103.

Tanner, W. F., 1974. Applications of the model. In: Sediment Transport in the Near-Shore Zone, Proc. 1st Symp. on Coastal Sedimentology, W. F. Tanner, ed.; Fla. State Univ., Tallahassee; p.104-141.

Tanner, W. F., 1975. Symposium on beach erosion in Middle America: Introduction. Trans. Gulf Coast Assoc. of Geological Societies, v.25 p.365-368.

Tanner, W. F., 1983. Field trip notes. In: Near-Shore Sedimentology, Proc. 6th Symp. on Coastal Sedimentology, W. F. Tanner, ed.; Fla. State Univ., Tallahassee; p.A1-A6.

Tanner, W. F., 1985. Framework for design of a Pleistocene Calendar. TER-QUA Symp. Series, Nebraska Acad. of Science, v.1 p.71-74.

Tanner, W. F., and F. W. Stapor, 1972. Accelerated crisis in beach erosion. Internat. Geography (Montreal, Canada), v.2 p.1020-1021.

Weller, Robert A., J. P. Dean, J. Marra, J. F. Price, E. A. Francis and D. C. Boardman, 1985. Three-dimensional flow in the upper ocean. Science, v.227 p.1552-1556.

STATISTICAL ANALYSIS OF STORM TIDE ELEVATIONS FOR NEW ENGLAND COASTAL COMMUNITIES

Y. J. Tsai
Environmental Consultant
Stone & Webster Engineering Corporation
Boston, MA 02107

ABSTRACT

Along the New England coast, most of the flooding damage is caused by winter storm or northeaster induced surges. Because the cause of flooding is unique, the Federal Insurance Administration has sponsored the development of a methodology to calculate water elevations using both the synthetic winter storm model and the storm surge model. Based on the methodology, determinations can be made of coastal flood levels having return intervals of 10, 50, 100, and 500 years. This paper describes in detail the statistical analysis adopted for the Flood Insurance Study using the results of storm surge modeling and the selected probability distribution model.

The Pearson Type III Distribution was found to be a good probability distribution for STEs at the tide gage stations in New England with all the fits accepted by the chi-square test at the 95 percent confidence level.

INTRODUCTION

No tide gage stations exist in most of the coastal communities in New England. In order to determine the flood frequency for each community under study, it is necessary to generate historical high water levels caused by either northeasters or hurricane events. Part of the Flood Insurance Study effort was to develop a synthetic northeaster model (S&W, 1978), modify a storm surge model (S&W, 1978), adopt a statistical method, and devise a wave crest profile including wave runup (S&W, 1981).

The method of joint probabilities is not applicable to the STE frequency analysis in New England for two reasons. First, some of the meteorological parameters used to model northeasters are dependent upon each other, making the use of this method difficult. The large size and slow speed of a northeaster also make this method time consuming. Second, the method of joint probabilities is not favored because it

V. P. Singh (ed.), Flood Hydrology, 249–256.
© 1987 by D. Reidel Publishing Company.

introduces a certain amount of error in assigning a probability to each class-representing value for all meteorological parameters.

For the direct method to be used, two areas must be addressed:

1. Generation of synthetic historical tide data for ungaged areas or areas with incomplete records.
2. Identification of a good probability distribution which adequately described the STEs' characteristics.

EVALUATION OF STORM TIDE ELEVATIONS BASED ON SELECTED STORMS

Selection of Storms

The first step in analyzing the 10, 50, 100, and 500 year coastal flood levels was to select severe storms that might have induced the annual maximum STEs from historical storm meteorological data.

The initial selection of northeasters was made from storm tracks published in the Monthly Weather Review (NOAA) from 1942 to 1958 and in qualitative description given in Storm Data (NOAA) from 1959 to 1976. National Climatological Data (USWB), HUR 7-97 (NOAA, 1968) and Technical Paper No. 55 (Cry, 1965) were used to select hurricanes that reached New England between August and October. The historical storms selected developed along the east coast of the United States within a band 1200 kilometers wide and moved through a circle around Portland, Maine of radius 600 kilometers. Figure 1 shows the designated area.

148 northeasters and 6 hurricanes were analyzed on the basis of severity (central pressure difference and storm size), their tracks, and the astronomical tides and observed STEs at the recorded gage stations during the storm period. After further analysis, 55 storms--54 northeasters and 1 hurricane--were selected to be used in the study.

Determination of Storm Surges

In order to perform the two-dimensional storm surge analysis, a grid system, as shown in Figure 2, was constructed to cover the study area. The nine communities under study are located between Portland, Maine, and Portsmouth. New Hampshire. Grid points were 4 nautical miles (7.1 km) apart in both x- and y- directions.

The meteorological parameters for the selected storms were used to generate the wind and pressure fields over the selected grid system at the specified time step. The northeaster model was used for the northeasters; a hurricane model derived by Graham and Nunn (1968) was used for the hurricane.

Using the input from the northeaster model (and the hurricane model), the storm surge model was then used to simulate the induced surge hydrograph along the coastline. Storm surges at the selected coastline grid points were calculated for each storm by assuming zero initial rise. An example of a storm tide hydrograph for one grid point, representing a study community, for the February 7, 1947 northeaster is shown in Figure 3. The figure also shows the computed astronomical tides and storm tide elevation.

Simulation of Astronomical Tides

The astronomical tide hydrograph was simulated using a computer program developed by Pore and Cummings (1967). The proper set of tidal standard harmonic constants and other tidal constituents constants for prediction at the major gage station, Portland, were obtained from the Manual of Harmonic Analysis and Prediction of Tides (USCG, 1940). The spatial variation in astronomical tide elevations and phases in the study area was accounted for by adjusting values predicted for Portland with the tidal differences published in the Tide Tables (NOAA).

Storm Tide Elevations Based on Selected Storms

The surge and astronomical tide hydrographs were superimposed to yield the STE peak value. At the recorded gage stations, the calculated values were compared to the observed STEs. Differences were obtained between the observed and calculated values at the gage stations that accounted for the initial rise random errors and the errors introduced by wind field, surge and astronomical tide predictions. These differences were then used to adjust the calculated surge tide hydrographs to yield the peak STE values at the selected locations for all storms.

EVALUATION OF STORM TIDE ELEVATIONS BASED ON CORRELATION ANALYSIS

Historical storm data are missing for some storms which generated the annual maximum STE at the recorded gage stations, Portland and Portsmouth. In these cases, correlation analysis was used to reconstruct corresponding STEs in the study areas.

Using h_i to denote STE at the i-th grid point, h_1 and h_{11} are STEs at Portland and Portsmouth, respectively. h_2 through h_{10} represent STEs at the intermediate communities. Using h_1 and h_{11} as independent variables and assuming h_2 through h_{10} to be functions of h_1 or h_{11}, the following two formulas are obtained:

$$h_i = a_i + b_i h_1 \quad (1)$$

$$h_i = c_i + d_i h_{11} \quad (2)$$

A linear regression analysis was performed for all the intermediate stations as related to Portland (h_1) and Portsmouth (h_{11}). The correlations were excellent. Correlation coefficients between stations 2 through 6 and Portland (station 1); and between stations 7 through 10 and Portsmouth (station 11) were equal to 0.99 or greater. Therefore for those years with recorded maximum STEs at gage stations but without storm data, Equation 1 was used for the prediction of STEs for the study areas between stations 2 and 6. For other study areas,

Equation 2 was used for the STE predictions. However, when h_{11} is not available, Equation 1 should be substituted.

Figure 4 shows the correlation of storm tide elevations between a selected ungaged location and recorded gage station with recorded values as the independent variables.

STATISTICAL ANALYSIS OF ANNUAL MAXIMA OF STORM TIDE ELEVATIONS

The annual maximum STEs for all stations were picked from the computed STEs, both those obtained from the storm surge modeling and those obtained by the correlation analysis.

The annual maximum STEs at each selected location were fitted by the Pearson Type III Distribution (HS, 1981). The magnitude of STE at selected exceedance probability is computed by the equation:

$$Y = \overline{X} + KS \tag{3}$$

where Y = magnitude of STE at selected exceedance probability
$\overline{X}$ = mean of the annual maximum STEs
K = frequency factor that is a function of the skew coefficient and selected exceedance probability
S = Standard deviation of the annual maximum STEs

The mean standard deviation and skew coefficient of data can be calculated using the following equations:

$$\overline{X} = \Sigma X/N \tag{4}$$

$$S = [\Sigma(X-\overline{X})^2/(N-1)]^{0.5} \tag{5}$$

$$G = N\Sigma(X-\overline{X})^3/(N-1)(N-2)S^3 \tag{6}$$

in which:
N = number of the annual maximum STEs
G = skew coefficient

A computer program (SWEC, 1980) was developed based on the above algorithm to compute the desired return period STE at each community location. The values of frequency factor, K, are stored in the program and used to interpolate the current frequency factor for a specified exceedance probability and computed skew coefficient.

All statistical fits are accepted at the 5 percent level of significance, which has a Chi-square value of 3.84 for 1 degree of freedom. The Chi-squares for all stations are less than 3.84. They are all excellent fits. Figure 5 depicts one such distribution. Goodness of fit can be recognized from the plot. The flood levels for 10-, 50-, 100-, and 500-year return periods at each of the selected locations were obtained from these frequency curves.

CONCLUSION

A methodology for flood level determination, consisting of northeaster storm modeling, storm surge simulation and statistical analysis, was developed and successfully applied in the Flood Insurance Study for the New England coastal communities. The Direct Method was used in which the annual maximum storm tide elevations for each community were obtained either from reconstruction of tide levels from storm data or from correlation of tide levels with the recorded gage data. Frequency curves based on the Pearson Type III distribution fitted for the station STEs were used to determine the desired return period flood levels for that location.

ACKNOWLEDGEMENTS

The methodology referenced in this paper was a part of developmental efforts sponsored by the Federal Insurance Administration, U.S. Federal Emergency Management Agency. The author acknowledges contributions by the Stone & Webster Engineering Corporation personnel who have been involved in the Flood Insurance Study projects and the technical staff of FEMA.

REFERENCES

1. Cry, G.W. 1965. 'Tropical Cyclones of the North Atlantic Ocean, 1871-1963.' U.S. Weather Bureau, TP-55.
2. Graham and Nunn 1968. 'Asymptotic and Peripheral Pressures for Probable Maximum Hurricanes.' Hydrometeorological Branch of the Weather Bureau, U.S. Department of Commerce.
3. Hydrology Subcommittee 1981. Guideline for Determining Flood Flow Frequency. Bulletin No. 17B. Revised September 1981. Editorial Corrections March 1982. Interagency Advisory Committee on Water Data, Office of Water Data Coordination, Geological Survey, U.S. Department of the Interior.
4. National Oceanic and Atmospheric Administration (NOAA) 1942-1958. Monthly Weather Review. U.S. Department of Commerce.
5. National Oceanic and Atmospheric Administration (NOAA) 1959-1976. National Climatic Center. Storm Data. U.S. Department of Commerce. Asheville, NC.
6. National Oceanic and Atmospheric Administration (NOAA) 1968. Memorandum HUR 7-97: Meteorological Characteristics of the Probable Maximum, Hurricane, Atlantic and Gulf Coasts of the United States. U.S. Department of Commerce.
7. National Oceanic and Atmospheric Administration (NOAA). Tide Tables. High and Low Water Predictions, East Coast of North and South America.
8. Pore, N.A. and Cummings, R.A. 1967. 'A Fortran Program for the Calculation of Hourly Values of Astronomical Tide and Time and

Height of High and Low Water.' Weather Bureau Technical Memorandum TDL-6.
9. Stone & Webster Engineering Corporation (S&W) 1978. Development and Verification of a Synthetic Northeaster Model for Coastal Flood Analysis. S&W, Boston, MA.
10. Stone & Webster Engineering Corporation (S&W) 1978. Determination of Coastal Storm Tide Levels. S&W, Boston, MA.
11. Stone & Webster Engineering Corporation (S&W) 1980. 'Frequency Analysis by the Pearson Type III Model.' SWEC Computer Program, EN-217. S&W, Boston, MA.
12. Stone & Webster Engineering Corporation (S&W) 1981. Manual for Wave Runup Analysis, Coastal Flood Insurance Studies. S&W, Boston, MA.
13. U.S. Coast and Geodetic Survey 1940. Manual of Harmonic Analysis and Prediction of Tides. U.S. Coast and Geodetic Survey Special Report No. 98, Revised edition.
14. U.S. Weather Bureau. National Climatological Data. U.S. Department of Commerce, Asheville, NC.

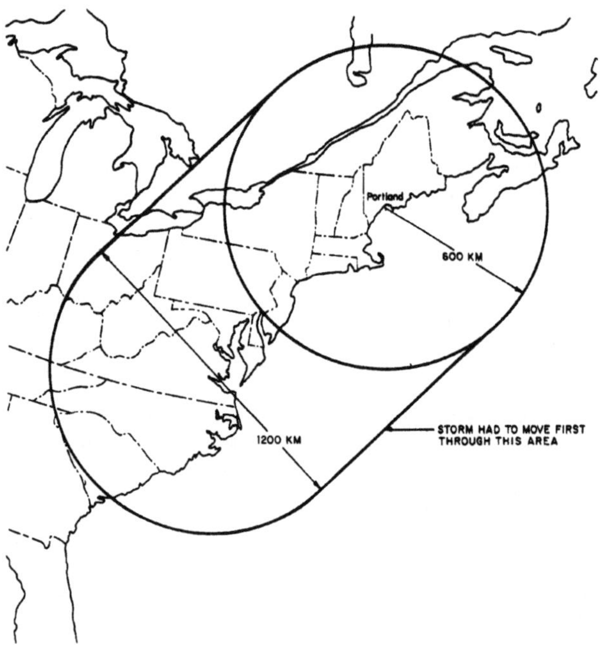

Figure 1. Storm Selection Criteria for Years Between 1942 and 1958.

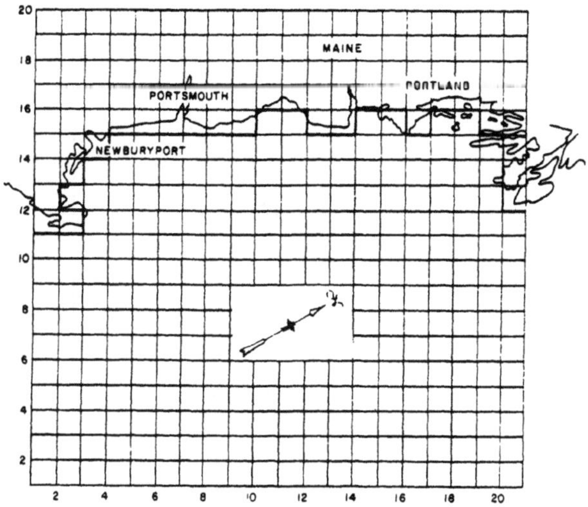

Figure 2. Grid System Off Southern Maine

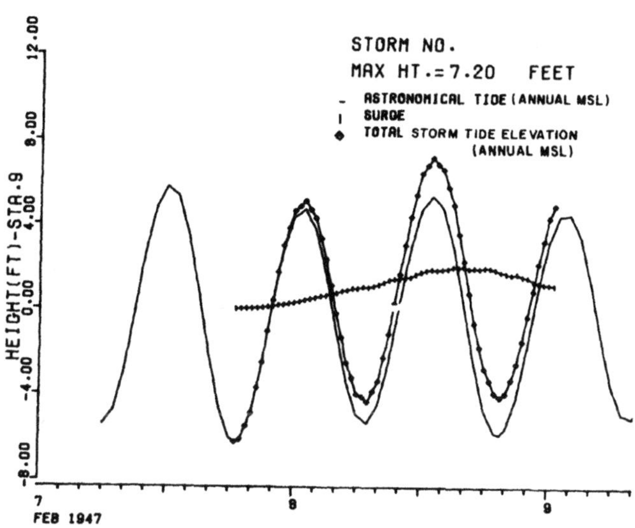

Figure 3. Computed Surge, Astronomical Tide and Storm Tide Elevations at Yotk, Maine - Storm of February 7, 1947.

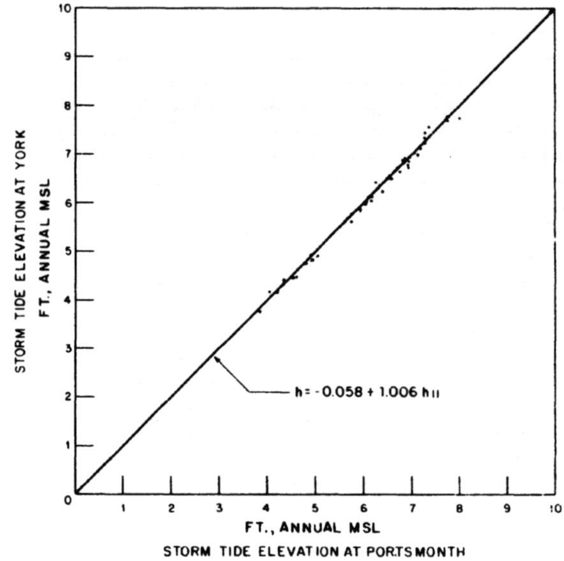

Figure 4. Correlation Between Storm Tide Elevations at York, Maine and Portsmouth, New Hanpshire.

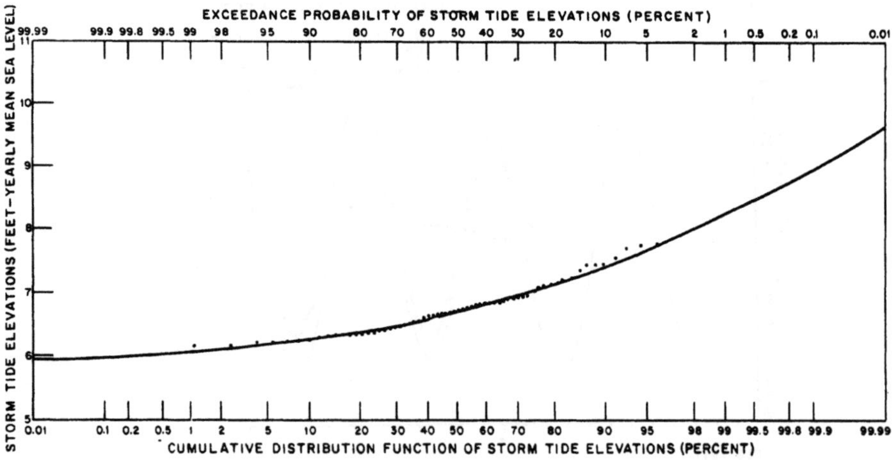

Figure 5. Frequency Curve of Storm Tide Elevations at York, Maine

ERRORS DUE TO LINEARIZATION IN TIDAL PROPAGATION

Panagiotis D. Scarlatos
South Florida Water Management District
3301 Gun Club Rd., P.O. Box 24680
West Palm Beach, Florida 33416, USA

Vijay P. Singh
Department of Civil Engineering and
Louisiana Water Resources Research Institute
Louisiana State University
Baton Rouge, Louisiana 70803-6405, USA

ABSTRACT. Propagation of tidal disturbance is simulated using the St. Venant system of equations which is a partial differential, nonlinear system of hyperbolic type and can be solved by means of numerical methods. However, certain assumptions can reduce the system into the linearized form of Telegrapher's equation, so that a closed-form solution is feasible. This linearization suppresses the tidal distortion induced by water shallowness, convectional velocities and bottom friction. Another source of error is the semi-empirical calibration of the linearized harmonic parameters.

Underestimation of tidal processes in an estuarial impact study may cause environmental and ecological problems. Flooding, sediment transport, salinity intrusion and pollutant diffusion are all closely related to tidal oscillatory flow. In this paper, the nonlinear distortion is investigated for a large number of numerically simulated cases. The independent variables are bottom friction, tidal amplitude, mean water depth and length of the waterway. Deviations of the harmonic solution from a finite element numerical solution are determined, and the risk of obtaining inaccurate results by using Telegrapher's equation is assessed.

1. INTRODUCTION

Most estuaries, rivers or canals directly connected to ocean waters are subjected to tidal action. The tides may be of astronomical or meteorological origin. Due to their long periodicity, tidal waves are considered as shallow waves, i.e., the disturbance reaches the bottom. As a result, the velocity can be considered as uniform along the water column (Dronkers, 1964).

Tidal oscillations in open water bodies can be predicted with remarkable accuracy. However, the same is not true for tidal oscillations within a channel. The geometry of the channel is much more

V. P. Singh (ed.), Flood Hydrology, 257–269.
© 1987 by D. Reidel Publishing Company.

limited than the oceans so that frictional effects influence the tidal motion considerably.

Mathematically, the tidal propagation along a channel can be described by the St. Venant system of continuity and momentum balance equations. Solution of this system in its complete form requires a numerical technique. For that purpose, there is a variety of numerical schemes based on either the finite differences or finite element methods (Mahmood and Yevjevich, 1975; Scarlatos, 1980). Analytical solutions of these equations are feasible only after certain simplifications are introduced into the system. The most important simplifications are small wave amplitude in comparison with water depth, linearization of frictional energy dissipation effects, and negligence of nonlinear convectional terms. The result of such treatment will be supression of nonlinear distortion, i.e., a sinusoidal wave might be amplified or damped but it will remain sinusoidal throughout the channel. Thus, there is a trade off between accuracy and easy access to a solution by using a numerical or analytical technique.

Utilization of the harmonic method for estuarine studies may lead to inaccurate information that can cause serious problems. Underestimation of water elevation might result in flooding of low-lands, while inaccurate estimation of current velocities might create shoaling or erosion within the waterway. Environmental studies of pollutant transport or salinity intrusion depend also on the accurate prediction of water velocities. The magnitude of the nonlinear distortion is a function of both physical and geometrical characteristics of the estuarine system and cannot be estimated easily. Therefore, before applying a simplified solution that neglects nonlinear effects, it is useful to know the risk of obtaining such inaccurate results that might jeopardize the estuarine system.

The purpose of this study is to determine nonlinear distortion of tidal oscillations within a waterway and accuracy of the harmonic method. Attention is given to both tidal amplitude and currents. Assuming that the St. Venant system of equations gives the best representation of estuarine tidal hydrodynamics, the system is solved by a finite element technique for a large number of cases (Scarlatos, 1980, 1982). Then, according to the numerical data, proper harmonic parameters are calibrated for utilization of linear Telegrapher's equation (Ippen and Harleman, 1966). Comparing the results of the two methods, the nonlinear tidal distortion can be easily assessed, and the risk of inadequate simulation of the estuarine ecosystem by using linearized approach can be quantified.

2. GOVERNING EQUATIONS

The one-dimensional, time-dependent, free-surface tidal flow can be successfully described by means of the St. Venant system of equations which are

Continuity of Mass:

$$A_{,t} + (uA)_{,x} = q \tag{1}$$

Momentum Balance:

$$(uA)_{,t} - (u^2A)_{,x} + gAH_{,x} = gA(S_o - S_f) + qu \tag{2}$$

where A is the wet cross-section area, u is the mean velocity of the cross-section, q is the lateral inflow, g is the acceleration due to gravity, H the water height above the reference datum, S_o is the bottom slope and S_f is the energy dissipation gradient. Distance and time are denoted by x and t respectively, and comma before a subscript expresses partial differentiation with respect to that variable.

For unit width flow with zero lateral inflow and horizontal bed, the system of Eqs. (1) and (2) becomes

$$h_{,t} + (uh)_{,x} = 0 \tag{3}$$

$$u_{,t} + uu_{,x} + gh_{,x} + gS_f = 0 \tag{4}$$

where h is the wave height above mean sea level. The energy dissipation gradient S_f is proportional to the second power of the velocity and can be expressed either by Manning's or Chezy's formula. Equations (3) and (4) are both nonlinear with respect to the dependent variables u and h. In case that the wave amplitude h is small in comparison to the water depth H, Eq. (3) can be written in linear form as

$$h_{,t} + Hu_{,x} = 0 \tag{5}$$

Linearization of Eq. (4) can be achieved by neglecting the velocity convection term and expressing the energy gradient as a linear function of the velocity, i.e.,

$$u_{,t} + gh_{,x} + gMu = 0 \tag{6}$$

where M is the linearized friction factor (Parsons, 1918) given as,

$$M = \frac{f}{3\pi} \frac{u^*_{max}}{gH} \tag{7}$$

In Eq. (7) f is the Darcy-Weissbach friction coefficient and u^*_{max} is an idealized maximum tidal velocity. Combination of Eqs. (5) and (6) leads to Telegrapher's equation:

$$c^2h_{,xx} = h_{,tt} + gMh_{,t} \tag{8}$$

or

$$c^2 u_{,xx} = u_{,tt} + gMu_{,t} \tag{9}$$

where c is the wave celerity. Double subscript after a comma denotes second partial differentiation.

By solving and comparing the results between the system of Eqs. (3) and (4), and Eqs. (8) and (9), the nonlinear distortion effects can be assessed.

3. METHOD OF SOLUTION

3.1 Finite Element Solution of Equations (3) and (4)

For solution of Eqs. (3) and (4), a finite element technique was utilized (Scarlatos, 1982). The x-t solution domain was discretized and each dependent variable was expressed as a weighted function of its parametric values at the nodes of a rectangular, space-time, finite element, i.e.,

$$u_{(e)} = \sum_{k=1}^{4} N_k(x,t)\ u_k \tag{10}$$

$$h_{(e)} = \sum_{k=1}^{4} N_k(x,t)\ h_k \tag{11}$$

where $N_k(x,t)$ are the weighting functions described as

$$N_k(x,t) = \frac{1}{4}\ (1 + \xi_0)\ (1 + \zeta_0);\ \xi_0 = \xi\xi_k,\ \zeta_0 = \zeta\zeta_k \tag{12}$$

$$\xi = 2[x - 0.5(x_i + x_{i-1})]/(x_i - x_{i-1}) \tag{13}$$

$$\zeta = 2[t - 0.5\ (t_{j+1} + t_j)]/(t_{j+1} - t_j) \tag{14}$$

The meaning of the subscripts i and j and parameters ξ_k, ζ_k are represented in Figure 1.

Substitution of the approximated variables $u_{(e)}$ and $h_{(e)}$ into Eqs. (3) and (4) resulted in residual values R_1 and R_2 that were generally different from zero. To minimize the error, the Galerkin technique was applied to the residuals and finally a system of four equations with four unknowns was obtained. The four unknowns were velocities and water elevations at the nodal points (i-1, j+1), (i, j+1). The variables at time level j were known either from the initial conditions or from the previous time step solution.

Since tidal oscillations are a slowly varying process, advantage was taken of this fact to obtain a linear system without sacrificing the nonlinear character of the phenomenon. This was accomplished by

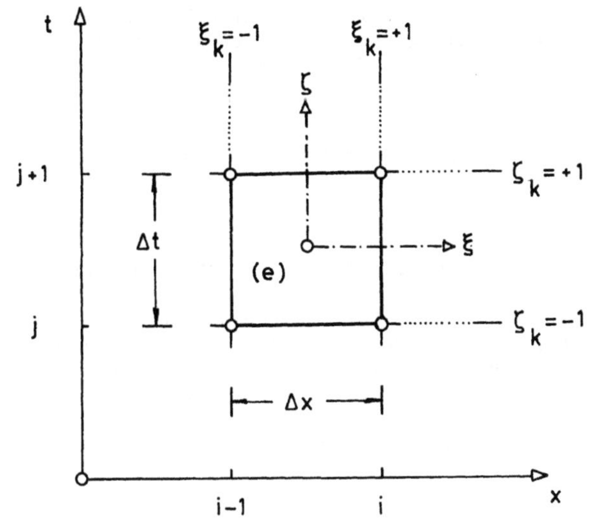

Figure 1. Space-time finite element discretization.

referring one of the two variables in nonlinear terms to the immediately previous time step.

Once the local system for each element was established, the entire strip between times t_{j+1} and t_j was assembled into a global system (Zienkiewicz, 1972), and it was solved by the conjugate gradient method. In that way the solution was implicit in space and explicit in time.

For an initial condition, a horizontal water surface with zero velocities was assumed. After five to six tidal cycles, a quasi-state condition was established. The boundary condition was sinusoidal water elevation disturbance at the open-sea end, and a reflecting wall at the other end. The scheme was free from instability or convergence problems for all practical purposes.

3.2 Harmonic Solution of Equations (8) and (9)

Telegrapher's equation accepts a sinusoidal type solution. For the case of a close-end waterway, the solution is the result of superposition of incident and reflecting waves, i.e.,

$$h = a_0[\exp(-\mu x)\ \cos(\sigma t - kx) + \exp(\mu x)\ \cos(\sigma t + kx)] \qquad (15)$$

where a_0 is half the wave amplitude at the closed-end boundary, μ is the damping modulus, σ is the tidal frequency and k is the wave number.

From Eq. (15), it is evident that a sinusoidal signal will remain sinusoidal during its propagation along the channel regardless of its amplification or damping due to various effects. Accordingly, the velocities are given as

$$u = \frac{a_0}{H} c \frac{k_0}{(\mu^2 + k^2)^{1/2}} [\exp(-\mu x) \cos(\sigma t - kx + \alpha) -$$

$$- \exp(\mu x) \cos(\sigma t + kx + \alpha)] \tag{16}$$

where k_0 is the frictionless wave number and α is the phase difference between water elevation and currents.

Both μ and k are not known apriori, but they should be calibrated from experimental data. This can be done by knowing the ratio between wave amplitude at certain station over the wave amplitude at the closed-end boundary, and the occurrence time of their maximum (Ippen and Harleman, 1966).

Once μ and k are determined, then the phase angle α can be estimated from the expression:

$$\tan \alpha = \mu/k \tag{17}$$

The linearized friction factor M is related to phase angle α as

$$M = \frac{\sigma}{g} \tan 2\alpha \tag{18}$$

From Eqs. (7), (17) and (18), dependency of parameters μ and k on the physical and geometrical features of the channel is evident.

4. NUMERICAL EXPERIMENTS

In order to investigate reliability of the linearized tidal equations, a set of numerical experiments have been performed under a wide variety of input data. The independent variables of the system were the length of the waterway ℓ, the mean water elevation H, the ratio between wave amplitude a_0 and water depth H, and the bottom friction given in term of Chezy's coefficient. The entire range of experiments results from combination of the variables given in Table 1. For all the experiments, the tidal period was kept constant and equal to 12 hours.

This problem was approached as follows. First, the velocities and water elevations at certain stations along the waterway were computed by the finite element method. The data were utilized for calibration of the harmonic parameters k and μ. Then the harmonic technique was applied to re-estimation of tidal water elevations and current velocities. Finally, the results of the finite element method and the harmonic analysis approach were compared to each other and the accuracy of the latter was quantified.

Table 1. Value of variables for numerical simulations.

Length of the waterway in km	45-60-75-90
Depth of the waterway in m	10-20-30
Ratio of amplitude over depth	0.025-0.050-0.075-0.100
Chezy's coefficient of friction in $m^{1/2}/s$	30-45-60-75-90-105

5. RESULTS

5.1 Tidal Distortion

The first quantity to be analyzed was nonlinear tidal distortion. This was done by introducing a normalized distortion quantity defined as

$$D_h = \frac{h_{max} - |h_{min}|}{\frac{1}{2}(h_{max} + |h_{min}|)} \times 100; \quad D_u = \frac{u_{max} - |u_{min}|}{\frac{1}{2}(u_{max} + |u_{min}|)} \times 100 \qquad (19)$$

where h_{max}, h_{min} refer to the water elevation at the closed-end boundary and u_{max}, u_{min} refer to the open sea boundary. Maximum conditions correspond to flood stage, while minimum to ebb. In order to determine the influence of each individual independent variable on tidal distortion, the results were plotted using a different parameter each time. For conservation of space, only a few representative figures will be given in this paper.

First, the distortion versus Chezy's coefficient of friction were plotted (Figure 2). From this figure, it can be seen that generally wave distortion increases for increasing tidal amplitude. The behavior for the case of current velocities distortion is more complicated. For 'smooth' estuarine beds, current distortion increases with increasing tidal amplitude, but for 'rough' beds distortion, it also seems to depend strongly on other quantities such as depth and/or length of the waterway.

Another interesting feature is that wave distortion attains a relative minimum while velocity distortion attains a relative maximum value when friction is the independent variable. The extreme values do not correspond to 'smoothest' estuarine beds but are linear functions of the a_0/H ratio. Indeed, from Figure 2 it is evident that the extreme values for various a_0/H values fall into a straight line. This is a proof that frictional energy dissipation is not the sole factor that has an important effect on the distortion processes.

Plotting tidal distortion versus ratio a_0/H (Figure 3), it can be easily seen that distortion is greatly affected by the magnitude of tidal range and/or mean water depth. The correlation, however, is not explicit because other nonlinear effects are present. From the entire set of data, it is concluded again that there is a tendency of these

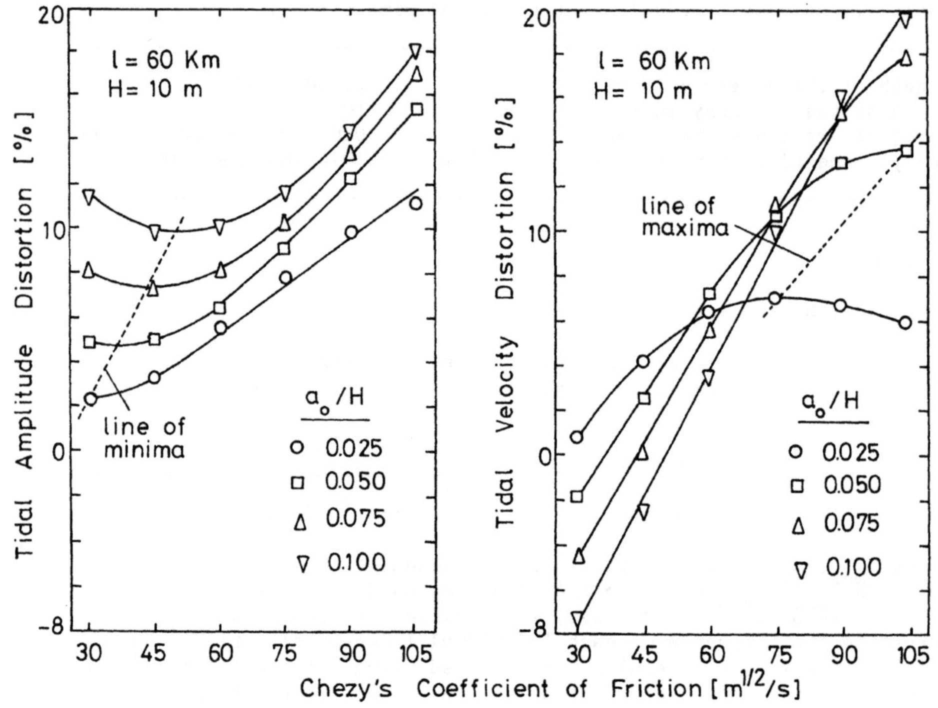

Figure 2. Nonlinear tidal distortion versus Chezy's coefficient of friction.

curves to obtain a relative maximum or minimum value. This is clearly demonstrated in the tidal velocity distortion curves of Figure 3. Since these curves should pass through (0,0) point, they were extrapolated accordingly (dashed lines of Figure 3).

In order to identify the effects of water depth H for constant value of the ratio a_0/H, in Figure 4 tidal distortion was plotted against mean water depth. Although the importance of the water depth to nonlinear distortion was evident, there was no indication suggesting any specific law of influence. From most of the cases, however, it seems that increasing depth decreases the water mass accumulation, i.e., decreasing high water elevation while at the same time supressing low water level.

Finally, distortion was plotted versus the length of the channel (Figure 5). For water depth H = 10 m, tidal amplitude distortion seemed to increase with increasing waterway length, while the opposite was true for current distortion. The situation was drastically

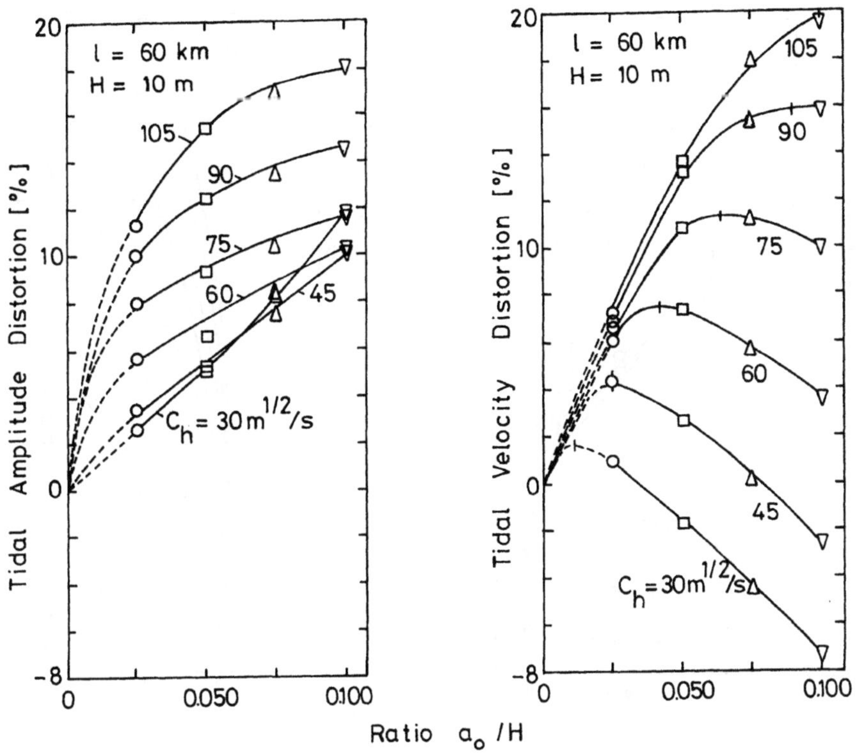

Figure 3. Nonlinear tidal distortion versus ration a_0/H.

altered for higher water depth (H = 20 m), where no specific tendency was identified.

Generally, tidal distortion appears to be a very complicated phenomenon depending on a variety of physical and geometrical parameters. From the entire spectrum of results, the importance of bottom friction, water depth, tidal amplitude and length of the channel was evident but assessment of their individual influence was not possible for explicit quantification.

As a qualitative remark, nonlinear distortion is higher for the case of tidal currents than for tidal amplitudes. The higher distortions that were recorded were 50 percent for the currents and 25 percent for the water surface elevations. Thus, negligence of the nonlinear terms may lead in certain cases to inaccurate information of tidal propagation processes.

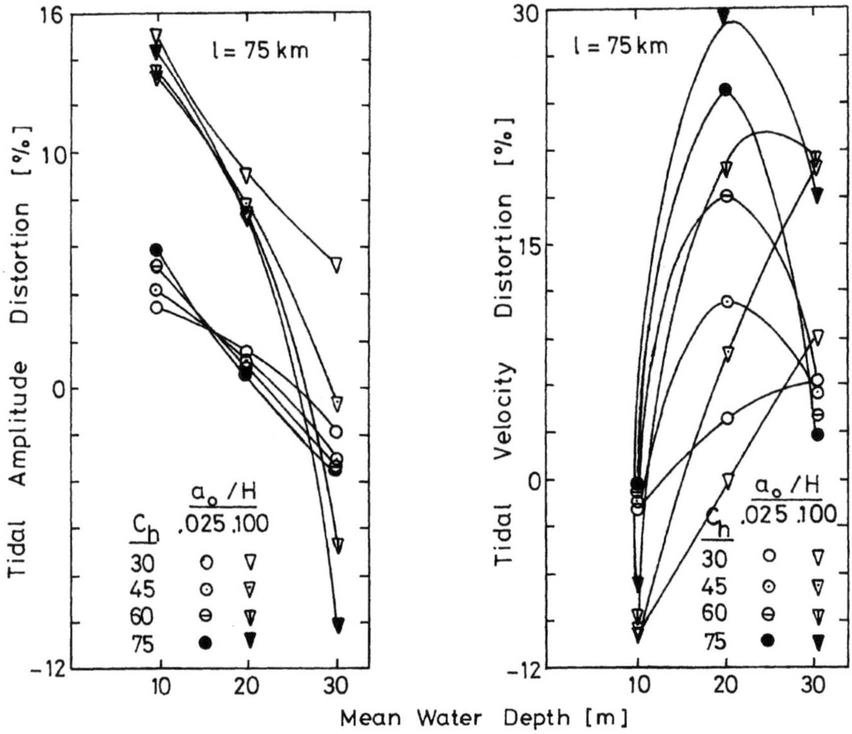

Figure 4. Nonlinear tidal distortion versus mean water depth.

5.2 Accuracy of Linearized Solution

In order to quantify the accuracy of harmonic solution based on Telegrapher's equation with regard to the results obtained by the finite element solution of the complete St. Venant system of equations, the following was done: First, an average value for tidal wave and current amplitude was attained by taking the arithmetic mean of their maximum and minimum values, i.e.,

$$h_m = \frac{1}{2} (h_{max} + |h_{min}|); \; u_m = \frac{1}{2} (u_{max} + |u_{min}|) \tag{20}$$

Then those mean values were compared with the harmonic analysis results. For convenience, an accuracy deviation quantity was defined as

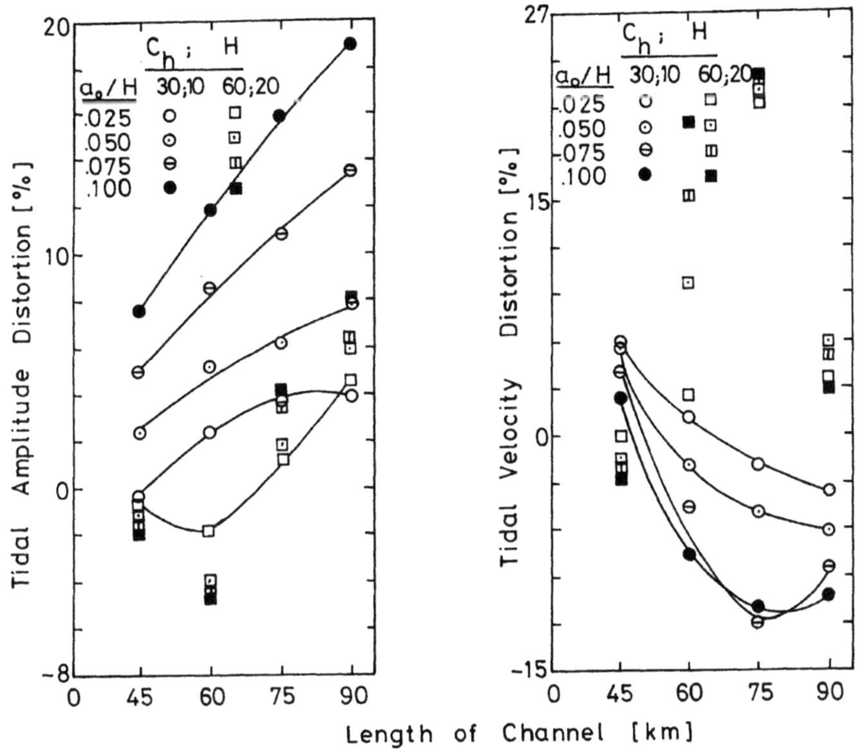

Figure 5. Nonlinear tidal distortion versus length of waterway.

$$D_h = \frac{h_m - h_{Hm}}{\frac{1}{2}(h_m + h_{Hm})} \times 100; \quad D_u = \frac{u_m - u_{Hm}}{\frac{1}{2}(u_m + u_{Hm})} \times 100 \tag{21}$$

where h_{Hm} is the wave amplitude at the closed-end boundary and u_{Hm} is the current amplitude at the open sea boundary. Both quantities were estimated by the harmonic approach.

From the entire data set, the probability of exceeding a certain accuracy deviation number was defined for tidal amplitude and velocity. These results are represented in Figure 6. Again, it is obvious that velocities are more sensitive to the computational technique than water elevations. Also, for both cases, the probability is exponentially distributed.

Concluding, it can be said that with a 90 percent probability, the accuracy deviation of the harmonic technique will be less than 4 percent for the water amplitude and 10 percent for the water

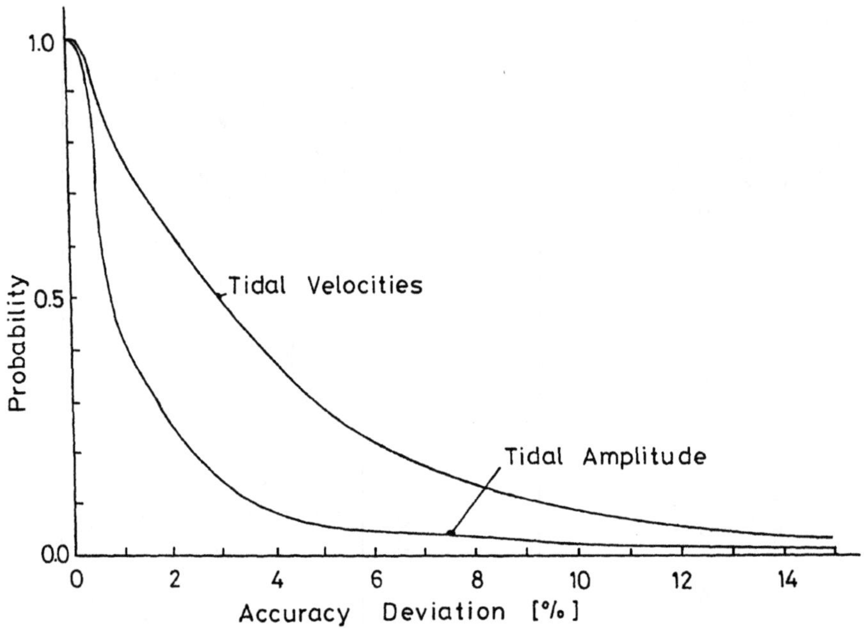

Figure 6. Probability of accuracy deviation due to linearization of tides.

velocities. Of course the real inaccuracy cannot be assessed unless the nonlinear distortion is incorporated.

6. CONCLUSIONS

The conclusions obtained from this study are as follows:

1. Tidal distortion is dominated by nonlinear effects such as high ratio of wave amplitude over mean water depth, convectional velocities and bottom friction.
2. Distortion depends on wave amplitude, mean water depth, length of the channel and bottom roughness. However, it was not possible to explicitly assess their individual importance.
3. Current distortion is higher than water amplitude distortion. The highest observed value of the former was approximately 50 percent, while for the latter, it was of the order of 25 percent.

4. The probability of obtaining inaccurate results with a relative error greater than 10 percent when using the harmonic technique is about 10 percent.

5. For estuarial impact studies, harmonic analysis must be used with care because it may occasionally give misleading information regarding tidal oscillatory behavior.

7. ACKNOWLEDGEMENTS

This study was partially funded by the Louisiana Water Resources Research Institute at Louisiana State University, Baton Rouge, Louisiana.

8. REFERENCES

Dronkers, J.J., 1964, Tidal Computations in River and Coastal Waters, North-Holland Publ. Co., Amsterdam.

Ippen, A.T. and Harleman, D.R.F., 1966, 'Tidal Dynamics in Estuaries,' in: A.T. Ippen (Editor), Estuary and Coastline Hydrodynamics, pp. 493-545, Engineering Societies Monographs, McGraw-Hill, New York.

Mahmood, K. and Yevjevich, F., Editors, 1975, Unsteady Flow in Open Channels, Vol. I, II. Water Resources Publ., Fort Collins, Colorado.

Parsons, W.B, 1918, 'The Cape Cod Canal,' Transactions of the ASCE, Vol. 82, Proc. Paper 1403, pp. 106-138.

Scarlatos, P.D., 1980, 'A Study of Tidal Disturbances in Channels of Finite Length,' Scientific Bulletin, Section 5, Vol. H, pp. 23-84, Aristotelean University of Thessaloniki, Thessaloniki, Greece.

Scarlatos, P.D., 1982, 'A Pure Finite Element Method for the Saint-Venant Equations,' Coastal Engineering, Vol. 6, Elsevier, Amsterdam, pp. 27-45.

Zienkiewicz, O.C., 1971, The Finite Element Method in Engineering Science, McGraw-Hill, New York, pp. 12-15.

FLOOD MANAGEMENT IN THE NETHERLANDS FROM THE MIDDLE AGES TO THE SPACE ERA

Johannes W. van der Made
Senior Engineer
Rijkswaterstaat
Tidal Waters Division
P.O. Box 20907
2500 EX THE HAGUE
Netherlands

ABSTRACT. After a general description of hydrological conditions in the Netherlands, with special reference to the threat of flooding from either the rivers or the sea, a historical review is given of the forecasting procedures for river floods and for storm surges. The present situation is described, highlighting the organization of the information and warning services and the methods used to assess forecasts. Finally a brief look is taken at possible future developments of modelling and observation techniques.

1. FLOODS IN THE NETHERLANDS THROUGH THE AGES

The Netherlands, which is partly formed by the Delta of the Rhine and Meuse rivers and which borders on the North Sea, has been faced with the danger of flooding since time immemorial. The country is threatened by floods from two sides: the sea and the rivers (fig. 1).

In the Middle Ages the first embankments were constructed to protect the population, which at that time was not very dense,and the land against flooding. But at the same time, owing to rises in sea level, tectonic soil subsidence and agricultural land drainage, water levels rose higher and higher in relation to the elevation of the land. This fact, sometimes combined with a lack of maintenace of the dykes, resulting from internal troubles, meant that the defence against flooding was not always successful. Besides, the rise in sea level not only increased the danger of flooding from outside. The drainage of excess water from local rainfall could no longer be effected by gravity alone, but had increasingly to be performed by pumping. This later led to the system of polders, windmills and pumping stations.

As stated earlier, the land is exposed to floods from two sides, i.e.

V. P. Singh (ed.), Flood Hydrology, 271–280.
© 1987 by D. Reidel Publishing Company.

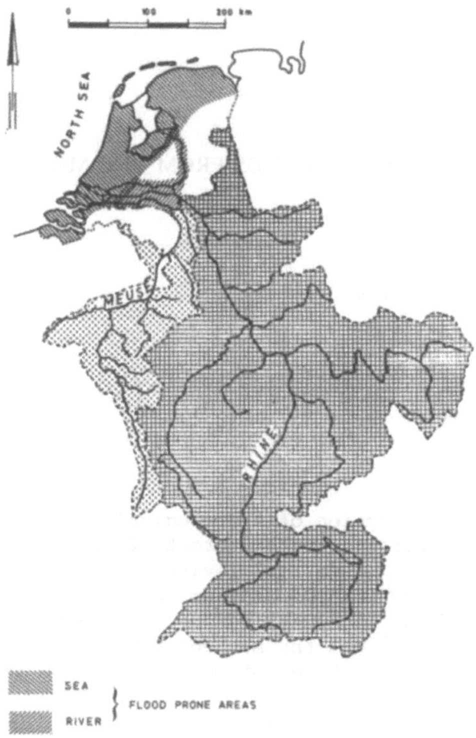

Fig. 1. The Netherlands between the rivers and the sea.

the sea and the rivers. Since the Middle Ages, about one hundred serious floods are known to have occured. Table 1 contains a list of years in which floods occurred, with an indication whether they were sea (S) or river (R) floods. This table includes 57 sea floods and 36 river floods. It is most likely that this list is not complete. Particularly in the more distant past, floods may have occurred of which no information is recorded in the old chronicles. So the peak for the 1500s should not be overestimated with respect tot earlier centuries. It is possible that in later centuries somewhat fewer floods occured, maybe owing to improved dykes, but the decreasing number of floods may be overshadowed by the extent of the damage. Higher dykes caused a feeling of safety and the population and investment had grown, as they continue to do today.

It is obvious that the list cannot indicate the frequency of storm surges and river floods, and, particularly not its trend over time. Recently, especially, many storm surges and river floods have occurred which did not cause flooding.

century beginning after:															
1200		1300		1400		1500		1600		1700		1800		1900	
yr	S/R	yr	S/R	yr	S/R	yr	S/R	yr	S/R	yr	S/R	yr	S/R	yr	S/R
19	S	15	S	03	S	02	S	10	S	02	S	09	S	16	S
48	S	22	R	04	S	07	S	12	S	09	R	20	R	26	R
87	S	62	S	11	R	08	S	16	S	15	S	24	S	28	S
		74	S	13	R	09	S	21	R	17	S	25	S	53	S
		77	S	21*	S	10	S	24	R	26	R	53	S		
		90	R	22	S	11	S	25	S	40	R	55	R		
		93	R	32	R	14	S	35	R	44	R	61	R		
		97	R	37	S	21	S	37	R	47	R	73	S		
				52	S	23	R	51	S	51	R	76	R		
				68	S	30	S	58	R	55	R	81	S		
				70	S	32	S	62	S	60	R	89	S		
				77	S	52	S	63	R	75	S	94	S		
				96	R	53	S	65	S	76	S				
				97	R	58	R	75	S	91	S				
						70+	S	77	R	92	S				
						71	R	82	S						
						72	R								
						73	R								
						75	S								
						95	R								
						99	R								
3 x S		4 x S		9 x S		14 x S		10 x S		7 x S		7 x S		3 x S	
		4 x R		5 x R		7 x R		6 x R		8 x R		5 x R		1 x R	
3		8		14		21		16		15		12		4	

*) St. Elizabeth Flood
+) All Saints Flood

Table 1. Sea and river floods in the Netherlands in course of the ages.
(taken from information of Fockema Andreae, 1953).

2. FLOOD WARNINGS

As early as the Middle Ages responsibility for water affairs was borne by "Water Boards", a typically Dutch administrative unit, being a kind of corporation of land owners in the area concerned. These Water Boards had sovereignty in their own territory with regard to flood protection and water management. It is possible that they were also responsible for providing information about floods which occurred and warning of danger. Most of the activities must have consisted of taking measures in advance or of repairing damage afterwards.

Over the centuries, several attempts were made to arrive at a certain degree of centralization, first by the Counts of Holland and later by the Dutch Republic, i.e. the federation of states founded in 1579, but progress was very slow. Central power did not extend to water affairs, which remained entirely in the hands of the Water Boards. Their interests were often in conflict with those of their neighbours and often disputes could not be settled. It was not until the establishment, under the impact of the French revolution, of the Batavian Republic in 1798, which took the place of the former Republic, that a cental agency for water affairs was created, the predecessor of the present-day "Rijkswaterstaat". From 1803 attempts were undertaken to make organisational arrangements for issuing messages about rivier floods and ice cover in order to take quick measures against impending disasters. The earliest regulations date from 1806. Then it was decided to establish two river services: one to administer rivers under normal conditions and the other to deal with emergencies. In times of river floods special couriers on horseback carried messages from upstream to the Water Boards concerned. Messages were also transmitted by flag signals and by firing guns. In the mid-19th century messages about the state of the river began to be transmitted by telegraph and later on by telephone, and since the 1930s the radio has played an important role, recently supported by systems like videotext.

The instructions for the "River Flood Emergency Service" were changed several times. In 1900 this institution was brought with the framework of the General Public Works Act. The latest general regulation dates from 1951. The River Flood Emergency Service has always been instructed to act by the Minister of Public Works. In emergencies Government engineers are directed to dyke posts and are invested with special executive power, which they take over from the Water Boards. The responsibilities of the Emergency Service do not include making and disseminating forecasts. The water level measured at Cologne (Federal Republic of Germany) was the official criterion.

Until the 1930s no institution existed to issue warnings about storms surges. Here the water boards were thrown entirely upon their own resources.

3. FORECASTS OF STORM SURGES AND FLOOD WAVES

In 1911 a storm surge which flooded a timber yard on the lower reaches of one of the rivers led to a request to the Government for to the possibility of a warning and forecasting system to be studied. Ten years later, in 1921, an experiment in this direction was started, leading, a further ten years later, to the establishment of a Storm Surge Warning Service, which was provided by the Royal Netherlands Meteorological Institute. However, this system was very limited and related only to the lower reaches region. The 1953 disaster, which hit the South Western part of the country hardest, drowning about 1900 people, was the motive for setting up a completely new system, in which the Rijkswaterstaat became involved.

The procedure followed by the Storm Surge Warning Service when storm surges are expected is as follows. The Meteorological Institute, which permanently monitors the weather, contacts the Rijkswaterstaat, i.e. the engineer in charge of the Storm Surge Warning Service, if the weather situation develops in such a way that a storm surge along the Dutch coast may be expected. After consultation, this engineer may decide to occupy the forecasting centre for monitoring the development of water levels. This forecasting centre is equipped with telemetering water level recorders.

If the situation becomes such that certain critical levels are expected to be exceeded around high tide, the forecasting centre issues telegrams warning the authorities responsible for the sea dykes. It is required that such telegrams be issued at least 6 h before the high tide concerned in order to give the responsible authorities the opportunity to take the necessary measures. There are two levels of warnings:

1) limited dyke watch
2) extensive dyke watch

In the case of the limited dyke watch only some of the offices are occupied, but in the event of an extensive dyke watch being declared continuous monitoring of the critical locations is organized; conditions in situ are then inspected continuously and organizations like the Red Cross, Civil Defence, Army and Navy will also be warned.

For the purposes of the Storm Surge Warning Service the Dutch coastline is divided into five sections as shown in Fig. 2, where the tidal conditions may be different. A sixth section covers the lower reaches of the rivers which may be attacked from two sides: storm surges from the sea and flood waves from the rivers.

The method which has so far been used to calculate the wind set up is a manual method, applied by experienced staff at the Meteorological Institute. It is based on a division of the North Sea into a number of sub-areas, as shown in Fig. 3. First a forecast is made of the atmospheric situation 7-10 h ahead, a so-called pre-barethic, leading to certain wind forces in the different areas. Each area contributes to a certain extent to the wind set up in a particular section. By adding together the influences of all areas a value is produced for the total wind set-up (Weenink, 1958). At the forecasting centre this information is considered together with the actual development of water levels, ultimately leading to a final forecast of the set-up, which is communicated in the warning telegrams.

The forecast made by the Meteorological Institute disregards astronomical tides. In about the last ten years a hydrodynamic-numerical model has been developed in which wind forces act on a water mass (Timmerman, 1977). Here too, tidal motion is not taken into account. Recently, the results of this model have also been considered when making forecasts.

An example of a storm surge is given in Fig. 4. The upper diagram shows the sea level actually recorded together with the computed astronomical tide. The black triangles indicate the provisional forecasts for the high tide level on 2nd February 1983, at 5 PM, calculated at different

Fig. 2. Sections of the Storm Surge Warning Service.

Fig. 3. North Sea sub-areas used to calculate wind set up.

times. The open triangle indicates the final forecast, assessed by the forecasting centre and given in the warning message which was sent tothe appropriate authorities. The levels for limited (L) and extended (E) dyke watches are also shown.

The middle diagram shows the difference between the two curves in the upper diagram, i.e. the wind set-up, together with forecasted curves produced by the computer model, starting at various times. In the lower diagram one finds information on wind speed and direction in the vicinity of the station concerned.

Table 2 shows a summary of the results of 30 years work by the Storm Surges Warning Service. The period (1954-1984) is divided into two sub periods, 1954-1973 and 1974-1984. In 1973 a reorganization of the system was carried out. Some improvement can be seen in the standard deviation, although the results vary. This indicates the complicated character of storm surge phenomena in the North Sea.

Forecasts for the river stages were first issued in 1937, when a flood forecasting service was established.These were based on correlation curves with the river stages in the reaches of the rivier Rhine in Germany.

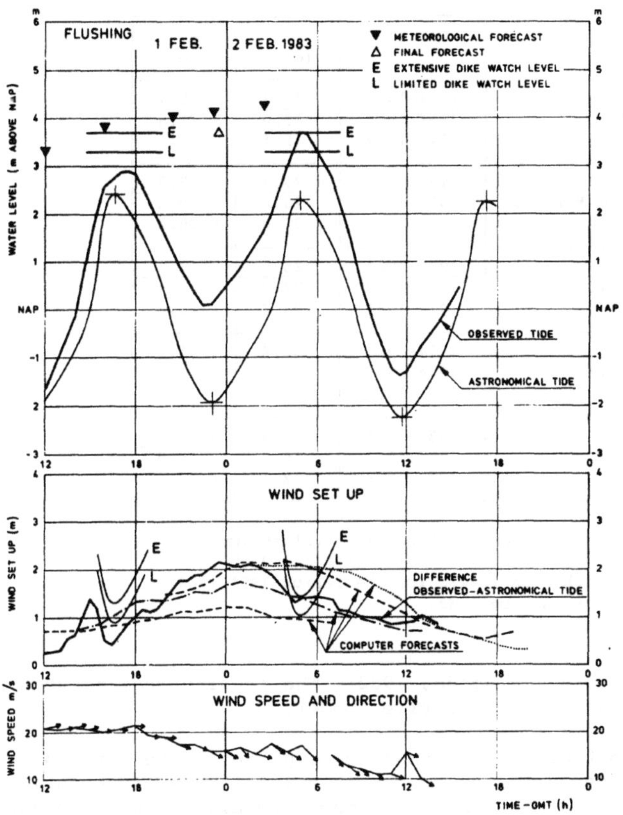

Fig. 4. Storm surges data for 2 February 1983 at Flushing (Netherlands)

In 1950 the "Wemelsfelder method" of river flow forecasting came into use (Wemelsfelder, 1972). This method involved adding the main river discharge to the discharges of the larger tributaries. Corrections were applied for flood wave propagation time and subsidence. Although this method was based on physical considerations the variable character of the flood wave was not adequately reproduced. The method was used with the aid of a standard table and graphs.

Recently this method has been replaced by a linear regression method, operated with the help of a desk computer. In this method the discharges of the main river and some tributaries act as input, together with precipitation data for a number of selected precipitation gauges in the drainage basin. The input consists of the data of the actual day and of

Station	Period	sd (cm)	mv (cm)
FLUSHING	1954-1984	22,2	- 16,1
	1954-1973	25,0	- 21,4
	1974-1984	19,4	- 12,1
HOOK OF HOLLAND	1954-1984	26,4	- 17,7
	1954-1973	28,2	- 9,2
	1974-1984	21,8	- 26,2
DEN HELDER	1954-1984	25,5	- 1,3
	1954-1973	27,3	5,4
	1974-1984	23,7	- 6,3
HARLINGEN	1954-1984	36,4	1.5
	1954-1973	44,1	7,6
	1974-1984	27,3	- 4,3
DELFZIJL	1954-1984	43,4	- 16,6
	1954-1973	55,9	- 6,6
	1974-1984	32,1	- 23,4

Table 2. Standard deviations (sd) and mean values (mv) of the differences between the observed and the forecasted high tide levels for the period 1954-1984 and for two sub periods.

the four days before (De Ronde, 1982). The results seem to show some improvement on those of the earlier model.

For the assessment of the forecast to be given in the warning telegram, which is to be issued to the responsible institutes, forecasts received from Switzerland and the F.R. of Germany are also taken into account. This may lead to some corrections to the results of the model.

4. FURTHER DEVELOPMENT OF FORECASTING MODELS

For the forecasting of the sea level data new hydrodynamic numerical models are being examined. In particular the continental shelf model should be mentioned, developed in cooperation between the Tidal Waters Division of Rijkswaterstaat and the Delft Hydraulics Laboratory.

The storm surge model is based on the WAQUA system in which the full non-linear shallow water equations are solved with an alternating direction implicit (ADI) finite difference scheme (Stelling, 1984). The equations, written in spherical coordinates, are discretized on a fine grid of 1/9° longitude and 1/12° latitude. This constitutes a model of over 20,000 active grid points. Along the open boundaries the water level is described in terms of 6 harmonic constituents(O_1, K_1, M_2, S_2, N_2 and K_2). These data originate from a coarser grid model that covers a large part of the north-east Atlantic Ocean. The external forces - wind speed and atmospheric pressure gradient - are provided by the Royal Netherlands Meteorological Institute on a 1/3° latitude grid at 3 hour intervals. For the tidal calibration and validation two 4-day periods were used which fell within the JONSDAP '76 and '73 campaigns, respectively. During these campaigns a lot of prototype data were collected. The model predictions are quite accurate. The storm surge modelling is being calibrated for a 9-day period in February 1983 and will be verified for 5 recent storms. The area covered by the model is shown in Fig. 5. Some results of the February 1983 storm are given in Fig. 6.

For river discharge forecasting too, new models are being developed, particularly based on flood routing models and on quantitative precipitation forecasting. The application of optimum estimation techniques like Kalman filtering is included.

For the North Sea model too, Kalman filtering is being considered. In view of the great number of elements in this model, the use of a super-computer is required. Otherwise more simple approaches are being examined like:
- the combination of a complete model and a sub-optimal filter;
- the combination of a reduced model and an optimal filter.

Another development is data acquisition through satellite observations. Although there are financial problems there are proposals to perform rainfall observations from geostationary satellites like METEOSAT. For sea level observations attempts are being made to use the ERS-1 satellite, to be launched by ESA in 1989, which will be equipped with a radar altimeter. There are also contacts with NASA with regard to the TOPEX/POSEIDON satellite, which will also have these facilities. Sea level data acquired in this way can be used as input data for the North Sea model. For real operational use in the distant future more than one satellite would be required since these would be polar orbiting satellites which cross the North Sea area only occasionally.

Of course all these possibilities are at a more or less advanced stage of development. The extent to which they will be able to be of use for real-time operational forecasting will depend on the experiences in the research phase, with regard not only to the accuracy and reliability of the results, but also to their practical applicability.

5. REFERENCES

Fockema Andreae, S.J., 1953, An outline of South Holland flood disasters in ancient times' (in Dutch), Zuid-Hollandse studiën deel III. Historische Vereniging van Zuid-Holland onder de zinspreuk "Vigitate Deo Confidences". Voorburg, Netherlands.

De Ronde, J.G., 1984, 'The forecasting and warning system of Rijkswaterstaat for the river Rhine', Report 6 , Dept. of Hydraulics and Catchment Hydrology, Agricultural University, Wageningen, Netherlands, pp. 325-336.

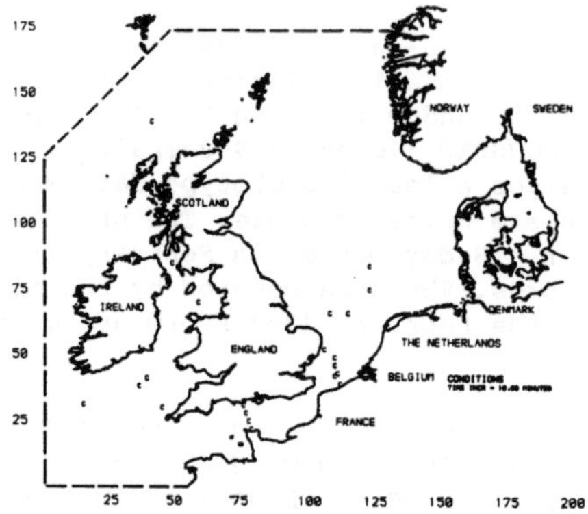

Fig. 5. Overall situation of the North Sea Storm Model.

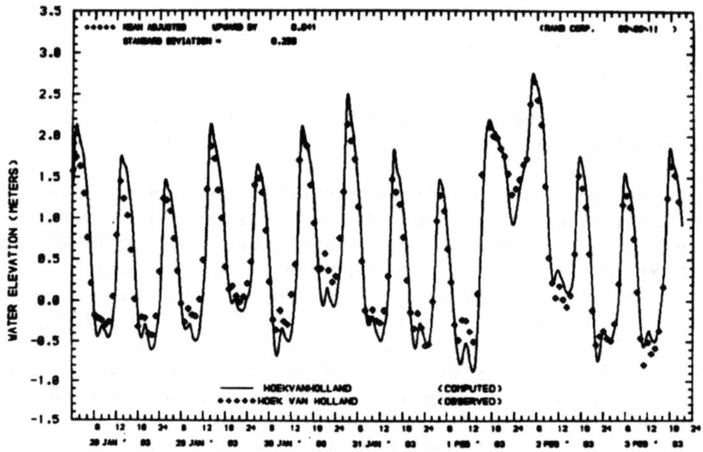

Fig. 6. Water levels at Hook of Holland station.

Stelling, G.S. (1984), 'On the construction of computational methods for shallow water flow problems', Rijkswaterstaat Communications 35 , Rijkswaterstaat, The Hague, The Netherlands.

Timmerman, H., (1977). 'Meteorological effects on tidal heights in the North Sea'. KNMI, Mededelingen en verhandelingen, 99 . De Bilt, The Netherlands.

Weenink, M.P.H., (1958). 'A theory and method of calculation of wind effects on sea levels in a partly enclosed sea, with special application to the southern coast of the North Sea'. KNMI, Mededelingen en verhandelingen, 73 . De Bilt, The Netherlands.

Wemelsfelder, P.J. (1972). 'Fighting floods and storm surges in the Netherlands'. Quaderno N. 1969, Accademia Nazionale dei lincei, Roma. pp 751-772.

OPTIMIZING FLOOD PROTECTION FOR CYPRESS CREEK, HARRIS COUNTY, TEXAS

Paul C. Wilson, Jr., P.E.
Assistant Chief, Planning Division
U.S. Army Engineer District, Galveston
P. O. Box 1229
Galveston, Texas 77553-1229, U.S.A.

Robert Pudlewski, P.E.
Project Engineer and Floodplain
Management Coordinator
U.S. Army Engineer District, Galveston
P. O. Box 1229
Galveston, Texas 77553-1229, U.S.A.

ABSTRACT. This paper describes the formulation of a flood damage reduction project for Cypress Creek in the San Jacinto River Basin in Harris County, Texas, U.S.A. The investigation was conducted in accordance with the planning guidance and hydrologic study guidelines of the U.S. Water Resources Council and the National Weather Service, supplemented by instructions and regulations of the U.S. Army Corps of Engineers. In developing the discharge-frequency curve, the computed Probable Maximum Storm discharge of 3,115 m^3/s was arbitrarily assigned a return frequency of once in 10,000 years. The optimum plan is an improved channel to carry 40-year frequency flood flows under 1990 conditions of watershed development. This level of flood protection is expected to degrade over time, unless county officials are successful in their efforts to control future runoff to the creek.

INTRODUCTION

In January 1985 the Galveston District, Corps of Engineers, completed a flood damage reduction study of Cypress Creek in northern Harris and eastern Waller Counties, north of the City of Houston, Texas. Cypress Creek is a tertiary tributary of the West Fork of the San Jacinto River, and drains the 319 square-mile (826 km^2) watershed shown in Figure 1. The ground surface slopes gently from 310 to 60 feet (94 to 18 m) above the National Geodetic Vertical Datum (NGVD) of 1929. The streambed slopes 2.7 feet per mile (0.5 m/km). The study area is about 8 percent urbanized with development concentrated along the attractively-wooded reaches of Cypress and Little Cypress Creeks downstream from U.S. Highway 290, which cuts diagonally through the upper portion of the

V. P. Singh (ed.), Flood Hydrology, 281–287.
© 1987 by D. Reidel Publishing Company.

watershed. A low southern divide separates the portion of the watershed upstream from U.S. Highway 290 from the adjacent Buffalo Bayou Watershed, and some of the Cypress Creek floodwaters flow southward across the divide into Buffalo Bayou. This basin overflow is expected to continue uninterrupted.

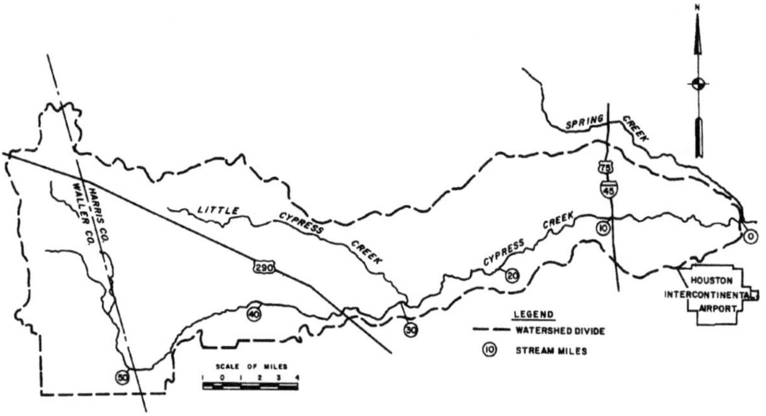

Figure 1. Cypress Creek Watershed map.

The principal water resources needs in the Cypress Creek Watershed are reduction of urban flood damages, improvement of water quality, satisfaction of outdoor recreation needs, and preservation of the existing fish and wildlife habitats. The study addressed all of these needs except improvement of water quality.

The primary purpose of the study was to formulate a plan to reduce urban flood damages in the watershed. Presently 3,400 structures are subject to flooding by a Standard Project Flood, and 2,400 by a 100-year flood. Most of these structures have been constructed since 1960. Average annual flood damages from the occurrence of the full spectrum of possible flood events over time total $9.3 million.

COLLECTION AND ANALYSIS OF HYDROLOGIC DATA

Since 1944 the U.S. Geological Survey has maintained a stream gage at the Interstate Highway 45 (IH 45) bridge across Cypress Creek, 11.0 miles (17.7 km) upstream from the mouth. The drainage area upstream from the gage is 285 square miles (738 km^2). The present datum of the gage is 63.89 feet (19.474 m) NGVD, unadjusted for land subsidence. Figure 2 shows the estimated peak annual flows corresponding to the peak water surface elevations recorded at the gage during the 40-year period 1945 through 1985. Flooding commences when the stream flow at the gage is between 4,000 and 4,300 cubic feet per second (cfs)(113-122 m^3/s). The maximum known flood elevation at the gage site, 97.89

feet (29.84 m), was observed in May 1929, indicating a peak flow of 26,000 cfs (736 m^3/s). A flood elevation of 95.89 feet (29.23 m) was observed in November 1940, indicating a peak flow of 15,000 cfs (425 m^3/s). Additionally, rainfall during the storm of December 6-8, 1935 ranged from 11 to 20 inches (28 to 51 cm) along Cypress Creek (Davis, 1953) and doubtless caused significant overbank flooding, although there are no recorded data concerning flood heights, flood volumes, or flow rates along the creek.

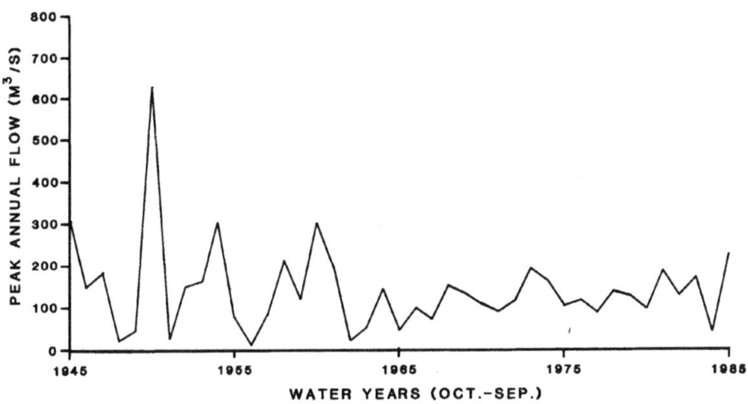

Figure 2. Peak annual flows, Cypress Creek at IH 45.

A chain of thunderstorms on October 25, 1984 caused the most severe flood since 1960. Intercontinental Airport, just south of the mouth of Cypress Creek, recorded 9.2 inches (23 cm) of rain in 24 hours, 7.5 inches (19 cm) of which fell between 10 a.m. and 4 p.m. on October 25, 1984. A peak flow of 7,920 cfs (224 m^3/s) resulted at the stream gage at IH 45. Several hundred homes along Cypress Creek were flooded, and approximately 500 persons were forced to seek emergency shelter.

Storm rainfall was determined in accordance with Hydrometeorological Report No. 51 (Schreiner and Riedel, 1978). The 10 square mile-72 hour Probable Maximum Storm (PMS) index used was 55.7 inches (141 cm). The Standard Project Storm (SPS) was taken as 50 percent of the PMS index. Rainfall depths and distributions were determined in accordance with Engineer Manual 1110-2-1411 (U.S. Army Corps of Engineers, 1952), supplemented by instructions from the U.S. Army Engineer Division, Southwestern. The peak PMS streamflow at IH 45, termed QMAX, was estimated to be approximately 110,000 cfs (3,115 m^3/s).

The statistical frequency curve for the stream gage at IH 45 was computed using Bulletin No. 17A (U.S. Water Resources Council, 1977) procedures with one modification. The skew factor used in the analysis was determined to allow the frequency curve to pass through the Probable Maximum Flood (PMF) discharge plotted at an assigned frequency of 0.01

percent (once in 10,000 years). The standard map skew was not used because of the unusual overflow characteristics of Cypress Creek. The estimated peak streamflows under existing conditions of watershed development in 1990 are listed in Table I.

TABLE I. ESTIMATED STREAMFLOWS ON CYPRESS CREEK AT IH 45 (EXISTING CONDITIONS OF WATERSHED DEVELOPMENT IN 1990)

Exceedence Frequency (Percent)	Recurrence Frequency (Years)	Percent of PMS Index	Peak Streamflow	
			(cfs)	(m^3/s)
0.01	10,000		110,000	3,115
(SPF)	(SPF)	50	49,400	1,399
1	100	30	27,700	784
2	50	24	21,500	609
4	25	19.5	16,600	470
10	10	14.2	10,700	303
20	5	10.9	7,310	207
50	2	6.9	3,650	103

The hydrologic analysis for Cypress Creek was performed before Bulletin 17B was issued. Subsequent reanalysis in 1985 indicates that for storms more frequent than once in ten years, Bulletin 17B procedures would result in slightly higher discharges than were used in the study. For storms occurring at frequencies between once in ten years and once in one hundred years, Bulletin 17B procedures would result in slightly lower discharges than were used in the study. Use of Bulletin 17B procedures probably would result in slightly higher average annual flood damages; however, the perceived impact is not significant enough to require revision of the report.

PLAN FORMULATION

The Corps of Engineers' study was commenced in 1976. The same year Harris County commenced land acquisition for the Cypress Creek Parks Project using $1.3 million in Federal and State grants matched with local funds and donated land. The Parks Project was conceived to preserve undeveloped land along Cypress Creek, both to serve as a natural flood absorber, and to provide open space and parkland for the enjoyment of future generations. This series of parks ultimately will consist of about 4,000 acres (1,619 hectares) of lands in the floodplain along Cypress Creek and the streams into which it flows, Spring Creek and the West Fork of the San Jacinto River. Over 1,400 acres (567 hectares) of land have been acquired to date.

As shown in Table II, anticipated watershed development between 1990 and 2040 is expected to increase the 100-year streamflow at IH 45 by 23 percent, with a further increase anticipated between 2040 and 2090. In 1984 Harris County adopted a policy of encouraging real estate developers to incorporate runoff limiting measures in their projects as a supplement to channel improvement (Harris County Flood Control District, 1984). It is still too soon to forecast the hydrologic impact of this new direction.

TABLE II. ESTIMATED STREAMFLOWS ON CYPRESS CREEK AT IH 45 (MIDLIFE CONDITIONS OF WATERSHED DEVELOPMENT)

Recurrence Frequency (Years)	Peak Streamflow	
	(cfs)	(m^3/s)
(SPF)	65,600	1,858
100	34,900	988
50	26,200	742
25	19,700	558
10	12,500	354
5	8,670	246
1	4,340	123

Both nonstructural and structural measures to reduce flood damages in the Cypress Creek Watershed were considered, singly and in combination. The final array of action alternatives are shown in Table III.

TABLE III. ALTERNATIVE ACTION PLANS FOR CYPRESS CREEK

Plan No.	Description	First Cost* ($1,000)	Annual Benefits ($1,000)	Annual Costs ($1,000)	Excess Benefits ($1,000)	Benefit-Cost Ratio
2	10_Yr Buy Out	141,056	8,025	10,556	--	0.8
3	10-Yr Channel**	96,540	14,799	11,133	3,666	1.3
4	25-Yr Channel**	152,613	15,425	17,035	--	0.9
5	50-Yr Channel**	207,081	15,672	22,902	--	0.7
6	5-Yr Channel**	73,986	9,042	8,550	492	1.1

*Exclusive of recreation and mitigation costs.
**Estimated degree of flood protection in 2090.

CONCLUSIONS

Plan 3 for channel improvements to carry future 10-year flows would provide the greatest excess of benefits over costs and thus offers the largest contribution to National Economic Development (NED). The most efficient means of providing flood protection was found to be Plan 3 for channel enlargement and straightening in combination with valley storage of flood waters on parks and undeveloped land, and floodplain management measures to prevent further unwise developments in the residual 100-year floodplain.

Channel excavation is designed to minimize disturbance to the Cypress Creek Parks Project. In reaches where only one bank is occupied by a park, the excavation would be done on the opposite bank. Where there are parks on both banks, the excavation would be done on the bank having the least valuable habitat, or on the bank adjacent to the larger park.

The recommended plan provides for channel improvements with a capacity of 23,000 cfs (651 m^3/s) to furnish 40-year flood protection under 1990 conditions and 10-year flood protection under estimated conditions of watershed development in 2090. If the Harris County Flood Control District can successfully reduce anticipated future increases in runoff from new residential and commercial developments, the degree of protection may not degrade this much. The channel would be trapezoidal with grass-lined side slopes of 1 vertical on 3 horizontal and a bottom width varying from 30 to 200 feet (9 to 61 m). The improved channel would be 29.4 miles (47 km) long and would extend from the mouth to U.S. Highway 290. The channel side slopes would be armored with stone riprap in reaches where stream velocities are expected to exceed 6 feet per second (1.8 m/s).

Additional project features include trails and other minimum basic recreational facilities, and habitat development to compensate for the project's significant adverse environmental impacts.

The estimated total first cost of the project is $108.4 million (1984 prices). The estimated annual benefits are 1.3 times the estimated annual costs. One-half the cost of the recreation features and one-fourth the cost of the flood control and mitigation features would be provided by Harris County. The proposed project was approved by the Board of Engineers for Rivers and Harbors of the U,S. Army Corps of Engineers at the October 1986 Board meeting.

ACKNOWLEDGEMENTS

The authors gratefully acknowledge the advice and editorial assistance rendered by Mr. William G. Wooley, P.E., Chief, Planning Division, during the preparation of the feasibility report on which this paper is based. Additionally, the authors wish to acknowledge the contributions of the following personnel of the U.S. Army Engineer District, Galveston: Mr. Gerald M. Dunaway, P.E., Hydraulic Engineer; Mr. Roy W. Different, P.E., Chief, Channel Design and Flood Control Section; Mr. John P. Babbitt, Regional Economist (Retired); and Mr. Frank G. Incaprera, Chief, Economics and Social Analysis Branch.

REFERENCES

1. Harris County Flood Control District. February 1984. Criteria Manual for the Design of Flood Control and Drainage Facilities in Harris County, Texas. Houston, Texas.
2. Rice University, School of Architecture. September 19, 1977. A Preliminary Investigation of Structural and Nonstructural Flood Control Alternatives for Cypress Creek, Texas. Houston, Texas.
3. Schreiner, L.C. and Riedel, J. T. June 1978. Probable Maximum Precipitation Estimates, United States East of the 105th Meridian. Hydrometeorological Report No. 51, U.S. Department of Commerce, National Oceanic and Atmospheric Administration, National Weather Service, Office of Hydrology. Washington, D. C.
4. U.S. Army Corps of Engineers. March 26, 1952. Standard Project Flood Determinations. Engineer Manual No. 1110-2-1411. Washington, D. C.
5. U.S. Army Engineer District, Galveston. August 8, 1977. Buffalo Bayou and Tributaries, Texas, Hydrology Report for Addicks and Barker Reservoirs. Galveston, Texas.
6. U.S. Army Engineer District, Galveston. January 1985. San Jacinto River and Tributaries, Texas, Flood Damage Prevention, Interim Report on Cypress Creek, and Supplemental Report Dated August 1986. Galveston, Texas.
7. U.S. Water Resources Council. June 1977. Guidelines for Determining Flood Flow Frequency. Bulletin No. 17A. Washington, D. C. (The current edition of this publication is Bulletin 17B, March 1982.)

EFFECT OF EMBANKING ON RIVER AND SEA FLOODS

Professor Adriaan Volker
International Institute for
Hydraulic and Environmental
Engineering
P.O. Box 3015
2601 DA DELFT
NETHERLANDS

ABSTRACT. The paper deals with the effects of embanking on floods in the nontidal reach of a river as well as in the tidal reach. These effects, often called "side-effects", are of various kinds: hydraulic effects, morphological effects and effects on the water management of the land areas.

In the nontidal reach the hydraulic effect generally consists of a rise in the maximum flood levels. This effects is most pronounced in the case of flash floods with sharp peaks, and may amount to as much as a few metres.

Before the introduction of physical and mathematical models the hydraulic effects of embanking were difficult to evaluate and unexpected difficulties occurred after construction. Classic examples are the Yellow River in China, the Red River in northern Vietnam, the River Rhine in the Netherlands and the Irrawaddy River in Burma.

Nowadays reliable predictions can be made but the costs of the models, including the extensive data collection, are high. In order to justify these expenses, especially in projects in developing countries, preliminary estimates are required showing the order of magnitude of the hydraulic effects, for which only a few data are needed.

Other side-effects of embanking are in the fields of river morphology, water management, cropping pattern and sanitary conditions. The most important effect on river morphology is a possible rise in the river bed, causing a rise in the flood levels in addition to the hydraulic effect. Furthermore, embanking may entail increased meandering of the river and increased bank erosion.

In the tidal reach, embanking of tidal marshes which are exposed to flooding during the tidal cycle involves a reduction of the tidal volumes. The tidal currents which govern the morphology of the tidal creeks decrease and silting-up may occur. This reduces the possibilities of tidal gravity drainage as experienced in Bangladesh.

V. P. Singh (ed.), Flood Hydrology, 289–297.
© 1987 by D. Reidel Publishing Company.

Enclosure of tidal embayments and estuaries may also modify the tides outside the enclosing dam. Under certain conditions the reduction of the area of the basin exposed to flooding during storm leads to a rise in the storm surge levels (1.2 metres in the case of the Zuiderzee dam in the Netherlands for durges of short duration).

Whereas at present the hydraulic effects of embanking can be predicted well, provided that adequate data is available on the original hydrologic regime, it is still much more difficult to make a quantitative prediction of the morphologic effects.

1. INTRODUCTION

Embankments are most ancient, the most simple and, in general, the most economic way of protecting land areas against floods. Whereas protection against river floods can also be achieved by upstream flood detention reservoirs, flood diversion and channel rectification, embankments form the only means of protecting land areas in coastal zones against flooding by storm surges at sea.

As experienced in many cases the erection of embankments produces changes of various kinds in the river and land system which were not intended and, often, not foreseen. These changes are mostly indicated as "side-effects", or environmental effects.

In the nontidal river reach hydraulic effects may occur in the form of a rise in flood levels and an increase in discharges. Embankments may affect the river morphology, resulting in a rise in the river bed and an increased tendency towards meandering and bank erosion. These hydraulic and morphological effects must be duly taken into account when designing embankments.

Embankments hamper the natural drainage of excess water from local rainfall. After embanking, a drainage system has to be installed consisting of outfall structures, conveyance systems and field drainage. Embankments also eliminate the beneficial effects of flooding such as supplemental irrigation of wet rice during local dry spells and the flushing of the land areas and channels after a long dry season.

In the tidal reach of a river embankments may change the vertical and horizontal astronomical tides which govern the morphology of the tidal creeks. Embanking or damming off of estuaries and other tidal embayments may produce a substantial rise in the maximum storm surge levels.

In the framwork of the International Hydrological Programma III of Unesco, a Working Group is active in drafting a manual on the reduction or elimination of negative side-effects of water resources development in which the issue of the side-effects of embanking will be duly considered. Any contribution to this activity will be highly appreciated.

2. EFFECTS OF EMBANKING IN THE NONTIDEL REACH

2.1. Hydraulic Effects

When the bank-full discharge of a non-embanked river which enters a flood plain is exceeded the water overtops the natural levels and flooding of the land areas occurs. This flooding means that a substantial volume of water goes into temporary storage and the discharge downstream is smaller than the discharge at the station where the river reaches the flood plain (Fig. 1). Because of the longitudinal slope of the flood plain in he downstream direction an overland flow develops, shunting the channel flow. Since the depth of flooding of the plain is relatively small and the rugosity is high because of vegetation, trees, roads, houses etc. the velocity of the overland flow is small (of the order of 0.1 to 0.2 m/s) but the width is considerable. The result is that the overland discharge can be of the same order of magnitude as the channel discharge.

Embanking changes the flow processes in two ways. The overland flow is confined to the cross-sectional area between the two embankments, resulting in an increase in the discharges compared with the original situation. Also overbank storage is eliminated, causing a further increase in the channel discharge.

The effect of the elimination of the overbank storage is most pronounced in the case of flash floods; in the case of gentle floods with slow rises and flat peaks the effect may be quite small. All the rivers of the large deltas in South-East Asia belong to this type.

Accurate predictions of the hydraulic effects of embanking in the nontidal reach can be made using physical analog models and, since about 10 years ago, mathematical models of a special type. In these models advantage has been taken of the fact the flow direction of the overland flow is determined in most cases by field irregularities and obstructions like roads and low inner dikes, or the presence of channels and small rivers in the flood plains. According to these features the flood plain (and the river channel) can be divided into compartments or "boxes" which are connected to each other by flooded low dikes and roads, sluices and channels. The boxes simulate storage, inflow and outflow, and the connections represent resistance effects. A special type of connection is formed by river reaches.

Both the physical and the mathematical models require extensive data on the original hydrologic conditions with respect to discharges at various stations and water levels, including water levels on the flooded land areas. This field survey often accounts for the greater part of the total costs.

2.2. Preliminary Estimate of the Hydraulic Effect.

In order to justify the high costs of an extensive field survey a preliminary estimate must be made which shows the order of magnitude of the hydraulic effect. For this estimate only a few data, which do not entail high costs, should be sufficient.

Such an estimate can be made by sondiering separately the effect of the flow confinement and the effect of the elimination of the overbank storage. The former effect can be found by applying the simple Manning equation for steady flow to the maximum value of the flood hydrograph both for the situation with flooding and the situation after embanking. This requires a flood hydrograph upstream and an estimate of the overland flow (Fig. 1).

The effect of the elimination of the overbank storage can be derived from the lateral flow Q1 from the channel into the flood plain. This flow cannot be measured but can be estimated from the rate of rise of the water level h on the flood plain, assuming that

$$Q1 = A\ dh/dt$$

where A is the area of the flood plain. If observations of the depth of flooding at a few points are available, Q1 can be computed and added to the downstream discharges.

2.3. Case Studies.

2.3.1. The case of the Rhine river

The deltaic reach of the Rhine river in the Netherlands was embanked on both sides in the period 1100 - 1200. Down through the ages the embankments were heightened periodically, which may be explained by competition between the water boards on either side of the river. However it seems that around 1750 a rise in the flood levels occurred which was ascribed to a rise in the river bed. There was pessimism as to the possibility of safeguarding the land in the long term. Around 1800 serious proposals were made to abandon certain areas which, it was hoped, could serve as flood detention areas and as areas for the deposition of sediments. Naturally, the decision was not an easy one to take, and the matter was left undecided. Shortly after 1850 the bucket dredger was introduced in the Netherlands. This dredger became the "deus ex machina" which solved the problem. Soon a need grose for huge amounts of sand for filling in of prospective sited for cities, industries and roads. Sand was borrowed from the bed of the river Rhine and its branches to such an extent that it was deepened and regulations had to be imposed to prevent undesirable effects.

2.3.2. The case of the Irrawaddy river (Burma)

The system of embankments in the deltaic area of this river provides a unique example of partial flood protection. The major dikes existing at present are shown in Fig. 2. They form a system of horseshoe-shaped embankments around the areas between the river distributaries, leaving the downstream ends of the compartments open. In this way, protection of the higher portion of the interdistributary areas is obtained, and the lands can drain freely in a downstream direction. In the event of extreme floods, the lower portions of the compartments act as a flood detention basin, thus slightly reducing the flood peaks.

Dike building started in 1861 and many dikes were built around 1880 and 1920. Up to about 1920 embanking was considered essential and did not meet much opposition. However in 1926 considerable flood damage occurred, mainly to the Burma railways. In the same year the Mississippi disaster occurred; it was attributed to the double embanking of the river. The authorities then began more and more to oppose ambankments on both sides of the Irrawaddy and further extension of the dikes in other areas.

In a report of 1929 of the Public Works Department it was even recommended that the embankments on the east bank be demolished at a rate of 16 km per year. Similar proposals were issued after the big flood of August 1939. The recommendations were never put into effect because the farmers did not want to be drowned out for the sake of future generations.

It is only recently that reliable information has become available as to the effect of complete flood protection by dikes on the flood levels. Since the floods of the Irrawaddy river are of the gentle type this effect is relatively small but still amounts to about 1 metre for major floods in the middle reaches of the river. This information was obtained through an extensive field survey an a mathematical model.

2.4. Other Environmental Effects.

Embanking may affect the morphology of a river. After embanking the river channel has to carry the silt which was formerly deposited on the flood plain. On the other hand the sediment-carrying capacity has increased because of higher velocities and it depends on the prevailing factor whether silting up or deepening will occur. Classic examples are the Yellow River in China and the Red River in Vietnam. Furthermore embanking may entail an increase in the meandering of the river and bank erosion. Also embankments halt the deposition of "fertile" silt on the flood plains and their further building-up.

Embankments hamper the drainage of excess water from local rainfall and an artificial drainage system with field drainage, conveyance system and an outlet structure are required to control the water levels.

In humid tropical areas, where wetland rice is grown, moderate floods are often beneficial in that they constitute supplementary irrigation during dry spells in the local rainfall. Embanking eliminates this beneficial effect and it may be necessary to install an irrigation system.

Embanking involves a complete change in water management and the agricultural pattern. If the floods are of the gentle type rice can be grown in the absence of flood protection and the local varieties are adapted to the prevailing local hydrological conditions. Also man has learnt how "to live with the floods". Embanking would completely upset this traditional pattern (delta of the Mekong river in Vietnam and delta of the Chao Phya river in Thailand).

3. EFFECTS OF EMBANKING IN THE TIDAL REACH

3.1. Effects on the Astronomical Tides.

The propagation of the astronomical tides in the creeks of low-lying coastal regions is affected by the extent of the areas which are exposed to flooding during the tidal cycle. Embanking to protect these areas from flooding modifies the horizontal and vertical tides in the creeks. The vertical tides are of significance for the gravity drainage of the land areas and the horizontal tides determine the morphology of the creeks. Where the tidal currents decrease, silting up may occur, hampering tidal drainage, as happened in a number of outfalls of drainage sluices in Bangladesh.

Enclosure of tidal embayments and estuaries in order to create freshwater reservoirs and to reduce the length of the lines of defence against storm surges may also modify the tides outside the enclosing dams. This is important for navigation and the maintenance of harbours in the remaining tidal area (Zuiderzee dam in the Netherlands, dam in the lagoon of Cochin, Kerala State, India).

3.2. Effects on Storm Surges.

Storm surges can be defined as abnormally high sea levels generated by landward winds. They may cause sea levels which are several metres higher than the high tide levels of the astronomical tides. Non-embanked areas are flooded to a considerable depth. Embanking of such areas eliminates the overbank storage and may cause higher maximum levels at the site, especially in the case of storm surges of short duration. There is an analogy with the hydraulic effect of embanking on the river flood levels in the nontidal reach. Thus the vast mangrove area in Bangladesh between the city of Khulna and the Gulf of Bengal provides efficient protection of the city against high cyclone levels at sea.

A rise in storm surge levels may also occur in the case of enclosure of estuaries and other tidal embayments when the dam is aligned upstream at some distance from the coast. The remaining water area between the dam

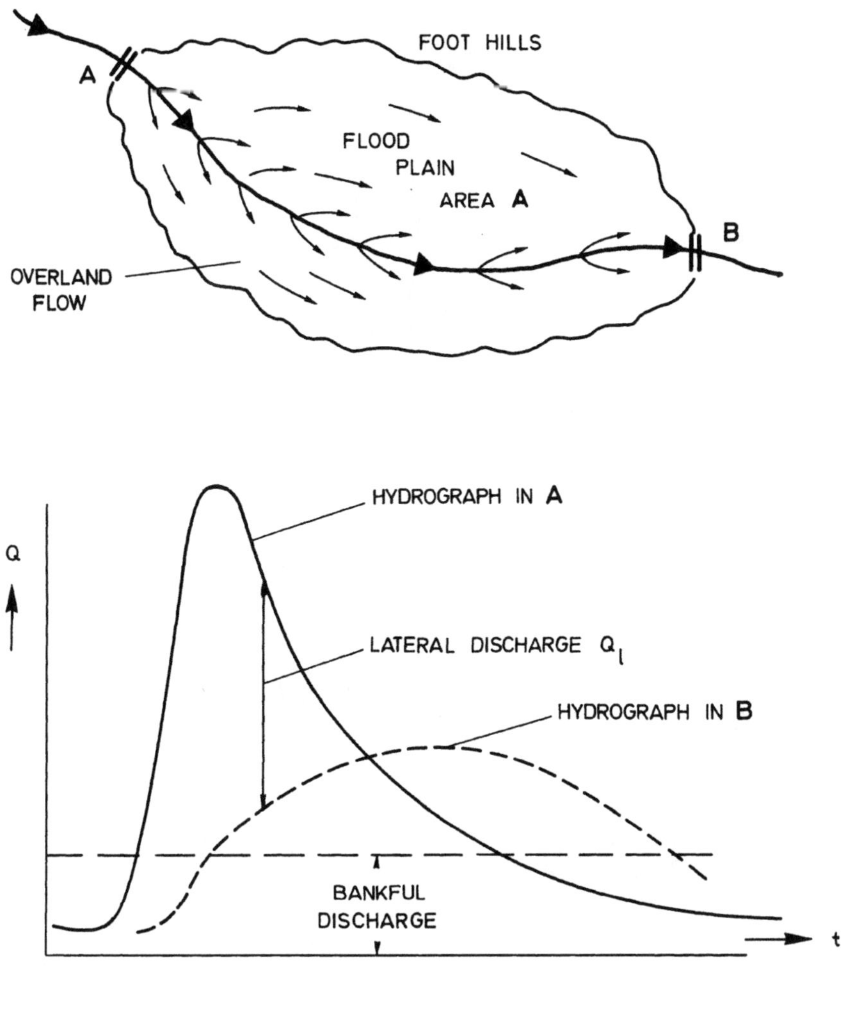

Fig. 1 EFFECT OF FLOODING ON FLOOD HYDROGRAPHS

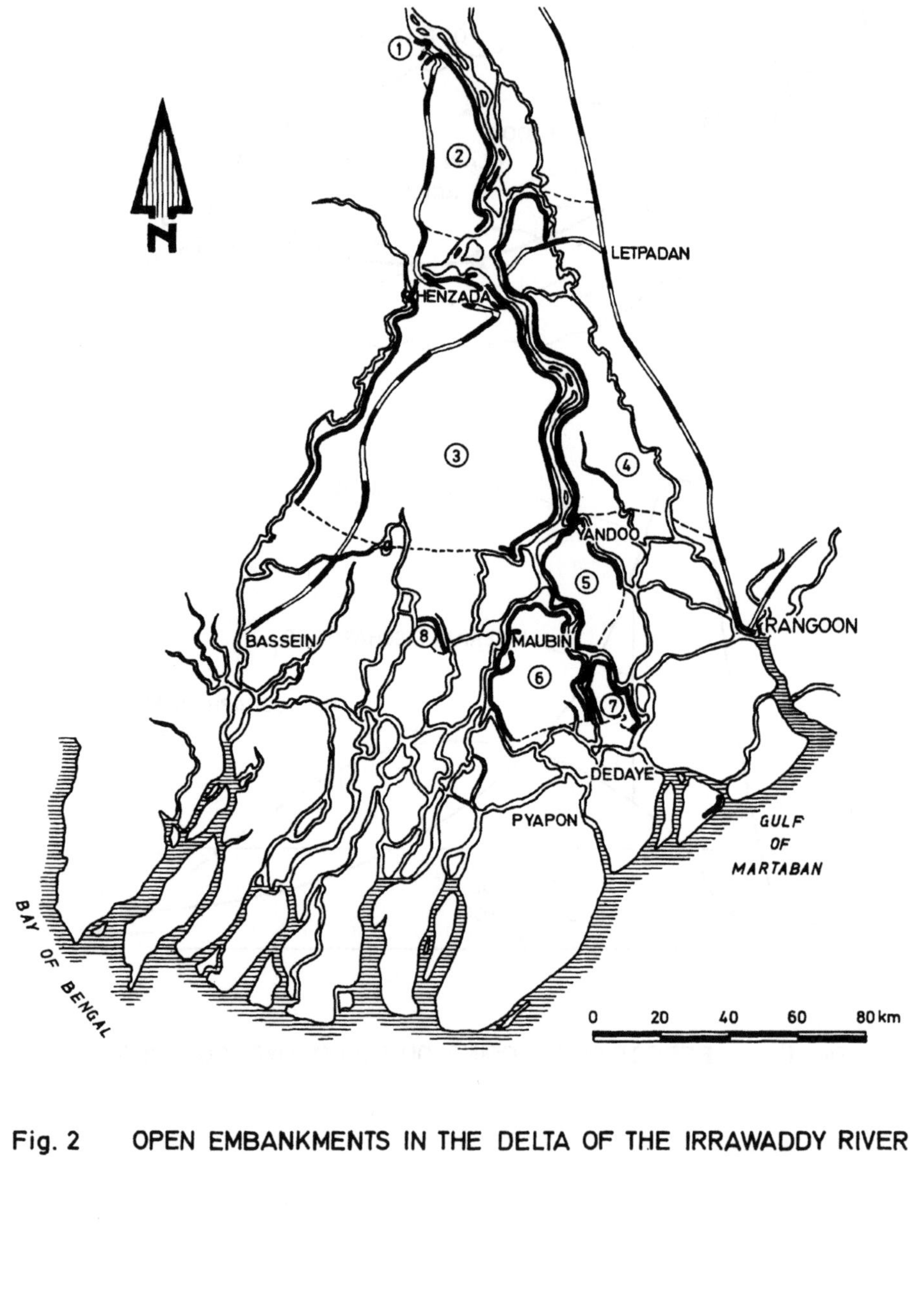

Fig. 2 OPEN EMBANKMENTS IN THE DELTA OF THE IRRAWADDY RIVER

and the coastline is filled more rapidly than before, leading to higher levels than occurred previously, especially in the case of storm surges of short duration (1.2 metres in the case of the Zuiderzee dam in the Netherlands).

Accurate prediction of the effects of embanking on the astronomical tides and storm surges is possible using physical or mathematical models provided that adequate information is available on the horizontal and vertical astronomical tides and on the storm surge levels under the original conditions.

4. REFERENCES

UN-ESCAP, 1963, Proceedings of the regional symposium on flood control, reclamation and utilization and development of deltaic areas. Water Resources Series, **25**, Bangkok.

UNESCO, 1964, Scientific problems of the humid tropical zone and their implications. Proceedings of the Dacca Symposium.

Volker, A., 1983, Rivers of south-east Asia: their regime, utilization and regulation. Hydrology of humid tropical regions with particular reference to the hydrological effects of agriculture and forestry practice. In: Proceddings of Hamburg Symposium, August 1983. IAHS Publication, **40**

A MODEL FOR THE OPERATION OF SPILLWAY GATES IN PLUVIAL FLOODS

Luis Arrau
Civil Engineer, Head of Projects, Irrigation Directorate
Ministry of Public Works
Morande 59, Fifth Floor
Santiago, Chile.

ABSTRACT. A methodology based on the Decision-Making Theory under conditions of uncertainty is developed for the operation of a gate spillway during floods of pluvial origin. An independent short-term forecast model for such type of floods is supposed to be available. Two types of uncertainty conditions are described: hydrological conditions based on the forecast and economic conditions due to the lack of knowledge concerning the actual damages that a given flow may cause downstream of the storage reservoir. These imply making two decisions: one regarding the maximum flow to be made when the flood wave reaches the reservoir. The ultimate goal is to minimize the economic damages caused by the discharged flow plus the damages caused by a deficient operation resulting in a smaller volume of stored water when the flood is over, compared to a normal operation condition.

1. INTRODUCTION

One of the most interesting aspects of actual time operation of spillway gates during a pluvial flood is the maximum use of the water volume available in the reservoir in order to minimize damages likely to be caused downstream by the spillway discharge. Such operation should be performed without reducing the total water resource available in the reservoir once the flood is over.

A methodology that enables the operation of the gates during the flood event mentioned above is developed in this paper, considering the existence of a series of uncertainty factors at the time of decision making. It is assumed that the basin dealt with presents a sufficient time lag between the rainfall in the upper zone and the flood wave flowing into the reservoir so that part of the decisions may be made on the basis of some type of forecast.

2. GENERAL DESCRIPTION OF THE PROBLEM

The peak flow discharged is usually used as an index to estimate the

V. P. Singh (ed.), Flood Hydrology, 299–308.
© 1987 by D. Reidel Publishing Company.

consequences of a flood wave. Said flow is the function of the stage reached by the lake, and this in turn is not only the function of the reservoirs' initial volume but also of the volume stored during the rise time of the outflowing hydrograph.

Briefly, there are three basic conditions that determine the situation under normal operation: the initial volume of the reservoir, the shape of the affluent flood, and the gate operation rules.

The situation is illustrated in Figure 1. In this figure, diagram 1 refers to the case in which there is no reserve volume available in order to reduce the flood; the maximum discharged flow is Qp. Diagram 2 refers to the case in which even with a reserve volume, given by V1+V2, it is insufficient to reduce the maximum discharged flow. Diagram 3 refers to a reserve volume V > V1+V2, which will reduce the peak flow that has been discharged to value Q'p. Now then, in the cases described above, curves Qe(1), Qe(2) and Qe(3) pertain to a given gate operation law, which is generally independent from the affluent flood, that is, it is assumed to be unknown (it is only a function of the lake stage). In other words, the three curves are different but they follow the same law. The alternative now is to operate the gate mechanically without regard to this law, which is the logical step if an estimate of the forthcoming maximum affluent flood is available, given a certain rainfall. An example of this situation is illustrated in Figure 2. In this figure, gates are operated in such a way that the discharged flow will not exceed Q"p value for which an additional reserve volume of the reservoir (Va) is required, compared to a normal operation meaning a discharged Qp peak flow. Theoretically, this means that the damage caused by the spillway operation is reduced. This is only possible if the affluent flood does not correspond to the maximum designed, and allows for the possibility of storing the Va volume. If the forecast is incorrect and the flood is larger, which is most likely because all forescasts imply a certain degree of uncertainty, the discharged peak flow value may not only be higher than Q"p but may even exceed Qp. On the other hand, whatever the discharged peak flow may be, there will always be an estimate on the damages likely to occur downstream, both in the reservoir structures and in the river course, but this is also uncertain. In other words, damage likely to be caused downstream also present a degree of uncertainty that must be considered.

Then, a decision making problem under uncertainty conditions is raised, basically of two types: hydrological and economical. Additionally there appears the difficulty of separating the hydrological uncertainties into two variables: the flood volume return period, and the shape of the hydrograph. The reason for this is that, generally, it is possible to forecast a volume with an acceptable probability, but the same does not occur with respect to the shape taken by the hydrograph. The latter is much more aleatory, and requires a different treatment.

3. STRUCTURING THE DECISION MAKING

The model developed is illustrated in the decision tree shown in Fig. 3.

There is a first information node, "initial conditions", which

FIGURE 1

EXPECTED OUTFLOW HYDROGRAPHS ACCORDING TO THE RESERVE VOLUME

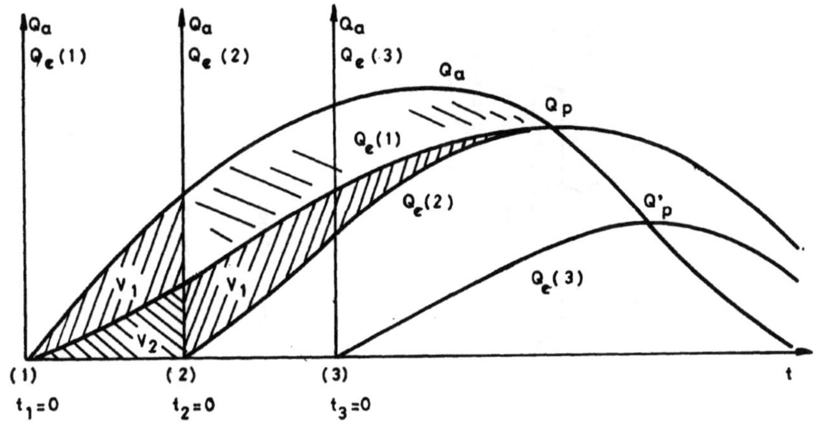

t_1, t_2, t_3 : TIMES OF GATE OPENING IN CASE (1), (2) AND (3), RESPECTIVELY.

FIGURE 2

LIMITATION OF THE OUTFLOWING WATER CASE

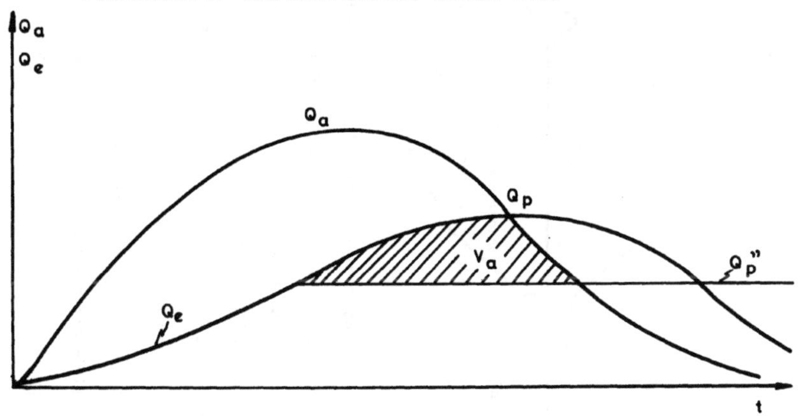

FIGURE 3

DECISION TREE

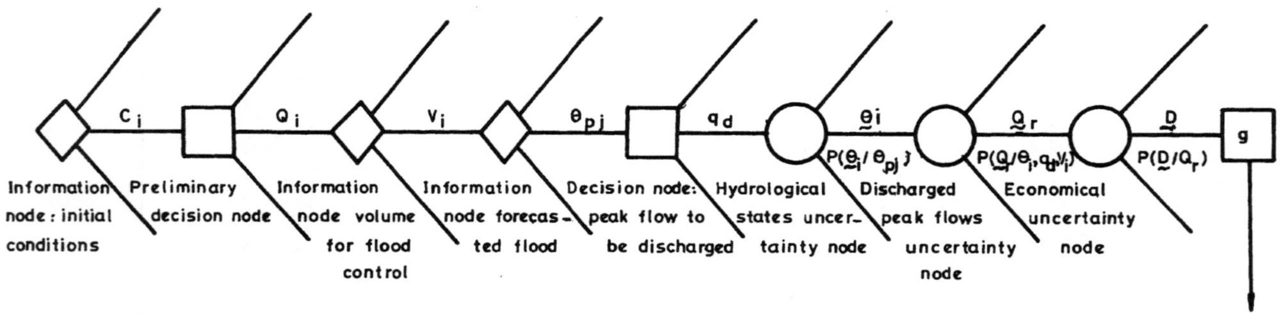

C_i = Status of basin at the beginning of the rainfall.

V_i = Reserve volume of reservoir at the beginning of affluent flood, $i = 1, -, n$

Q_i = Initial decision for operation

θ_{pj} = Forecasted hydrological state (return period of the forecasted total volume of flood)

q_d = Decision alternative peak flow to be discharged

θ_i = Natural and actual hydrological condition (return period of the actual total volume of flood)

Q_r = Peak flow actually discharged.

D = Possible damage variable for a given discharged peak flow

g = Global consequence function as a result of the complete operation process $= g(Q_i, q_d, \underset{\sim}{\theta_i}, \underset{\sim}{Q_r}, \underset{\sim}{D})$

refers to the condition of the basin at the beginning of the rainfall. Such a condition (Ci) refers to an initial condition index given by various factors according to the specific problem: obviously, the storage volume at that moment; state of the reservoir structures that may exist upstream; state of the basin soil since it might have rained only a few days earlier and, therefore, an almost total runoff of the rainfall could be expected; restrictive situations for immediate discharge given by the conditions downstream of the reservoir, etc.

On the basis of the initial conditions, it is necessary to make a first decision whether to open the gates mechanically. The alternative decision in this case would be either to discharge a given Qi flow through the spillway in order to wait for the affluent flood wave with a lower stored volume, or to wait until a forecast is available (in this case Qi =0).

As a result of the decision made, and of the initial conditions of the reservoir, there will be a reserve volume that will be used to reduce the flood wave at its onset. In the "decision-making tree" such a situation is represented by an information node, "reducing volume", which is subdivided into a series of values that a Vi variable may take.

Then we have an information node: the affluent flood forecast that, as mentioned above, is normally suggested to be made in terms of the return period of the flood volume, with a certain estimate of the possible peak flow, and of the flood shape. The latter data is important for decision making since without it the outflowing hydrograph cannot be estimated with certainty. However, in order to overcome the unreliability of the curve, the analysis is mainly performed in volumetric terms based on considerations of this type mentioned in the preceding paragraph. The forecast will be subdivided according to θpj different hydrological states, depending on the total volume forecasted.

On the basis of the θpj forecast and the Vi volume, it will be necessary to make a decision on gate operation: whether the gates are left to work their normal law, or the discharge flow is limited to a certain qd value. Such a situation is illustrated in the tree as the decision alternatives for the peak flow to be discharged.

At this point the final decision has been made, and we only have to wait for the results delivered by nature. Two hydrological uncertainty nodes and one of economical uncertainty are defined for this case. The first one represents the actual hydrological state that took place in terms of the total volume of the affluent flood. Then we have the node representing the actual state from the point of view of the flood shape or the affluent hydrograph, which consequently results in the Qr discharged peak flow being larger than the one which was decided to be discharged, or else, in very special cases, being smaller.

Finally, the economical uncertainty node refers to the lack of accurate knowledge on the D damage that a given flow may cause. Here, it is even possible to illustrate the duration of the maximum discharged flow, on which the consequence function or the damage also depend.

As a final result of the process, a g consequence function is shown, which should include not only the damage caused downstream by a given peak flow but also other consequences of the decisions made, such as the opening of the gates when rainfall starts to occur.

Furthermore, through this consequence function it is necessary to evaluate the case in which at the end of the flood the stored volume may be smaller than the stored volume available under "normal" operation of the gates, that is, according to their general rules not supposing a previous estimate of the affluent flood.

For each of the three uncertainty nodes defined, the corresponding conditioned probabilities of occurrence will have to be estimated, as described below:

-P (θi/θpj) is the probability of occurrence of an θi hydrological state since the θpj hydrological state was forecasted.

-P (Qr,θi,qd,Vi) is the probability of the discharged peak flow being Qr since the θi hydrological state occurred, it had been decided to operate by limiting the peak flow according to qd alternative, and a Vi reserve volume was available.

-P (D/Qr) is the probability of a D damage being caused since the discharged peak flow takes Qr value.

4. DECISION MODEL AND ESTIMATING METHODS COMPONENTS

According to the structure of the model illustrated in the decision tree in Figure 3., the following basic components may be distinguished:

- state variables: Ci, Vi, θpj
- uncertainty variables: θi, Qr, D
- decision variables, or alternatives: Qi, qd
- consequence function: g.

Following is a detailed explanation of each model component and the appropriate estimating procedures.

4.1.- State Variables

4.1.1. Initial Conditions. This variable has been included in the decision tree, and it has been subdivided into a specific number of possibilities (Ci) which actually have to include a series of conditions of various types and are not easy to summarize in a single index. It is unpractical to establish a general criterion to evaluate the initial conditions index. Furthermore, its definition is also subjective if one bears in mind that the general criteria for the operations must be maintained, that is, the definition of the index must be useful for decision making that has to take into account the interests or the ultimate objectives of the work. However, a list of all the parameters that exert an influence upon them should be made, for each problem, as well as their relative importance, so that they may be evaluated and the index defined according to previously established guidelines.

4.1.2. Reserve Volume. For the model, the reserve volume is defined as a subdivided number of value ranges, Vi, not necessarily of the same magnitude. Thus we could have:

V1 = reserve volume between 0 and 5 x 10^6 cubic meters.
V2 = reserve volume between 5 and 20 x 10^6 cubic meters.
V3 = reserve volume between 20 and 50 x 10^6 cubic meters., etc..

The following procedure is presented in order to establish the exact limits of each Vi state:

1st. Determine the affluent flood volumes for various return periods and their typical distribution in time (hydrograph).

2nd. Simulate, with the gates under normal operation rules, the flood flowing through the spillway for the largest number of possible reserve volumes, and for the various return periods.

3rd. Define the Vi states, in accordance with the discharged peak flows so that any value the reserve volume may take in a specific Vi range does not alter the discharged peak flow significantly.

4.1.3. Forecast States. They are given by an independent flood forecast model with respect to a given rainfall. Such a forecast provides the flood return period in terms of its total direct runoff volume as information for the decision model. The said volume determines a forecasted θpj hydrological state.

4.2. Variables of Uncertainty (or Random)

4.2.1. Hydrological States. The classification of hydrological states is performed in accordance with the relative importance of the various flood waves, according to the damages likely to occur. Thus, for example, the θ1 hydrological state may pertain to those cases in which the probability of exceedance of the flood's direct runoff volume exceeds 90%; the θ2 state, to those cases in which the probability of exceedance ranges between 70% and 90%, etc.

Having forecasted a θpj state, the P(θi/θpj) probability of the actual occurrence of a θi hydrological state is defined. This conditioned asignment of probabilities can be calculated in terms of frequency, comparing the number of times these θi states have occurred in flood statistics when the forecast model has forecasted θpj. Given the objectives of the decisions model, the lack of major flood data is not too relevant because in these cases it can be assumed that the forecast model keeps as a maximum the error trend of historical data.

4.2.2 Discharged Peak Flow. As mentioned above, each θi hydrological state is defined by a range of direct runoff volumes. When processing the various floods in this range, given also a certain Vi reserve volume and a qd alternative decision, a series of values of the discharged peak flow is produced: Qr min i to Qr max i with a characteristic Qri value. On the basis of this information we can define a range of peak flows that may be discharged for each θi, Vi hydrological state, and qd alternative, which can be subdivided into a series of Qr volumes. However, considering the actual information generally available, assignment of P(Qr/θi,Vi,qd) probability is difficult to perform. Since from the point of view of the solution of the decision model it is sufficient to evaluate only the expected value of this distribution, the problem is focused directly and in a simplified way, supposing that P(Qr/θi,Vi,qd) corresponds to Beta distribution limited between Qr min i and Qr max i. The corresponding estimate is shown below when dealing with the resolution method of the decision tree (paragraph 5).

4.2.3. Damages. The economic uncertainties are specified in terms of the randomness of the D damage variable. Its evolution is quite difficult since all the possible damages likely to occur both in the reservoir's structures and downstream from it for various Qr peak flows have to be identified. This can be closely achieved by calculating the river's hydraulic axis and the flooded area for each flow. However, only likelyhood can be estimated. Considering the foregoing, it is suggested to assume, together with the discharged peak flow, that P(D/Qr) corresponds to a Beta distribution limited between an estimated minimum value of the damages, D min r, and a maximum one being D max r; such a distribution is characterized by a better estimate, Dr, of the damages. The latter also allows for a consideration of the different durations the discharged flow may have in the corresponding estimates (point 5).

4.3. Decision Variables or Alternatives

4.3.1. Preliminary Decision. The primary decision to be made by the reservoir operator when rainfall starts is to define the reserve volume in the reservoir with which the flood wave will be expected, which means that the gates could be opened and a Qi flow could be discharged before the flood wave reaches the reservoir. Such a decision will basically depend upon: (a) the initial conditions index, specially the stored volume; (b) the possibility and/or need of an immediate and gradual opening of the gates; and (c) the reserve volume reached through this operation. In several cases perhaps the modification of the reserve volume reached will not suffice to achieve a practical reduction of the discharged peak flow (see Figure 1.), and therefore it would be preferable not to discharge water until the flood wave reaches the reservoir, and take advantage of the time lag to alert the people in order to prevent greater damages. In other cases it might be preferable to proceed to a discharge before the flood wave arrives, which is possible to foresee in the decision tree through the mathematically expected consequences for the various reserve volumes.

4.3.2. Post-Forecast Decision. Once the forecast and the available reserve volume are known, the decision is either to let the gates operate normally, or to limit the peak flow to a lower value than the one predicted. It is suggested that this qd variable be subdivided, based on an analysis of the required additional volumes in order to achieve such a limitation of the flow (see Figure 2.), and considering the subdivision performed for the hydrological states. The maximum value of qd in each node, as for normal operation, is defined by the simulation of the typical flood that defines each hydrological state, given the Vi reserve volume.

4.4. Consequence Function

The g consequence function is structured as the addition of the series of damages likely to occur as a result of the operations performed. Among them we can quote:

- d1 = damage to spillway structures and/or reservoir wall.

- d2 = damage caused downstream due to Qi discharge.
- d3 = damage caused downstream due to Qr flow.
- d4 = damage due to a smaller stored volume than the one under a normal operation, at the end of the flood.

Thus, in a general case we have:

$$g = \sum_i d_i$$

Then, the approach or analysis of the problem can be stated with various objectives or approaches, being possible to establish α_i ponderability factors, or relative importance of the total damages,resulting in:

$$g = \sum_i \alpha_i d_i$$

From the irrigators' point of view, for instance, $\alpha 3$ would be nil, unless Qr is lower than or at most equal to the normal qd of the operations. The general definition of the α i ponderability factors is then conditioned to the general structure operation patterns.

Nevertheless, for every specific case the g function will depend on the decisions to be made and on the values that uncertainty variables may reach, that is :

$$g = g\ (Vi,\ qd,\ \theta i\ ,\ Qr,\ D)$$

5. DECISION MODEL RESOLUTION

According to the usual methodology and hypothesis employed in Decision Making Theory, the optimum gate operation alternative will be the one that minimizes the expected value of the cosequence function. Such a process consists in successively substituting each node for the expected value of the function with respect to the random variable that corresponds to such node, until reaching the alternative node, where we proceed to the minimization process.

Mathematically, the regressive induction process mentioned above is expressed through the following sequence of relations (see Figure 3.)

(a) Substituting the economic uncertainty node for its expected g1 value, where:

$$g1 = E_D\ \ g(Vi/Qi,\ qd,\ \theta i/\theta pj,\ Qr/\theta i,\ D/Qr)$$

$$g1 = \int_D g\ (Vi/Qi,qd,\theta i/\theta pj,Qr/\theta i,D/Qr)\ P\ (D/Qr)\ dD$$

Supposing that P(D/Qr) corresponds to a Beta distribution limited between D min and D max, the best estimate of which being $\hat{D}$, then:

$$g1 = 1/6\ (g\ min + 4g + g\ max)$$

$$g\begin{cases}min\\max\end{cases} = g\ (Vi/Qi,\ qd,\ \theta i/\theta pj,\ Qr/\theta i,\ D\begin{cases}min/Qr)\\max\end{cases}$$

$$g = g\ (Vi/Qi,\ qd,\ \theta i/\theta pj,\ Qr/\theta i,\ \hat{D})$$

After these computations, g1 value is assigned to each branch of the uncertainty node of the discharged peak flows.

(b) Substituting the uncertainty node of the discharged peak flows by g2, expected value of the g1 consequence assigned to its branches.

$$g2 = g2\ (Vi/Qi,qd,\theta i/\theta pj) = \underset{Qr/\theta i}{E}\ g1\ (Vi/Qi,qd,\theta i/\theta pj,Qr/\theta i)$$

The process is similar to the calculate of g1.

(c) Substitution of the hydrological uncertainty node by g3, expected value of the g2 consequences assigned to its branches.

$$g3 = g3\ (Vi/Qi,qd) = \underset{Qi/\theta pj}{E}\ g2\ (Vi/Qi,qd,\theta i/\theta pj)$$

$$g3\ (Vi/Qi,qd) = \sum_{\theta i=\theta 1}^{\theta i=\theta n} (g2\ (Vi/Qi,qd,\theta i/\theta pj)\ P(\theta i/\theta pj)$$

(d) Optimization Process. After the foregoing reduction process a reduced decision tree can be designed consisting only of the data nodes and the decision nodes. At the end of the branch of each node of the discharged peak flow, the corresponding g3 value is assigned as a consequence. Therefore, the optimum q*d alternative will be the one able to minimize g3 . That is to say:

$$\underset{qd/Vi,Qi}{\min}\ (g3\ (Vi/Qi,\ qd)) \longrightarrow q^*d$$

Therefore, the decisions model designed enables the selection of an optimum alternative for spillway operation, q*d, given a preliminary operation and an initial volume for each situation of flood forecast.

6. REFERENCES

6.1. Analysis of Decisions under Uncertainty, R. Schlaifer, Mc Graw Hill Book Co., 1969.

6.2. On the Consideration of Variability in Cost Estimating', E. Case Kenneth, IEEE, Transactions of Eng. Management, November 1972.

6.3. Decisions Model for the Regional Assignment of Irrigation Water in a Basin, B. Espildora and L. Arrau, Department of Hydraulic Resources, University of Chile, Santiago, 1977.

6.4. Analysis of the Duration of Winter Floods at Paloma Reservoir Spillway, L. Arrau, Ministry of Public Works, Irrigation Directorate, Santiago, 1983.

FLOOD CONTROL WITH THE USE OF AN IRRIGATION STORAGE RESERVOIR

Enrique Kaliski
Luis Arrau
Civil Engineers, Irrigation Directorate
Ministry of Public Works,
Santiago, Chile.

ABSTRACT. This paper describes a methodology developed with the aim of analyzing the incidence on irrigation of an irrigation reservoir additionally serving flood control purposes. The analysis is basically carried out in economic terms, evaluating the expected net benefit in the longterm operation of a project used for both purposes.

The expected benefits from irrigation, owing to the use of a portion of the reservoir's storage volume for flood control, are obtained through a monthly simulation model, incorporating the useful capacity of the reservoir to store irrigation water as variables.

The expected benefits from flood control are obtained through a mathematical treatment of a series of land data related to the damage resulting from the inundations caused by floods in various return periods. These inundations are determined according to the characteristics of the reservoir, the river bed and flood wave.

Subsequently, an application of the methodology to the " Paloma System " and the results achieved in the 1984 storm are included.

1. INTRODUCTION

Multi-purpose dams have become increasingly widespread throughout the world in recent years, one of their uses being flood control. This use allows the storage of large affluents in the reservoir in order to reduce the damages that would result mainly downstream.

There are various reasons for the upsurge in flood control, but the major disasters that have occurred in some countries in the past is the most important of all.

A revision of the operational characteristics of some reservoirs in Chile has proved the need to carry out studies in order to quantify the convenience of having storage reservoirs serve flood control.

The purpose of this study is to develop an evaluation methodology for the additional use of an existing irrigation reservoir to serve flood control, and justify its convenience. This methodology includes the principal features of Chilean irrigation systems and, at the same time, is comprehensive enough as to make it independent from the

V. P. Singh (ed.), Flood Hydrology, 309–323.
© 1987 by D. Reidel Publishing Company.

specific conditions of the reservoir under study and other conjunctural situations.

Therefore, the methodology is not only valid for the Chilean case but can also be applied to other cases. However, it shall have to be adapted and changes will have to be introduced into certain basic assumptions according to the characteristics of each case and the data available.

2. DESCRIPTION OF THE PROBLEM

Flood control in a reservoir generally consists in leaving a portion of the reservoir's useful capacity during a certain period of the year as a reserve space to store the entire flood volume or part of it, and thus reduce the effects of floods both in the reservoir's structure and downstream of it.

If a reservoir has been originally designed for irrigation purposes only, the additional service of flood control may imply a reduction in the volume of the target or guaranteed water supply throughout the year, and, therefore, may result in less irrigation benefits. On the other hand, the damages caused downstream of the reservoir by flood discharge may be reduced to a large extent if at the beginning of a flood event the reservoir has an adequate reserve volume to control it.

Furthermore, flood control provides an additional benefit related to higher dam safety, considering that the spillway may release the floods of a higher return period than originally designed for.

Since flood control decisions have certain side effects, such as smaller employment opportunities if agriculture decreases and others, the conclusion is that the major complexity of the problem lies in the need to make compatible interests of various types and origins. In fact, general criteria of an economic nature at global basin level have to be harmonized with specific decisional aspects, frequently subjetive and not easy to evaluate.

The methodology described aims at using to the largest extent possible the data available for each case, providing a series of results that can be used as decision making elements by the appropriate managing bodies.

As a means of becoming independent from subjetive and/or incidental aspects, the procedure is developed in such a way that decisions on flood control may only be made on the basis of economic effects likely to occur in the long-term.

3. DESCRIPTION OF THE METHODOLOGY

3.1. General Development

The evaluation of the effects of irrigation and flood control in economic terms, according to the proposed methodology, may be illustrated by means of the scheme shown in Figure 1, in which effects are quantified through the mathematical values expected from annual agricultural benefits and flood damage. It is estimated that these magnitudes represent an average value of the reservoir's response and are independent from the randomness of annual hydrological resources and eventual conjunctures.

Therefore, a magnitude representing the combined effect of irrigation and flood control, as an average in the long-term, is defined as Net Expected Benefit (NEB), and is determined as follows:

$$NEB = EAB - ED \qquad (1)$$

In equation (1) EAB is the Expected Agricultural Benefit, and ED is the Expected Damage.

Graphically the condition is represented by the curves in Figure 2.

The correct estimation of each of the curves in Figure 2 requires the definition of certain mathematical models quoted further on, which implicitly carry the consideration of all the aspects at stake, both subjetive and objective. As a means of ordering them for their correct use, the overall system has been divided into five sub-systems described below.

3.2. Sub-systems

The overall system at basin level includes all those factors that are either cause or effect, both direct and indirect, of the additional use of an irrigation reservoir for flood control. Figure 3 shows a diagram of these factors in the Chilean case, as well as the mathematical models used.

According to the characteristics of these factors, the System has been divided into five sub-systems: administrative, agro-economic, physical, operational, and hydrological sub-systems.

3.2.1. Administrative Sub-system. In Chile it would be constituted by the institution responsible for the management of large projects and by irrigators' associations who distribute the water at farm level. These bodies are usually responsible for the decisions concerning the operation of the structures in each season, and all the subjective aspects related to operation should be considered through them.

3.2.2. Agro-economic Sub-system. Is directly related to agricultural revenues. All types of inputs that allow the evaluation of income in relation to the availability of water for irrigation purposes are included here: crop distribution and rotation, prices, yields, agricultural costs, irrigation rates and, generally, all agricultural and market factors.

3.2.3. Physical Sub-system. Basically comprises land infrastructure, related both to irrigation and other type of infrastructure that might be affected by inundations caused by floods. Dams, rivers, channel systems with their conveyance capacity and efficiency, types of soil, irrigated areas, flood areas, water distribution techniques at farm level, etc.

3.2.4. Operational Sub-system. Mainly comprises the project's design features as from the operational point of view, particularly the basic criteria related to the reservoir's delivery and operation rules for the spillway gates discharge curve.

3.2.5. Hydrological Sub-system. Concerns the system's water resources. Studies on floods likely to occur and monthly affluents to the reservoir, records of historical floods, dam evaporation, forecasting models that might exist in the future, either of monthly discharge or thawing, etc. are included in this sub-system.

3.3. Developing and Obtaining Expected Benefit

3.3.1. Agricultural benefits and reservoir water delivery relationship (curve 2 (a) in Figure 2). In order to establish a graphical or analytical relationship between Agricultural Benefits and the Annual Water Delivery of a reservoir (Figure 2 (a)) a resource distribution model for each irrigated area is proposed, enabling the evaluation of the agricultural production for each sector.

This model determines the way in which the resources are distributed for various annual volume supplies, according to preestablished criteria. We intend to analyze the two following cases with the model:

(a) Conditions under which water is supplied to the respective irrigated areas according to the standards of the appropriate administrative bodies. The crop area in each sector is obtained through an agricultural survey.

(b) Optimal or future situation, in which water and crop distribution are determined in order to maximize a benefits function. Linear, dynamic and other programming techniques can be used in this case.

After determining how to represent resource and crop distribution, the annual agricultural benefit is determined as:

$$B_K = \sum_i \sum_j S_{iJK} \left(P_{iJ} R_{iJ} - C_{iJ} \right) \qquad (2)$$

where:

B_K is the annual agricultural benefit for a "K" supply from the reservoir in monetary units.

S_{iJK} is the crop area "i" in the J zone, in hectares, for a "K" supply from the reservoir.

P_{iJ} is the price of crop "i" in the J zone, in monetary units per ton.

R_{iJ} is the agricultural crop yield "i" in the "J" zone, in tons per hectare.

C are the agricultural costs of crop "i" in the "J" zone, in monetary units per hectare.

All the abovementioned magnitudes may vary depending on the volume of the annual water supply and throughout the years. Therefore, market conditions for the area must be known a priori, in order to obtain an expression as representative as possible of these conditions for the long-term.

Thus, evaluating Equation (2) for the reservoir's various supply levels a curve for the relation sought is obtained, as from which it is also possible to obtain an analytical expression through the correlation.

In case of estimating distribution through optimization models several restrictive aspects must be taken into account, such as channel capacity and efficiencies, balances in agricultural productions, maximum arable areas, and others, depending on the zone under study.

3.3.2. Supply target and Reserve Volume relationship (Curve 2 (b) in Figure 2). In Chile, annual supply targets for irrigation are usually defined as those targets that can be supplied with an 85% guarantee. Hence, the possibility of a reserve volume for flood control, even if it is only for a few months of the year, reduces the effective capacity of irrigation reservoir, with the subsequent decrease in the supply target (Figure 2 (b)).

The foregoing is quantified through an operational simulation method that takes into consideration two types of inputs:

- the economic, physical, logical condition of operation and hidrological data features, that do not vary from one simulation to another; and
- the structure's operation rule, mainly considering the annual delivery level and the reserve volume for floods in the reservoir.

Once the parameters of the specific problem are entered, the monthly simulation is performed for all the selected statistical years. Finally, the curve in Figure 2 (b) is obtained through the successive adjustment of parameters.

Essentially, due to the large amount of data and computations required, the evaluation process of the monthly operations are generally performed in a computer. The answers obtained for a specific operation rule are, inter alia, annual supplies, annual benefits, evaporations, spillovers, and all other factors interesting to quantify.

3.3.3. Expected Agricultural Benefit and Reserve Volume Relationship Curve 2(c) in Figure (2). The curve in Figure 2 (c) is obtained by combining the results of curves 2 (a) and (b). This process is directly carried out with the operational simulation model described in 3.3.2, which incorporates the computation formulae and required data.

3.3.4. Physical evaluation of the damages for various maximum flow discharges. The determination of the effects of flood on reservoir structures and on the areas downstream of the reservoir requires the physical evaluation of the assumed damages for various maximum flow discharges.

The procedure to obtain a curve as that shown in figure 4 (a) may be summarized in the following stages:

(a) Analysis of hydraulic axis in areas downstream of the reservoir, determining the flooded sectors for each flow discharge.

(b) A cadaster of the eventual flooded areas and evaluation of losses.

(c) The damages in the discharge structures and in the reservoir in general are added to the damages occurring in the areas likely to be flooded.

The values obtained enable the establishment of a graphical relation of flood damage in function of the maximum flow discharge, as well as analytical relationships through correlations.

3.3.5. Flood Study. Flood hydrographs representing the potential events that may occur in the long-term have to be considered in this stage, such as floods in the return period (T) of 100, 200, 500 and 1,000 years, for instance. For every flood considered, reservoir operation during the flood wave can be evaluated and the maximum flow discharge determined. The discharge will mainly depend on the reservoir's initial volume (reserve volume for flood control), and on how discharge structures are operated. Thus, performing simulated operations of the reservoir during each flood which take into account the various conditions, maximum flow discharge curves are obtained in function of the reserve volume, as shown in Figure 4 (b).

Applying the relationships between flood damages and the maximum flow discharged, flood damage curves may be obtained in function of the reserve volumes in the reservoir for various return periods, as shown in Figure 4 (c).

3.3.6. Determination of Expected Damage (Curve 2(a) Figure 2). A series of flood damage values and its appropriate probability of exceedance rate (1/T) are obtained from the previous stages for each VR value, thus enabling the determination of the mathematical expected damage, defined as:

$$\text{Expected Damage} = \int_{-\infty}^{+\infty} x \ f(x) \ dx \qquad (3)$$

where:

f(x) is a distribution of the x random variable (Ln damages)

It is considered appropriate to use a lognormal distribution function in principle, to easily adapt to the problem's characteristics in a large part of the cases. In an application, however, the existence of an adequate data correlation for this distribution should be checked.

The Expected Damages in this distribution are defined with the expression:

$$ED = e^{u - \frac{\sigma c}{m}} \qquad (4)$$

-u and are lognormal, average and standard deviation distribution parameters, respectively, of variable x (Ln damages).

-c and m are correction coefficients of the distribution parameter, obtained from a linear correlation between variables t and t*.

-t is a typified variable corresponding to a specific probability of exceedance of x, and takes the value (x-u)/

-t* is the mathematical value of the lognormal distribution, associated to a certain exceedance probability of x in order to check:

$$t^* = mt + c \tag{5}$$

Therefore, Equation (4) is applied to the reservoir's various flood reserve volumes, finally obtaining the desired curve 4 (d).

3.3.7. Net Expected Benefit and Reserve Volume for Flood Control. With the results shown in the curves in Figures 2 (c) and 2 (d), the curve in Figure 2 (e) can be infered. Only the difference between Expected Agricultural Benefits and Expected Damages has to be calculated for this purpose.

4. APPLICATION TO THE PALOMA RESERVOIR

4.1. Description of the Paloma Reservoir

The Paloma Reservoir has a 750 million cubic meters capacity and is located some 400 km. north of Santiago, Chile, in the Limari river basin (Figure 5). The reservoir is part of the Paloma irrigation system, additionally consisting of the Recoleta (100 million m3.) and Cogoti (150 million m3.) reservoirs, and a network of main and lateral canals (with capacities between 3 and 8 m3/s. with a total length of 590 km.).

Their physical disposition is such that approximately 40% of the 40,000 hectares benefited by the Paloma system can be supplied by two reservoirs alternatively (Paloma and Cogoti, or Paloma and Recoleta).

The Paloma reservoir has a gate spillway designed to discharge a flow up to 4,000 m3/s. in the millenary flow.

4.2. Results

Results obtained with the application of the proposed methodology at the Paloma Irrigation System are shown in figures 5(a), (b), (c), (d) and (e).

4.3. Operation of the Paloma Revervoir during the July 1984 Storm

In July 1984 there was a very heavy storm in the area of an 80 year return period, with total precipitation reaching nearly 300 mm. On that opportunity the Paloma reservoir was operated in accordance with the methodology presented herein, and excellent results were achieved.

In fact, the snow conditions in the Cordillera allowed for a reserve volume over 50 million m3 in the winter season, but under 100 million m3. Thus, at the onset of the event the gates were gradually opened, bearing in mind the risk of large damages occurring downstream and the rainfall in the area. The final results of the operation show that the reservoir was used at its full maximum eventual capacity, and the peak affluent flow of 1,800 m3/s. was reduced to 700 m3/s. at the reservoir outlet. This flow was sufficient enough to control the situation and prevent major damages (Figure 6).

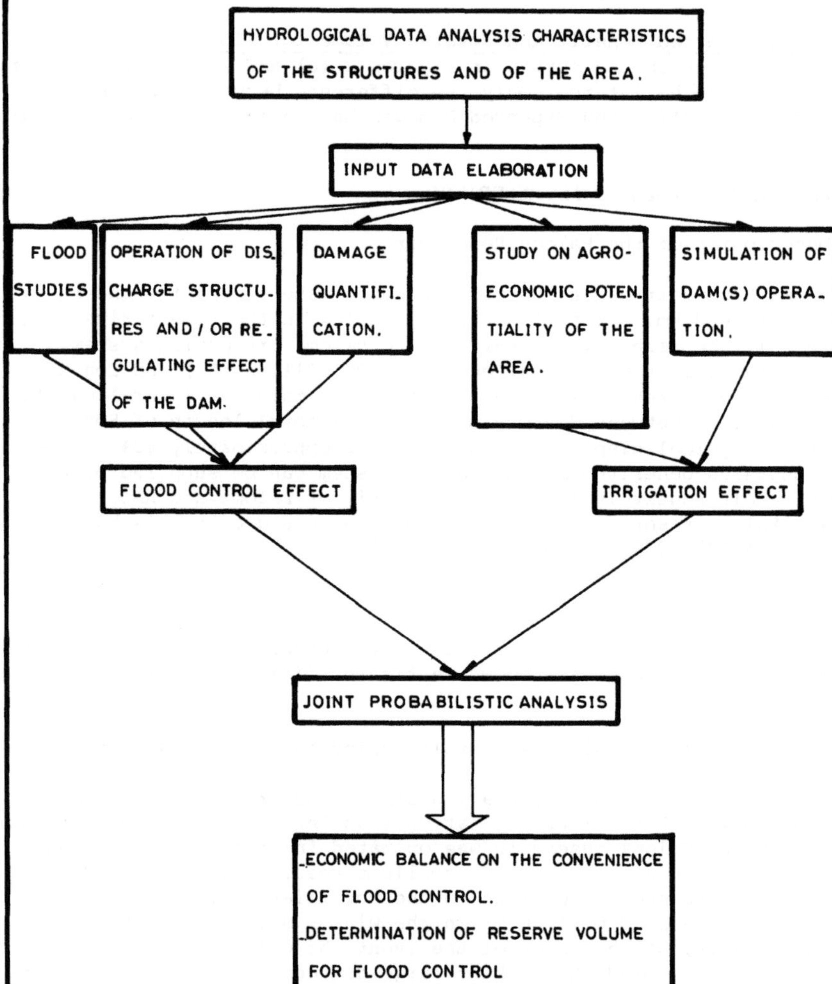

FIGURE 1

GENERAL METHODOLOGY FOR THE EVALUATION OF FLOOD CONTROL IN A DAM SCHEME.

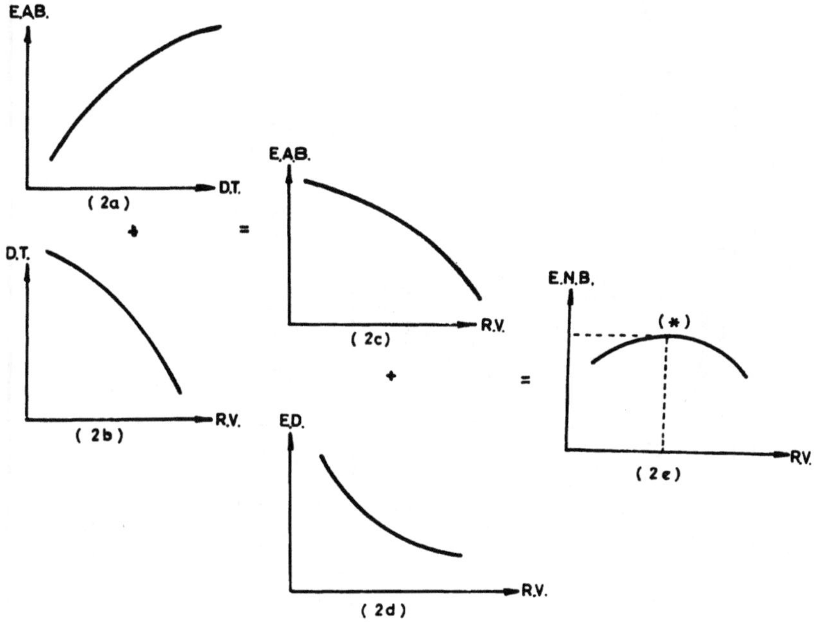

NOMENCLATURE : DT = Annual supply target for irrigation

RV = Reserve volume for flood control

EAB = Expected agricultural benefit

ED = Expected damage

ENB = Expected net benefit

(*) = Optimal condition

FIG 2.

EXPECTED NET BENEFIT EVALUATION CURVES

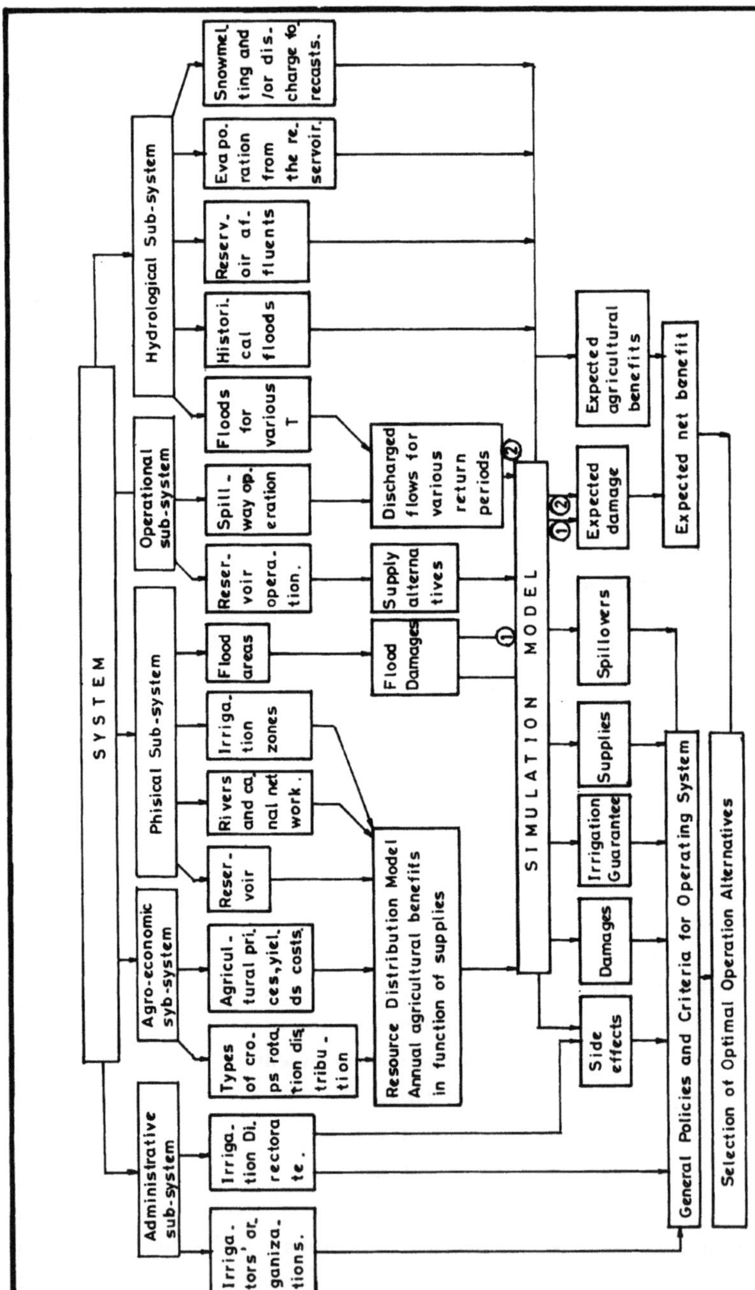

FIGURE 3

Sub-systems involved factors and general scheme of the methodology

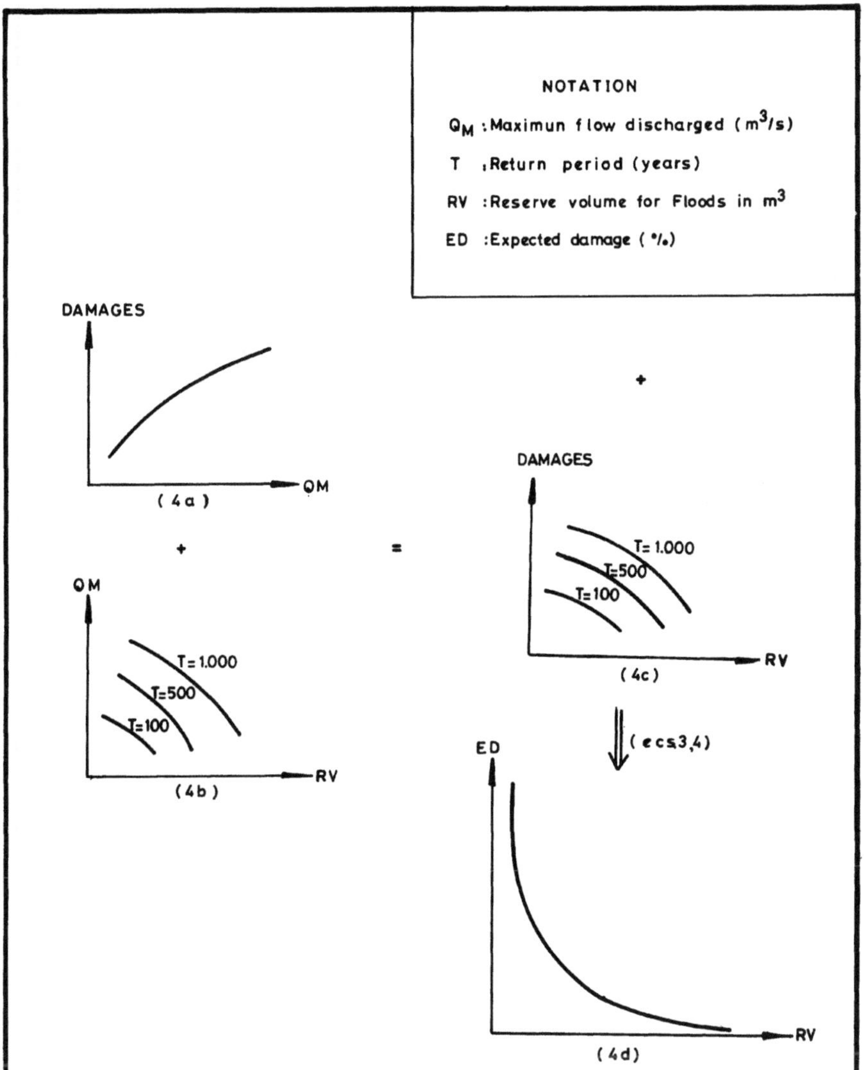

FIGURE 4, DETERMINATION OF EXPECTED DAMAGE

(4a) Results of the cadaster of areas inundated by floods

(4b) Reducing "effect" of the reservoir for various floods

(4c) Effect of the reserve volume on the damage caused by floods.

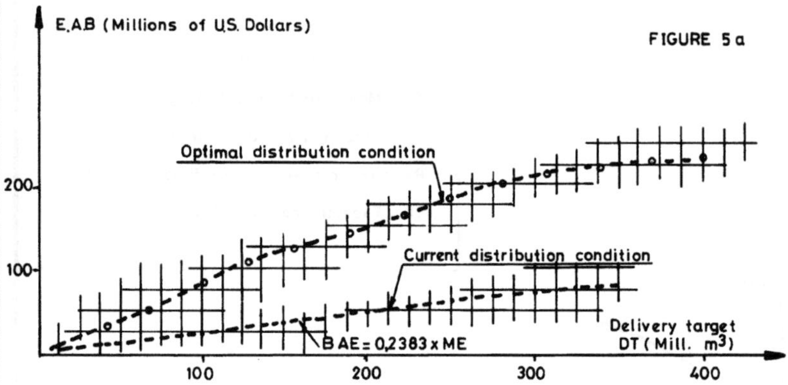
E.A.B (Millions of U.S. Dollars)
FIGURE 5 a
Optimal distribution condition
Current distribution condition
BAE = 0,2383 x ME
Delivery target
DT (Mill. m3)
200
100
100
200
300
400

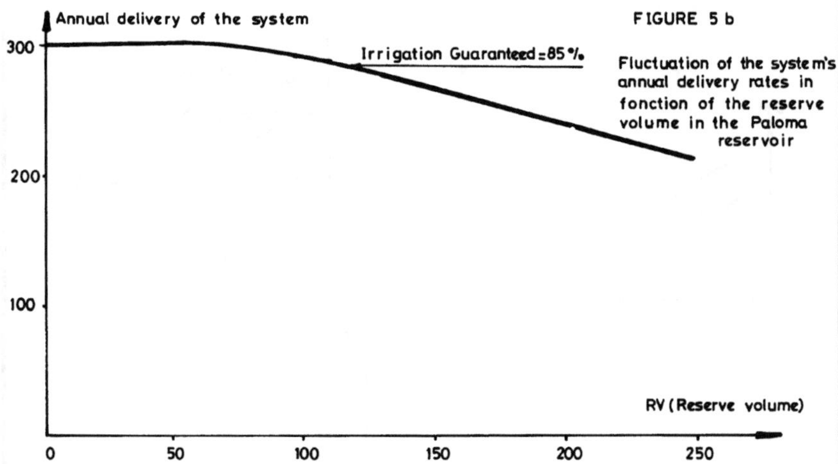
Annual delivery of the system
FIGURE 5 b
Irrigation Guaranteed = 85 %
Fluctuation of the system's annual delivery rates in fonction of the reserve volume in the Paloma reservoir
RV (Reserve volume)
300
200
100
0
50
100
150
200
250

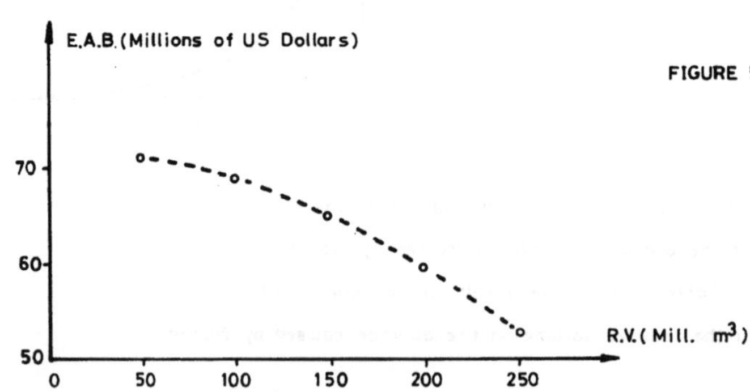
E.A.B. (Millions of US Dollars)
FIGURE 5 c
R.V. (Mill. m3)
70
60
50
0
50
100
150
200
250

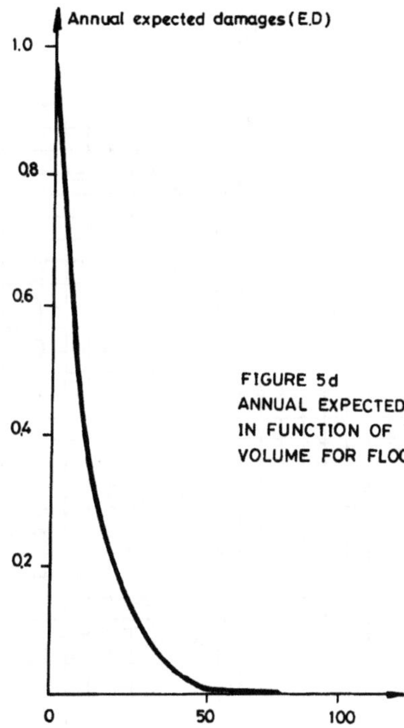

FIGURE 5d
ANNUAL EXPECTED DAMAGES IN FUNCTION OF THE RESERVE VOLUME FOR FLOOD CONTROL

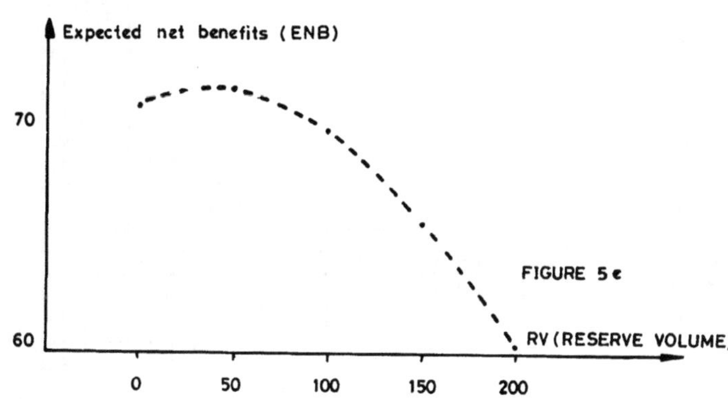

FIGURE 5e

FIGURE 5 a, b, c, d, and e:

Representation of the results of the application of the methodology, at the Paloma reservoir case

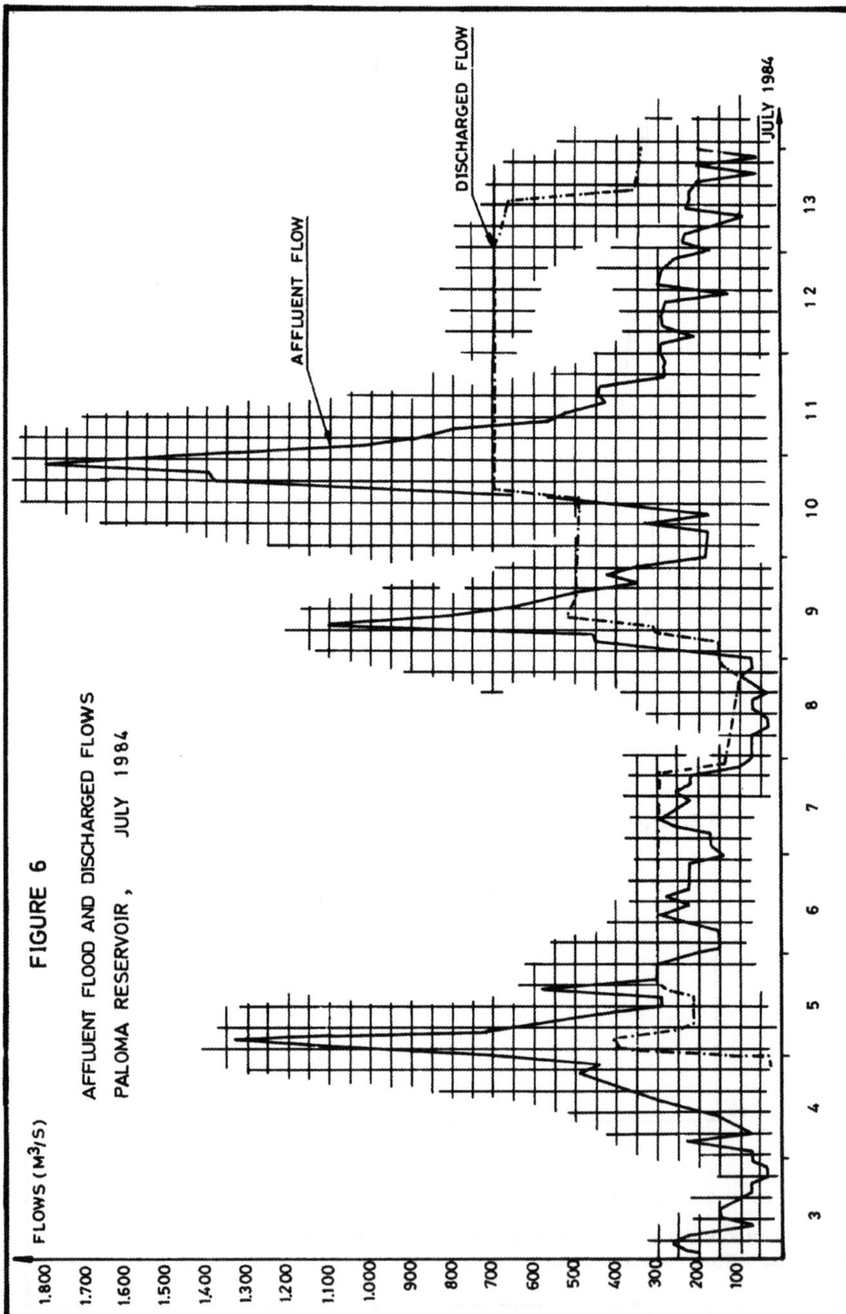

FIGURE 6

AFFLUENT FLOOD AND DISCHARGED FLOWS

PALOMA RESERVOIR, JULY 1984

5. CONCLUSIONS

(1) The methodology developed enables the evaluation of the operation of a reservoir serving both irrigation and flood control purposes, as well as the determination of the convenience of such uses.

(2) The methodology described herein is both relatively simple and effective, since it uses fairly common modeling techniques which are flexible enough to be adapted to the diverse situations that may arise in each case.

(3) Despite the foregoing, the application of the methodology is a long and sensitive process requiring a series of specific analyses, such as the analysis of floods, hydraulic axis of the river downstream of the reservoir, flood damages, determination of agro-economic factors, and others.

(4) The methodology developed was applied successfully to the Paloma system using the largest amount of data available. Analysis of the results shows that an adequate representation of the operational characteristics of the structures was achieved.

(5) Considerable damages are caused by floods in the areas adjacent to the Grande and Limari rivers, that can be reduced to a large extent with a slight decrease in the Paloma reservoir's useful capacity during certain months of the year.

(6) Operation of the Paloma reservoir with a reserve volume for flood control in the winter months of May through August, up to 100 million cubic meters, does not decrease substantially neither the target nor the guaranteed water supply of the system nor its safety.

6. REFERENCES

5.1. Estudio hidrológico y operacional del Sistema Paloma, Brown, Ernesto., and Ferrer, Patricio, 1976, Santiago, Chile, Ministry of Public Works, Irrigation Directorate, pp. 4.29-4.38, and 5.12-5.24.

5.2. Determinacion del Eje Hidráulico del Rio Grande-Limari, Hidroproyectos Ltda., Civil Engineers, 1980, Santiago,Chile, Ministry of Public Works, Irrigation Directorate, pp. 14-21.

5.3. "Water Resources Engineering", Linsley, Roy K., and Franzini, Joseph B., 1972, pp. 119-143.

5.4. Report on the Irrigation and Development Project, Rendel, Palmer and Tritton, 1967, Republic of Chile, Ministry of Public Works, Irrigation Directorate, Vol. 1, pp. 110-130.

ST. HELENS BLAST RELATED SEDIMENTATION MODELING AND PLANNING REALITIES IN THE TOUTLE/COWLITZ/COLUMBIA RIVER SYSTEM

William A. Rabiega
Urban Studies and Planning
Portland State University
P.O. Box 751
Portland, OR 97207

ABSTRACT. Following the Mount Saint Helen's eruption of May 18, 1980 a major planning effort was started by the United States Army Corps of Engineers to deal with potential flood and sedimentation problems related to that event. This paper reviews the four major reports to come out of that effort with emphasis on the role of sedimentation, frequency, and risk modeling in the planning process. New types of information concerning the assumptions of supporting models and the positive and negative consequences of selecting various management strategies were incorporated into the documents. The inherent stochastic nature of the modeling process emerged as an important issue in evaluation of recommendations from planners and engineers instead of being masked by single conclusions based on average conditions. A higher public profile for models and modelers and the attendant demands this new role implies is likely if the planning process exemplified in these documents is generally adopted.

1. INTRODUCTION

On May 18, 1980 Mount Saint Helens in Southwestern Washington State of the United States of America erupted, depositing 2.9 billion cubic meters of avalanche, pyroclastic, and ash materials in the Upper Toutle River Valley. A mudflow with 21 million cubic meters of sediment rolled through the Toutle and Cowlitz Rivers eventually dissipating in the Columbia River. In a few hours the entire North Fork Toutle River, Lower Toutle River, and Lower Cowlitz Rivers stream geometries where radically modified, and the deep-water navigation channel of the Columbia River was blocked. The United States Army Corps of Engineers (COE) responded with emergency actions, primarily dredging programs, to re-open navigation and provide minimal flood control to the communities of the Lower Cowlitz Valley and with a major research and planning program to provide for a long term program to alleviate potential flooding and navigation problems related to the eruption.

It is this latter effort on which this paper focuses. Four major planning documents and a multitude of support studies were produced

V. P. Singh (ed.), Flood Hydrology, 325–331.
© 1987 by D. Reidel Publishing Company.

over the next five years. The planning effort is ongoing at this time. The major documents, A Comprehensive Plan (1983), Spirit Lake Decision Document (1984), St. Helens Feasibility Report (1984), and St. Helens Decision Document (1985), provide a wealth of material and are the principal sources for this analysis. What will become evident is that planning for the dynamic system which followed the eruption was a unique challenge to the Corps. It ultimately forced the recognition that the risks for which corrective measures were eventually to be implemented where not an immutable reality, that they were founded in the perception of reality reflected in the sedimentation and economic modeling selected for the planning effort. An innovative type of planning document emerged which articulates where subjective, qualitative decisions enter into the modeling process and which pinpoints their effects on the judgment as to the significance of problems and the reasonableness of proposed solutions to the problems.

2. PLANNING AT THE CORPS OF ENGINEERS

Planning and implementation of programs is basically carried out at the District level within the COE. The Districts are the first level in the administrative hierarchy of the organization. Above them, respectively, are the Divisions, the Office of the Chief Engineer, the Assistant Secretary of the Army for Public Works, the Secretary of the Army, the Secretary of Defense, and, finally, the President of the United States. Each district has a system of watersheds over which it has responsibility. The Oregon District carried out these planning studies and will implement the subsequent programs, because Mount Saint Helens is included in its watersheds. In this instance the planning effort was initiated under a personal directive from the President of the United States, a highly unusual circumstance which underscores the high public profile of the activities.

Once a planning process is set in motion, it proceeds according to standards and protocols set down within organizational regulations. Basin hydrology and sedimentation is modeled, leading to forecasts of the frequency and water surface elevations of various floods. This information is used in conjunction with inventories of current and projections of future land use to calculate possible flood losses with no action. These losses are averaged, with specified discount rates, over the planning period. Then losses which are avoided are conceptualized as benefits. Ultimately, the benefits of any proposed program must exceed its expected costs for action to be taken by the Corps. If more than one possible program is cost beneficial, then the alternative showing the greatest net benefits is preferred. It must be recommended by the District as the "National Economic Development Plan" unless there are grievous problems of local acceptability. Thus the recommendation of the District usually falls out from a routine and fairly straightforward set of activities. Recommendations at higher levels virtually always follow that of the District.

3. THE MAJOR PLANNING DOCUMENTS

3.1. The Comprehensive Plan

The Comprehensive Plan (Portland District, Corps of Engineers, 1983) predicted a total erosion of 764 mcm of sediment from the debris avalanche during the period 1980 to 2030. Using the Corps' HEC-6 model, patterns of net deposition of the bulked sediment in the Toutle, Cowlitz, and Columbia Rivers for that same period were also forecast. Water surface levels in the Lower Cowlitz Valley were projected for selected points using that deposition forecast and verified with the Corps HEC-2 model. This "no action" scenario resulted in a prediction of cumulative deposition of sediment in the Lower Cowlitz River of 38 mcm of material by 1987. While deposition in the Columbia River navigation channel was not explicitly predicted, sand yield to that river was seen to peak at around 14 mcm by 1995 and taper down to .76 mcm per year by around 2020. It was inferred that a proportion of this sand would deposit in the channel. Increases in water surface levels in the Cowlitz Valley would result in yearly flooding of three of the towns along the river, Castle Rock, Lexington, and Kelso, by 1990 and thus, their abandonment by or prior to that year. Likewise a major Interstate highway and a mainline railroad bridge over the Toutle River would be inundated by 1990. Sedimentation in the Columbia navigation channel was expected to diminish its capacity to allow the passage of deep-draft maritime ships. An additional catastrophic flood hazard was posed by unstable debris dams left by the eruption in the outlets of upland lakes, most notably Spirit Lake in Washington State, and the consequent possibility of dam failure and lake "breakouts."

Thus four situations, 1) flooding of towns, 2) inundation of land transportation corridors, 3) lake breakouts, and 4) disruption of Columbia River navigation, were determined to be the problems for which the Corps had to plan. A number of alternatives, including partial evacuation of the Cowlitz Valley, various dredging strategies, levee raises, and sediment retention structures (dams) were preliminarily considered as solutions to the problems other than a lake breakout. This latter problems was dealt with in the subsequent Spirit Lake Decision Document (1984).

To this point, the planning analysis had followed the established regulations of the Corps. Alternatives which were not cost beneficial were eliminated and, of those remaining for further consideration, ones with the best net benefits were identified as the favored strategies.

3.2. The Spirit Lake Decision Document

The Spirit Lake, Washington Decision Document (Portland District, Corps of Engineers, 1984) followed the established procedures. The sediment budget for the Comprehensive Plan was employed in the analysis and the alternative with the highest net benefits, an outlet tunnel in hard rock, was recommended. With the implementation of this tunnel and minor works for the smaller lakes, the problem of lake breakouts was met.

3.3. Feasibility Report

The Mount Saint Helens, Washington Feasibility Report (1984) marks the beginning of a change in content of the planning analysis in response to modeling uncertainties. This report departed from a revised sedimentation forecast using the same modeling procedures as the Comprehensive Plan. The total sediment erosion estimate was revised downward to 500 mcm over the next fifty years. Cumulative sediment deposition in the Cowlitz River was projected to peak around 60 mcm by the year 2005 and scour out after that time. Annual sand deposition in the Columbia navigation channel was expected to decline from 6 mcm in 1985 to .76 mcm by 2010 and maintain that yearly rate through 2035. While the essential problems of flooding and navigation disruption were not altered by these changes in the forecast, the timing and severity of the problems were. Consequently, the efficacy of the preferred solution was called into question.

This document responded to that question by introducing a sensitivity analysis to the report (Portland District, Corps of Engineers, 1984b, pp. IV-14 to -27). As conceptualized in this instance, the sensitivity analysis considers the effect of increasing or decreasing the sediment forecast by 50 percent on the economic benefits of the two best solutions. Here those alternatives were 1) a sediment retention structure on the North Fork Toutle River and 2) dredging in the Toutle and Cowlitz Rivers. Only under the 50 percent decrease did dredging exhibit the best benefit/cost condition. This analysis was done first, and in retrospect, for the Comprehensive Plan (1983) forecast and preferred management strategies and later on the national economic development plan of The Feasibility Report (1984) with its updated sediment budget and more detailed alternative designs. Again the sediment retention structure remained the better alternative with 100 percent or 150 percent of the predicted budget. The result of this analysis was to focus on a single sediment retention structure, dredging, and a limited set of levee alternatives for detailed study.

3.4. The Mount Saint Helens Decision Document

The Mount Saint Helens Decision Document: Toutle, Cowlitz, and Columbia Rivers (Portland District, Corps of Engineers, 1985), produced yet lower forecasts, 420 mcm total, and some radically different estimates about how sedimentation would be distributed in the Cowlitz and Columbia Rivers. Deposition in the Cowlitz was depicted as increasing through the year 2035 to somehwat over 90 mcm. Thus no natural scour was seen as likely to occur in the project life and substantially greater volumes of sand requiring dredging were seen to deposit. For the same period, no deposition was forecast in the Columbia navigation channel. Again, these are no action forecasts.

Thus, if one accepted the new forecasts, different planning problems were implicit than in previous documents. Since lake breakouts had already been independently handled, only the problem of Cowlitz River Valley flooding remained. Except of a highly improbable scenario involving two rare flood events back-to-back, sediment from the May 18th blast would not deposit in the navigation channel. The model had, thus, effectively eliminated the threat of navigation disruptions as a possible planning future.

These changes brought two points into clear focus: 1) that the planning process was itself model driven, and 2) that because of this, decision-mrkers needed a more explicit treatment of the inherent uncertainties in the evaluation process and the possible consequences of wrong decisions. These realizations are particularly reflected in the letters of transmittal of the Mount Saint Helens, Washington Decision Document (1985). The result is a chapter concerning "premise sets" (Portland District, Corps of Engineers, 1985, pp. VI-1 to VI-8), not previously seen in this series of planning documents, in addition to the sensitivity analysis of the two better management strategies.

A third substantial change in the nature of these reports is also evident, but not highlighted. To this point, various measures had been analyzed as discrete alterantives. Dredging itself, levee raises in themselves, and a sediment retention structure in itself were compared. Chapters III and IV are organized to consider these measures not only as discrete alternatives but in selected combinations. Hence, the eventual recommended plan is a combination of dredging, levee raises, and a sediment retention structure. This strategy evolved in the document rather than being an explicit alternative the District was charged to investigate. This innovation implies substantial ramifications to the planning process and modeling with regard to the economics of flood risk and frequency.

4. IMPLICATIONS FOR PLANNING WITH RISK/FREQUENCY MODELING

In his letter of transmittal to the Office and Management and Budget, Robert K. Dawson, the Acting Assistant Secretary of the Army (Civil Works), states that the 1985 Decision Document, "sets a new standard for Corps of Engineers planning reports and will serve as a model for future efforts. State, local, and Federal decision makers will find useful the explicit treatment of uncertainty and (it will) aid them in their appreciation of the trade-off between risk reduction and cost (1985, pp. 1-2)." Assuming that future documents of the United State Army Corps of Engineers will follow the format laid down in that document and that other agencies and institutions may follow, what the model is and what its implications are for researchers contributing risk and frequency modeling to the overall planning effort is of interest.

First the planning model underlying the 1985 document is inherently probabilistic. While risk/frequency modeling is itself probabilitic, only the average, i.e., most probable state, has been considered in previous planning analyses. Thus one best plan emerges for dealing with a single expected future. That is not the case in this instance. While the most probable state is used for the principle forecasts and the ultimate recommendation follows from it, recommendations for lower probability high and low estimates are also made. It follows that explicit treatment of those estimates, and assumptions about those estimates, is mandated.

Thus, and second, this document has elaborate treatment of the assumptions of the erosion and sedimentation modeling. The general veracity of the recommendations is seen to directly flow from the reasonableness of these assumptions. Previously, such treatments were more likely a part of the technical appendices to a report if they were done at all.

Third, and following from this treatment, is an assessment of the risks to decision-makers of accepting management strategies for futures characterized as having a higher or lower probability. This is done in the premise set analysis. Here the potential decision-maker is apprised of the premises, i.e., the model assumptions and analytical conclusions, s/he must accept to support a certain plan. Then what outcome is maximized and what future actions are foregone by that support is specified. This type of analysis of ramifications to and for the decision-maker has not previously been an explicit function of planning analysis.

Fourth, recommended management strategies in this report are optimized over many combinations of abatement measures, rather than rationalized versions a single measure such as dredging. This type of optimization involves many more variables than rationalization within a given measure. Hence the generalized planning approach which optimizes by rationalizing within directed alternatives for management is superseded and complicated.

The minimum implications for the modeling which supports this type of planning is that the assumptions of all modeling, sedimentation, stage-frequency, and economic, will receive greater scrutiny. The models themselves may require extremely detailed elaboration. Modelers will be asked to be their own critics, to explain where their work may go wrong and to assess the effects of it going wrong.

The maximum implications are more profound. An acknowledgement that planning processes are model driven is likely to directly involve modelers who have been somewhat insulated from the politics of decision-making in the political arena. While there has always been some level of implied pressure to serve the agendas of officials, the technical aspects of modeling have served to limit the degree to which it is open to modification from interest-based compromise. This relative separation between the political and technical is likely to fade. Certainly, the planners who use these models are brought into the political arena by the use of premise sets. Premise sets require the planner to anticipate the political implications of a decision to an official and that the planner enumerate those implications.

There are also purely technical implications. Demands for accurate performance from models are likely to increase. At the same time simple models which are easily explained are likely to be favored. Modelers may be faced with the problem of producing enhanced outputs from retrograde algorithms. Sophisticated and accurate black boxes will not be acceptable. Operating those models with a range of possible inputs and reporting the various outputs is also implicit. Taken to its extreme, modelers may be required to attach probabilities to a wide range of possible outcomes and identify any subjective element of those probabilities.

At the same time, some of the submodels in this total process may be complicated. In the 1985 Decision Document an optimization across dredging, levee, and sediment retention dam configurations is induced. The costs and benefits of a large number of possible combination strategies were considered. The eventual recommended plan was indeed the one which had the greatest net benefits of the set considered. However, there is no certainty that the solution was the global optimum, or the feasible engineering solution that most closely approximated the global optimum. If trade-offs between costs and protection are to be evaluated with the greatest accuracy, a modeling process which first identifies the global optimum and is then constrained by engineering and/or political feasibility is necessary. Such models are not currently implemented at the practice level.

If even the minimum of these implications becomes a reality, the demands placed on the modeler and models themselves will be intensified. Most evident will be calls for better communication and higher public profiles. Less evident will be the basic research to produce the algorithms to support the type of planning document for which the 1985 Decision Document is a prototype.

5. REFERENCES

5.1. Portland District, United States Army Corps of Engineers. 1983. Mount Saint Helens, Washington: A Comprehensive Plan.

5.2. __________. 1984a. Spirit Lake, Washington: Decision Document.

5.3. __________. 1984b. Mount Saint Helens, Washington: Feasibility Report and Environmental Impact Statement.

5.4. __________. 1985. Mount Saint Helens, Washington: Decision Document.

COST-BENEFIT ANALYSIS OF A PROPOSED STORM SURGE BARRIER ON THE RIVER SCHELDT (BELGIUM)

J. Berlamont, Professor of Hydraulics
M. Sas, Research Engineer
University of Leuven
de Croylaan 2, B-3030 Heverlee, Belgium

P. Van Rompuy, Professor of Economics
University of Leuven
Van Evenstraat 2 B, B-3000 Leuven, Belgium

ABSTRACT. At the request of the Belgian government, the University of Leuven carried out a multi-disciplinary evaluation study of the proposed storm surge barrier on the river Scheldt, near Antwerp (Belgium).

After assessing the frequency of occurrence of storm surge levels at Antwerp and at any other station along the river, a number of inundation scenarios were simulated to estimate flooded area, and thus the damage as a function of the return period. It was found that the cost of the storm surge barrier was six to seventeen times higher than its benefit : it only offers a very expensive marginal extra safety. In consequence of the study the barrier project has been abandoned.

The study has proved that a multi-disciplinary approach is the best way to evaluate flood protection alternatives in a systematic way, within the wide range of socio-economical constraints.

1. INTRODUCTION

After the disastrous storm surges in 1953 and 1976 the great number of claims for damages, and the fact of 150000 people being at risk, forced the Belgian Government to launch a flood protection program.

This program (Ref. 3), called the SIGMA-plan, consisted of three components. The first was a flood banks improvement program. The second was the designation of three small flood basin sites. The third component offered two alternatives. Alternative 1 was the construction of a storm surge barrier with three navigable passes downstream of Antwerp. Alternative 2 was the setting up of three major controlled flood basins downstream of Rupelmonde.

The storm surge barrier would protect the whole river Scheldt basin against storm surges up to 9.05 m + datum. This level has a frequency of occurrence, estimated at 1/2,000 to 1/10,000 years, depending on the method used (Ref. 4). The project costs of alternative 1 were estimated at 300 million US dollars, including construction, maintenance and exploitation during 100 years.

V. P. Singh (ed.), Flood Hydrology, 333–345.
© 1987 by D. Reidel Publishing Company.

Alternative 2, consisting in the organization of three major flood basins, would provide a lower protection level. The frequency of flooding was estimated at 1/200 to 1/500 years. The investment, however, would be considerably smaller than for the barrier.

In view of these data, the University of Leuven was asked in 1981 by the Ministery of Public Works to evaluate both alternatives, not only in a purely economical way, but also with respect to ecological, environmental and even political aspects (Ref. 1).

The Hydraulics Laboratory, together with the Department of Applied Economics, coordinated the research by a team of about 30 scientists : hydraulicians, economists, agricultural engineers, ecologists, psychologists; medical, sociological and political scientists, and specialists in risk analysis and insurance policies. The study was to be realized within a period of less than 1 year. The budget was approximately 85,000 US dollars.

2. STORM SURGE IDENTIFICATION

The probability of occurrence of storm surges was first to be determined at Antwerp, where a long water level record is available (Fig. 1).

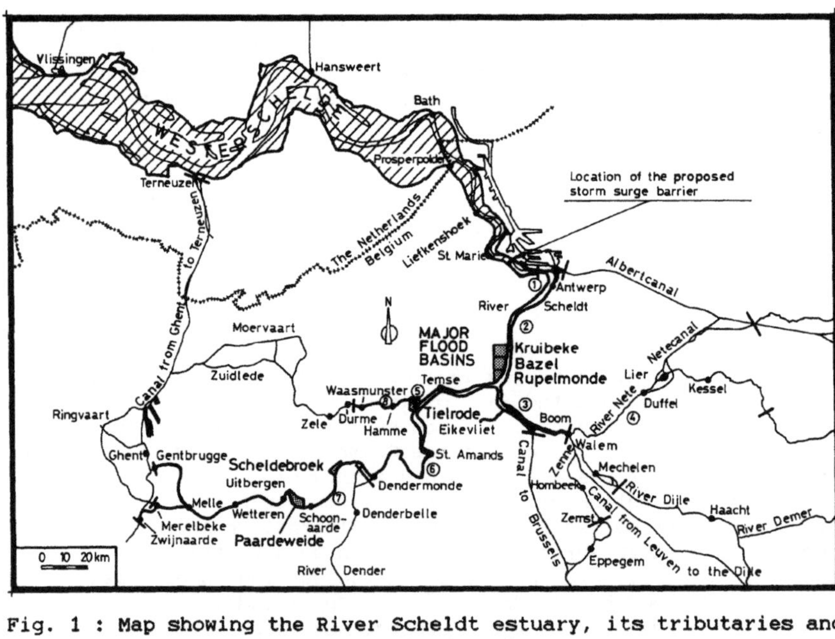

Fig. 1 : Map showing the River Scheldt estuary, its tributaries and the location of the flood basins

The storm surge with a return period of 10,000 years was of particular interest, since it was the standard indicated in the

SIGMA-plan. Earlier studies of the Ministery of Public Works (Ref. 6) indicated 9.05 m + datum for a 10,000 year surge level. This study considered the maximum high water levels recorded between 1901 and 1977 as a homogeneous sample. Actually, there is in the data a marked trend of increasing high tide levels, tidal range, and mean tidal level, due to the endiking of flood plains, dredging works and the gradual increase of the mean North Sea level (Ref. 4, 5). A Gumbel extreme value distribution adjusted to the data set for different time periods confirms this trend (Table I).

TABLE I

Maximum surge level (+ datum)	Return period (years) 10	100	1000	10000
(Theuns, Ref. 6)	7.16	7.78	8.42	9.05
Gumbel 1901-1970	7.05	7.80	8.53	9.27
Gumbel 1941-1970	7.24	8.07	8.90	9.72
Gumbel 1951-1970	7.33	8.21	9.07	9.94
Monte Carlo	7.29	8.06	8.85	9.58

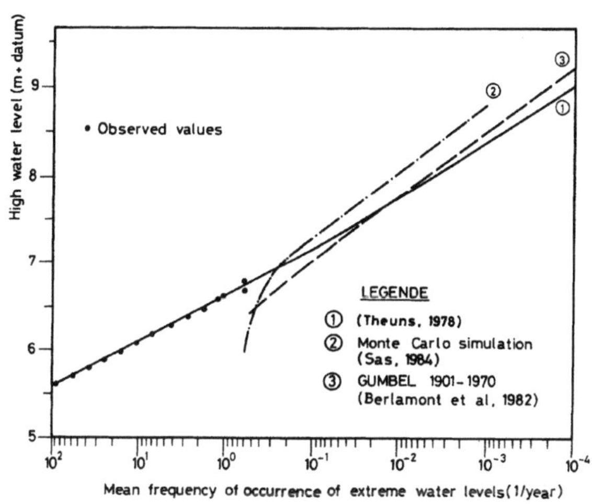

Fig. 2 : Mean frequency of occurrence of extreme water levels at Antwerp

A Monte Carlo simulation method (Ref. 4), which takes account of the trends in determining the frequency of occurrence of extreme high water levels, yielded relatively high values (Fig. 2).

For the cost-benefit analysis of the storm surge barrier, all storm surge levels causing damage were important, not only the one associated with the nominal return period of 10,000 years. In the present study it was assumed that the "nominal values" from the Ministery of Public Works study (Ref. 6), and the return periods obtained from the Monte Carlo simulation (Ref. 4) provided respectively a lower and an upper limit for the frequency of occurrence (Fig. 3). From this uncertainty interval on the surge levels associated with different return periods, the uncertainty interval on the return period associated with a given high water level could be derived (Fig. 3a).

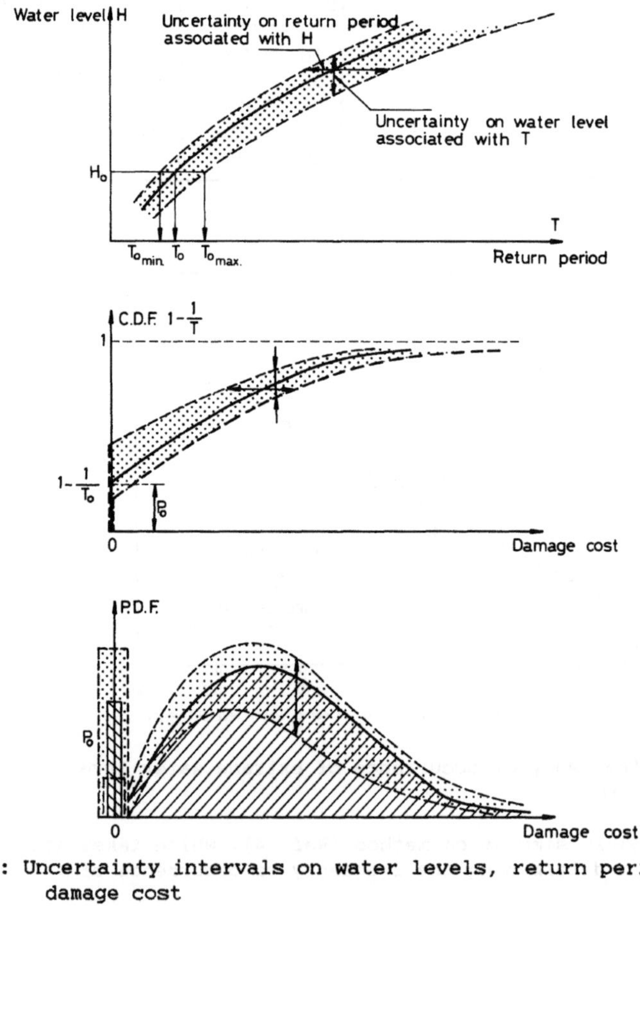

Fig. 3 : Uncertainty intervals on water levels, return period and damage cost

3. HYDRAULIC ASPECTS

The hydraulic information needed in the cost-benefit analysis was :

a)i) the safety provided by the improvement and raising of the flood banks, and the construction of the three minor flood basins;

ii) the additional safety provided by the three major flood basins;

iii) the frequency of inundation of the proposed major flood basins;

b) the amount of damage as a function of the return period for each of the two alternatives (i.e. storm surge barrier or major flood basins);

c) the effect of proposed dredging and regulation works in the river Scheldt.

This information was obtained by using a 1D mathematical model of the river Scheldt, bounded by controlled weirs upstream the river and its tributaries, allowing lateral inflow, river junctions and weirs connected with the flood basins (Fig. 1) (Ref. 1).

3.1. Major Problems

The major problems associated with the numerical modelling of the river Scheldt, its flood plains and, eventually, the inundated areas, were :

a) the lack of a coherent set of accurate topografic data :

i) not all available data were recent;

ii) since all surveys were carried out in view of navigation on the river and of dredging works, no data were available about the flood plains.

b) the different roughness of the river and the flood plains. The flow resistance in the flood plains with bushes, hedges, dikes, etc., is much larger than the resistance in the main bed. Therefore, the part of the wetted section, associated with the flood plains, was only taken into account in the continuity equation (storage) and not in the equation of motion : i.e. the flow velocity in the flood plains was estimated to be very small.

c) The variation of the fresh water discharge and the uncertainty about the operation strategy of the weirs, bounding the river system upstream.

The model was calibrated and verified for different tides, both with and without storm surges. For the river Scheldt (without tributaries) the following Manning coefficients were accepted :

(i) between the Sea and Prosperpolder : $n = 0.0325\ sm^{-1/3}$

(ii) from Prosperpolder up to Ghent : $n = 0.035\ sm^{-1/3}$

Relatively large errors on the estimated water levels appeared along the upstream tributaries (river Nete). A small, but systematic overestimation was found between St.-Amands and Wetteren. They were attributed to the uncertainty with which the channel geometry and the fresh water discharges were known, and to unrecorded flooding of some flood plains.

3.2. Assumptions

a) The frequency of occurrence of a certain water level at any station along the river was assumed to be the same as the frequency of occurrence of the water level at Antwerp, which corresponded to it through water surface profile calculations (Fig. 4, Fig. 5).

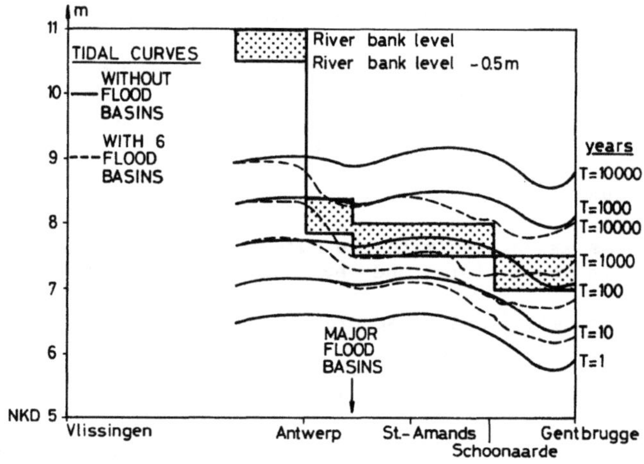

Fig. 4 : Tidal curves in the River Scheldt with, and without flood basis (calculated, no flooding)

b) It was postulated that the new and/or renewed flood banks would be stable and well maintained in the future. Then, a local flooding could only be due to the overflowing of the banks. As the actual crest level of the banks is not always the nominal level, due to subsidences and the limited accuracy of the topographic surveys, and because of the possible presence of wind waves, it was suspected that flooding could occur when the water level exceeds the nominal crest level of the flood banks minus 0.5 m, which is called inundation level (Fig. 4, Fig. 5).

3.3. Safety

The degree of safety at every station along the river Scheldt and its tributaries was determined by simulating storm surge penetration into the basin from the North Sea. The safety at a particular station corresponds to the frequency of occurrence of the surge for which the inundation level is reached at that station.

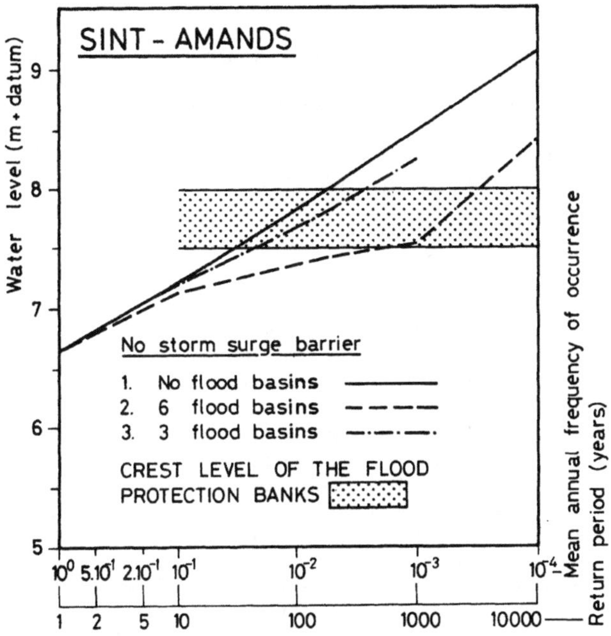

Fig. 5 : Water level at a typical station (St. Amands) along the River Scheldt versus return period for the three cases

The results of this study are :

a) In the absence of the three major flood basins, the safety in the river Scheldt basin varies greatly from place to place : the annual inundation risk is of the order of
   2 to 4 % along the river Rupel and its tributaries, along the tributary Durme and along the middle reach of the river Scheldt (between Rupelmonde and Dendermonde),
   0.8 to 2 % at Antwerp, left bank,
   0.15 to 0.4 % at Antwerp, right bank.
   The lowest values are associated with the nominal bank levels and a conservative estimate of the storm surge level at Antwerp.
b) In the presence of the three major flood basins (alternative 2), the largest part of the river Scheldt basin has an annual inundation risk of 0.2 tot 0.6 %. In the most upstream part of the river Scheldt and the tributaries of the river Rupel, the risk

is 0.4 to 1 %. Since the flood basins are situated upstream of Antwerp, the inundation risk at the city of Antwerp remains unchanged at 0.15 to 0.4 %. It may thus be stated that the six controlled flood basins provide the upper Scheldt basin with a safety against flooding which is the same as at Antwerp, i.e. an annual risk of flooding of roughly 0.5 %.

c) The storm surge barrier (alternative 1) would offer a protection level of 0.01 to 0.05 % annual risk of flooding (depending on the hypothesis about the frequency of occurrence of level 9.05 m + datum at Antwerp).
d) From the calculations, it could also be concluded that the three major flood basins would be used on an average 3 times every 10 years.

3.4. Damage

In order to make the calculation of the "benefit" of the construction of the barrier possible, the probable amount of damage, associated with all storm surges causing flooding, had to be evaluated. Therefore, a number of inundation scenarios were simulated for several storm surge levels.

The potentially inundated land was split up into eight zones, in each of which the inundation risk is almost constant.

Providing a side weir, with a crest level at nominal bank level minus 0.5 m, in the downstream part of a particular zone, and calculating the stored volume of water during the inundation, it could be checked whether the upstream parts of the zone were protected against flooding, for that specific surge, with a known frequency of occurrence. If they were not, either more or longer side weirs must be placed, and accordingly, larger inundation volumes must be reckoned with (Fig. 6).

With the local topography known, the inundation volumes can be transformed into inundation areas. Since the exact location of the flooding within a particular zone is obviously not known, a number of different possibilities must be tried out.

The resultant volumes and areas are a direct measure for the expected damage and thus for the benefit of an alternative, which avoids the damage. Some results are found in table II.

TABLE II : Flooded area and associated probability

Probability (%)	Flooded area (ha, 10^4 m^2)
0.4 to 1	500
0.1 to 0.3	1200
0.04 to 0.2	3200
0.02 to 0.1	catastrophic

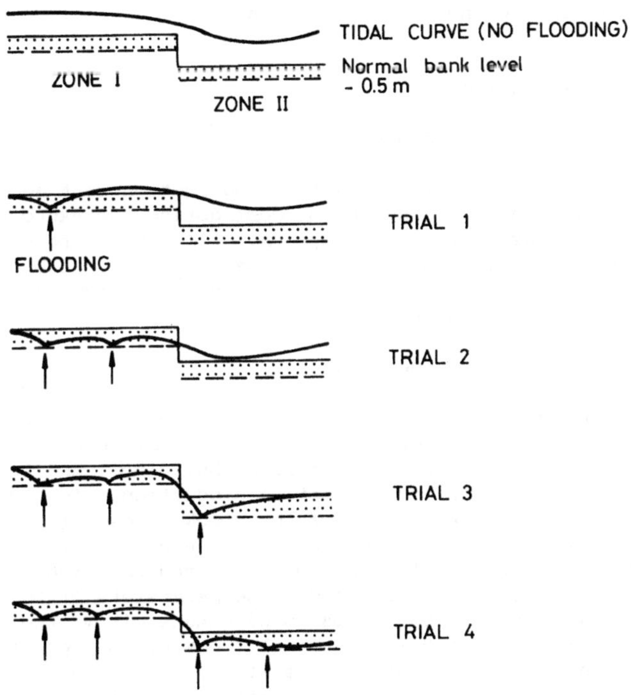

Fig. 6 : Schematic representation of the method for determining the inundation volumes

All estimated volumes (and damage) depend upon the storm surge levels, which have an estimated probability of occurrence. Therefore, the expected risk (damage * probability) must be calculated from the probability density function (PDF) of the estimated damage. This PDF is only the best estimate, since different procedures result in different return periods for the same water level, and different scenarios result in (slightly) different damage. In this way, the calculated risks are probabilistic variables (Fig. 3).

3.5. Effect of proposed works

By using the mathematical model, it was shown that neither the planned regulation works between St. Amands and Wetteren, which are intended to allow navigation with 1350 tons ships, nor the proposed dredging works, increasing the waterdepth upstream of Antwerp from 43 up to 48 ft, reduced the global safety of the river Scheldt basin in a substantial way. The expected increases in high water levels are situated within the uncertainty interval.

4. COST-BENEFIT ANALYSIS

4.1. Social cost of the project

The cost of the barrier includes the construction cost and the recurrent maintenance and operation cost during the expected lifetime of the project, estimated at 100 years. All costs were actualized by means of a social discount rate of either 4 or 8 %. Labour cost was adjusted to take into account the expected high unemployment rate during the construction period.

4.2. Social benefits of the project

The social benefits of the construction of the barrier are related to the costs avoided, i.e. the damage to buildings, such as houses, factories, administrative and commercial buildings, and to industrial capital, as well as the income lost in production. In order to evaluate these costs, estimates of the stock of dwellings and of industrial capital were made for each of the potentially flooded areas. Next, damage coefficients were defined as a function of the water level in the case of a flood. The parameters of this damage function were estimated on the basis of damage claims registered in 1976 following on a flood in the Scheldt basin. By applying the estimated damage coefficients to the stock of buildings and to the industrial capital in the critical areas, expected values for this benefit component could be obtained.

As for the income lost in production, estimates of the value added, lost in consequence of a flood, in agriculture, manufacturing and the service sector, were obtained for each of the potentially flooded areas by using data on employment in these areas, and nation wide sectorial characteristics.

All benefits were assumed to grow, in real terms, at an annual rate of 1.2 or 3 %.

4.3. Results of the Cost Benefit Analysis

If the flood banks are raised and improved, and only the three minor flood basins are built, the safety along the river varies from place to place; the expected yearly flood risk is 2 to 4 %. Assuming the most favourable conditions of economic growth, interest rates and return periods for storm surge levels, the barrier (alternative 1) leads to a cost six times higher than the actualized benefits, as compared with the solution with three major flood basins (alternative 2). By using the worst (or most conservative assumptions), that ratio increases to 17.

The economic evaluation of the barrier thus resulted in a substantial negative value of the expected net benefit. Therefore, alternative 1 could only be selected on other grounds than those taken into account in the cost-benefit analysis, e.g. extreme risk aversion or political reasons. It must be stressed that some costs, e.g. the loss of human lives and the value of environmental damage, were not evaluated. However, even on the basis of the most favourable assumptions, the cost of the barrier outweighs its benefit in such a way, that all plausible estimates of these unaccounted costs could not change the cost-benefit ratio substantially.

On the other hand, national or regional employment and income effects could speak in favour of the barrier. Although these effects should be excluded from a cost-benefit analysis, it may be said that estimates of these derived effects indicate that the employment and income multipliers of the barrier project compare favourably with multipliers obtained for an average public work project.

From an economic point of view, it is evident that the storm surge barrier is an expensive protection measure for the Scheldt basin : the provided extra safety as compared to alternative 2 is marginal but extremely costly. This is mainly due to the high maintenance cost of the barrier and the a priori condition that it must have navigable passes with a width of 80 m. If this condition were relaxed, the concept of the barrier and consequently its cost might be completely different. This is conceivable because all the harbour infrastucture of Antwerp is situated downstream of the barrier site and there is only one major ship yard upstream of the barrier.

4.4. Other aspects

The damage to agriculture, caused by a regular inundation of the flood basins, is negligible, especially if these areas are used as grazing land and hay fields, and provided that the quality of the river water were improved. Ecologists concluded that the environmental impact of the barrier would be small, if appropriate precautions were taken during the construction. The ecosystems in the flood basins would not suffer from the inundations if the drainage systems could prevent prolonged flooding and if the quality of the water were sufficiently improved.

On the basis of the damage estimates, an annual insurance premium of 625,000 US dollar has been calculated. Because of the specific factors involved, i.e. a very low probability of occurrence and a high loss value, the insurance for flood damage has to be placed in the category of catastrophy risks. These ought to be covered by an international pool of insurance companies.

5. EVALUATION

The results of all disciplines were compared to evaluate both alternatives, within the given constraints. A fairly strong constraint was the environmental impact of both solutions. The barrier would be constructed near Antwerp and be taller than the medieval cathedral, and moreover part of the recreational zones would have to disappear. For these reasons politicians and ecologists in the city of Antwerp rejected alternative 1, although it provides additional safety in Antwerp, whereas alternative 2 only provides protection more upstream. On the other hand, farmers and local politicians reject alternative 2, because they expect damage to crops and hay fields.

6. CONCLUSION

The technique of simulating inundation scenarios with a mathematical model has proved to be a useful tool for estimating damage and costs due to inundations caused by storm surges. The multidiciplinary approach emphasized the need of clear definitions by all teams, strict appointments and a fixed time schedule. It proved to be the best way to evaluate in a systematic and comprehensive way the protection projects for the river Scheldt basin, within the wide range of constraints, imposed by society.

It is stated by the authors that a cost-benefit analysis does not replace the political decision making; it is, however, very valuable quantitative information for the decision makers, although other arguments and motives may also be taken into account.

As a result of the study, the storm surge barrier project has been (temporarily ?) rejected. In 1985, the works were started to make the flood basins operational.

7. REFERENCES

1. Berlamont, J., Van Rompuy, P., 1982, 'Multi- and disciplinary evalutation study of the proposed storm surge barrier at Antwerp'. (in Dutch : Multi- en interdisciplinaire evaluatiestudie betreffende de stormvloedkering te Antwerpen).
Study performed by the Katholieke Universiteit Leuven, by request of the Belgian Covernment (Ministry of Public Works).
a-Summary (Berlamont J., Van Rompuy P.)
b-Hydraulic aspects (Berlamont J., Sas M., Van Langenhove G., Thienpont M.)
c-Medical and hygienic aspects (Van de Voorde H., Appeltans H.)
d-Agricultural aspects (Feyen J., Boon W., Couderé K.)
e-Psychological and sociological aspects (Smets G., Broux J.)
f-Ecological aspects (De Bont A., Van Assche J., Elen H., Geerardyn B., Konings J.)

g-Risk analysis (De Meester P., Poucet A.)
h-Cost-benefit analysis (Berlage L., Van Rompuy P., Vertongen R., Halsberghe K., Van Elewijck P.)
i-Insurance aspects (De Pril N., Goovaerts M.J.).

2. Cunge, Holly, Verwey, 1980, <u>Practical aspects of computational river hydraulics</u>.
3. Ministry of Public Works Belgium, 1977, 'the SIGMA-plan for the protection of the river Scheldt basin against storm surges on the North Sea' (in Dutch : Het SIGMA-plan voor het beveiligen van het Zeescheldebekken tegen stormvloeden op de Noordzee).
4. Sas, M., 1984, 'Frequency of occurrence of storm surges in an estuary : a stochastic approach'. 19th International Conference on Coastal Engineering. ASCE, Houston.
5. Sas, M., Berlamont, J., 1982, 'The effect of the mean sea level variation on the frequency of occurrence of the high water levels at Antwerp'. Hydraulics Laboratory Report 10-HY-06 K.U.Leuven (in Dutch : De invloed van de verandering van het gemiddeld zeepeil op de overschrijdingsfrekwentie van hoogwaterstanden te Antwerpen).
6. Theuns, J., 'The occurrence of storm surges at Antwerp', (in Dutch : De overschrijdingskromme van hoogwaterstanden te Antwerpen), Report AZ-78-1, Ministry of Public Works Belgium.

FLOOD PLAIN MANAGEMENT – THE NZ EXPERIENCE

Judy H Lawrence
Senior Water & Soil Planner
Water & Soil Directorate
Ministry of Works & Development
P O Box 12-041
Wellington
New Zealand

P "Takis" Koutsos
Chief Engineer
Hawkes Bay Catchment Board
P O Box 233
Napier
New Zealand

ABSTRACT. The New Zealand experience with flood plain management is described in terms of the nature of the flood problem, the features of New Zealand's flood plain management system which are unique, because of the direct links between government technical and planning agencies and the strong ties between central and local government.

The future directions are set out as a unified flood-loss reduction programme based on New Zealand conditions and needs and what has been learnt from the experience of other nations.

1. INTRODUCTION

The management of flood plains in New Zealand has great economic and social significance due to our dependence on such land for settlement, communications and primary production.

Currently nearly 100 communities in New Zealand are at risk from flooding, comprising over 1.64 m people (67% of New Zealand's total population) directly or indirectly affected by damaging events. Between 1920 and 1953 New Zealand experienced 820 damaging floods, both large and small. In the next 30 years, 115 large damaging floods occurred in spite of larger scale investment in structural protection works since the 1940s. Refer figure 1 for a comparison of protection and flood loss costs.

Ericksen (1985 in press) has calculated that the seven major floods that occurred during the 1968-81 period cost New Zealand $1.75 billion or $135 million/year (1984 $ value). Figure 2 shows that central government, local government and insurance have been picking up 50% of the losses - those taking the risk of locating in flood prone areas or approving such location, are bearing only part of the cost.

The New Zealand experience with flood plain management has some unique features which make it different from approaches adopted elsewhere. The country is serviced by 20 regional organisations set

V. P. Singh (ed.), Flood Hydrology, 347–368.
© 1987 by D. Reidel Publishing Company.

up to manage water and soil resources which have strong ties to the central government agency, the National Water and Soil Conservation Authority (NWASCA), in the form of finance and technical advice.

These organisations are special purpose regional catchment authorities managing both water and soil resources together and carrying out a wide range of functions, including the prevention of damage by floods. This activity is very closely linked with the objective of utilising land in such a way as to promote the prevention of damage by floods; flood management in New Zealand has always been linked with land management.

Flood plain management has reached a turning point in New Zealand; much can be learned from other nations, towards the development of a more unified approach complementing the traditional flood control measures with other options that modify the use made of the flood plain.

This paper will highlight the nature of the New Zealand flood problem, the features of our flood plain management system, review the adequacy of our approach by identifying what can be learnt from other's experience so that future directions can be set out.

2. NATURE OF THE FLOOD PROBLEM

The flood problem in New Zealand has two facets, physical and perceived characteristics.

To date, flood plain managers have largely concentrated on modifying the physical characteristics by building stopbanks, flood detention dams and other structures, as a defence against flooding.

Only recently has the issue of people's perception of the flood risk and hazard been addressed as part of the flood problem.

2.1 The Physical Characteristics

New Zealand is an island nation situated between latitudes 34°S x 48°S, comprising 270 000 sq km. Refer figure 3. Its climate is dominated by a succession of easterly moving anticyclones and depressions associated with cold and warm fronts.

High rainfall is the main cause of dominantly riverine flooding, topography causing wide regional rainfall variations. Intensity, duration, areal extent and distribution are the most important rainfall indices for flood analyses in New Zealand.

The most serious flooding is associated with high intensity long duration rainfall periods. However, two other situations are also common. In small catchments short duration, high intensity storms cause flood damage while in large catchments low intensity, long duration rainfall can also cause considerable damage.

The most intense 24 hour rainfalls can be expected to occur in the north and east of the North Island and to the west of the South Island associated with convection storms and active troughs of low pressure crossing the southern axial mountains.

Catchment characteristics such as size and shape, surface conditions and flood plain topography all have an important part to play in flooding.

Despite its small size New Zealand covers a wide range of river regimes and geomorphic features. Because it is geologically young, there is an abundance of vigorous, high energy rivers, the flows of which issue from rocky narrow gorges to steep, active alluvial fans where the deposition and reworking of gravels is the normal process. Only about 25% of New Zealand is below 200 m above sea level where more mature, slow, meandering rivers carry only silt loads, that flow inside defined flood plains.

Because of the wide variation in river type and regime, the flooding problems created by these rivers can be quite different. Flood damage in more mature topography is mainly caused by overflows, of the regime channel onto the flood plain. The risk from this type of flood can be related to flood levels and hence to a return period or statistical probability for different areas in the flood plain.

The threat imposed by the steeper rivers perched on alluvial fans is often connected to velocity, and the direction of the flood path. In these rivers, damage is caused not only by inundation but also by gravel and other debris deposition. (Refer figure 4.) Figure 5, Wright et al, 1969 shows an alluvial fan at Kaikoura any part of which can be subject to flood damage, depending on the direction of the flood path.

Flood debris, apart from covering fertile land and causing loss of production, can also affect flood levels by blocking bridge waterways and culverts thus creating much more severe ponding upstream of these structures, with consequent higher flood damage. Such impediments to flood flows are a major contributor to flood damages in New Zealand. Refer figure 6.

In recent years another flood problem has arisen. The standard of protection afforded by stopbanks has deteriorated over the years in some of the major river systems due to the accumulation of gravel in the bed of the river between the stopbanks. Reductions from 100 year frequency flood to 20-50 year protection has resulted.

The speed of onset of floods in New Zealand is never greater than a few hours, hence a sophisticated system of flood forecasting and warning has been set up to increase the lead time and so reduce the damage potential. The incidence of flood events in New Zealand is depicted in figure 7.

2.2 The Perceived Characteristics

Despite the high incidence of floods in New Zealand, the flood problem is typically perceived by flood plain dwellers as the extraordinary event not likely to recur in 100 years - the design standard most frequently adopted for structural flood control works.

Consequently, actions taken following major floods have frequently increased the hazard and hence the risks in the long term. Actions such as flood plain zoning so that the hazard may be identified and avoided and land acquired to remove dwellings at risk, are not widely used or indeed supported.

Flood plain occupancy is reinforced also by the institutional responses such as flood relief, and up until recently, a government administered damages fund. These responses occur despite the fact that many urban communities have been affected by floods on more than one or two occasions.

The presence of stopbanks creates a feeling of confidence in urban communities protected by them, which is increased with the length of time between flood events. The reality that a bigger than designed for event will occur does not seem to register. (Ericksen 1985 in press)

In addition Ericksen's work also indicates that there is an erroneous understanding amongst flood plain dwellers of the "return period" concept, as well as a different perception from scientists or technical experts of the main characteristics of the flood event.

There is not a straightforward relationship between information available on floods and the public assimilation and action on it. This was graphically illustrated during the Southland 1984 floods, where flood information had been provided but not made readily available to the public and when finally it was, it was not acted on to avert damage.

These perceptions markedly affect several aspects of flood plain management:

- the degree of population and assets at risk
- behaviour in response to floods
- the understanding and acceptance of flood information
- the choice of flood-loss reduction measures.

It is the linking together of the physical characteristics of flooding with the perceived characteristics that is the challenge for flood plain managers.

This paper will outline what New Zealand has done in the past and where is is going in the future, based on experience in New Zealand and other countries.

3. HISTORICAL DEVELOPMENT OF FLOOD PLAIN MANAGEMENT

3.1 Flood Control

The effects of massive deforestation carried out largely during the late 19th century were beginning to be felt by the 1920s. The following results were reported at the time;

> "greatly increased sheet washing of soils; great increase in the number of slips; slumps and rain gullies; aggradation of stream beds, wandering of streams over valley bottoms, lateral erosion of river-banks; burying of culverts and bridges; ... and more severe and frequent floods." Henderson, J. and Ongley, M. J., 1920.

As the national population increased, more farms and towns developed under the protection of stopbanks, and more roads and railway bridges were located alongside and across rivers.

Despite strong evidence of the link between catchment condition and flooding, all investigations into flooding problems up to the 1930s assumed that flood control was simply a question of works on rivers. It was not until disastrous floods on the east coast of the North Island in 1938, with loss of life and economic livelihood, that the need for catchment control linking soil conservation and flood control was recognised.

After much debate, the 1941 Soil Conservation and Rivers Control Act was passed, laying the foundations of flood prone land management for New Zealand. This Act has remained largely unaltered to the present time. Its essential elements provide a means by which soil conservation and rivers control works can be carried out on an individual farm or catchment wide basis with a funding system based on contributions by both government and those benefitting from the scheme; a successful central/local government linkage between NWASCA and catchment authorities.

Wide powers were given to the regional catchment authorities. These include power to construct, reconstruct, alter, repair and maintain works, including breaching any stopbank, controlling or regulating the flow of water towards, into and from watercourses, preventing or lessening any likelihood of the overflow or breaking of riverbanks, and preventing or lessening any such damage, preventing or lessening erosion and its likelihood, and promoting soil conservation.

Structural works have taken pride of place as the prime means of flood control since the 1941 legislation. Meanwhile, assets have blossomed behind the stopbanks.

In the 20 years to 1962, the government subsidy of flood control works averaged around 70% of the total cost. Grant rates are graduated according to their community benefit and comprehensiveness.

The most common type of flood control works in New Zealand are stopbanking, flood detention and channel improvements. Since 1945, 115 major flood control and drainage works have been funded which represents works on all the major rivers with flooding problems in New Zealand. In addition, many hundreds of isolated works have been completed throughout the country. Refer figures 8 and 9.

Twenty years of works to 1973 give the following statistics:

8 800 km	channel improvements	43	flood detention dams
480 km	diversion channel cuts	983	major groynes
300 km	river mouth openings	1 500	flood gates
30 400 km	channel maintenance	506	drop structures
5 600 km	bank protection	177	water level recorders
3 040 km	stopbanks	139	pumping stations
940 km	fencing	4 190	culverts and bridges.

Refer Ericksen 1976 from Parliamentary records, and Ericksen 1985 (in press).

It should be noted that since the mid 1970s there has been a reduced level of government spending due to the state of the economy which has slowed the pace of river control schemes. In addition, several of the major works in the last decade have been upgradings of previous schemes, due to increased settlement behind the stopbanks which became threatened by floods larger than those designed for. The consequence of this type of action is only now being witnessed, for example, the disastrous Invercargill flood of January 1984, where stopbanks were breached and parts of the city flooded.

The government costs of control and related works since the 1941 Soil Conservation and Rivers Control Act represents over 1 billion (1984 $ value) or about $20 million per year. If one adds the local contributions to this of about $1 for every $3 of government grant, which represents about $250 million (1984 $ value), the total cost to New Zealand exceeds $1.33 billion about $38 million per year.

These figures do not include the considerable amount of unsubsidised work carried out by other agencies. Neither do they include the costs to the nation of failed or exceeded flood control works. Restoration of flood damaged works between 1963 and 1972 alone amounted to $33 million (1984 $ value). (Cost figures from Ericksen 1985 in press.)

The cost to the nation from flooding are in fact much greater than these figures show. Flood relief from central government, costs to local authorities in repair works, costs to the insurance industry and the New Zealand Earthquake and War Damages Fund, and of course the social costs to those unfortunate enough to be flooded.

That large benefits accrue to the nation from adopting flood control schemes and from locating populations in the flood plain is certain from economic analyses of several schemes. However there are no estimates of the overall benefits published, thus the savings from flood control works that have prevented losses during past flooding remain unknown.

Some of these wider costs began to be recognised in legislation during the 1960s, but it was not until the Town and Country Planning Act 1977 and the Local Government Amendment Act 1979 that specific provisions were made for the avoidance of situations where flooding and other hazards occur.

3.2 The Role of Planning

The Town and Country Planning Act requires local authorities to prepare district schemes which promote the health, safety, convenence and general welfare of its inhabitants. In particular, the schemes must provide against land being used for unsuitable purposes including landuses in flood prone areas.

Under the Local Government Act local authorities must refuse subdivisions and buildings in areas subject to flooding.

Over the years both pieces of legislation have been amended to generally strengthen the provisions as they relate to flood prone areas. However by 1971 (Ericksen 1976) few local authorities had adopted measures in their planning schemes to avoid damage from flooding. Perceptions were directed to tangible structural works; there was little information which quantified the flood hazard sufficiently for adoption by planners who had a weak image of the problem. Many engineers did the planning for local authorities and did not perceive development control as an option. Continued use of the flood plain ensured non-interference in private property rights and as flood control works were subsidised, there was an incentive to a continuance of flood plain occupance and reliance on structural works.

In addition, the 1959 amendment to the Soil Conservation and Rivers Control Act had given catchment authorities wide powers to control land use practices which were likely to cause flooding, erosion and sedimentation.

Land use restrictions however have been met with resistance from local authorities and developers, especially in areas where buildings already existed and owners wished to alter or replace a dwelling. Local authorities found themselves liable in event of allowing development in such areas, if subsequent damage ensued.

Further amendments are currently being discussed which will provide local authorities with some flexibility for additions and alterations in floodable areas, yet at the same time discourage long term development in those areas.

3.3 Planning and Flood Control Linked

As all these legislative changes were taking place the NWASCA began to question the support of flood control schemes, especially the upgrading of schemes which had proven inadequate.

In 1982 a requirement was introduced which limited financial assistance to all schemes protecting urban areas from flooding. Catchment authorities were actively encouraged to take all possible steps to see that new developments were planned to include adequate safeguards against flooding. And so started the catchment authorities' active involvement in the district planning scheme process - providing flood hazard information, recommending raised floor levels, suggesting that particular land uses locate well away from the flood plain, that secondary flow paths be provided, recommending other more suitable land uses on flood plains and following these recommendations through to the appeal stage if necessary.

NWASCA was beginning to see that there were limits on river control works as a solution to flood problems. Other complementary actions were also taken. Urban land development guidelines were prepared to provide assistance to developers and local authorities to avoid flooding and siltation from occurring during the construction stages.

Provisos were placed on the grant of funds for flood protection schemes. Appropriate provision had to be made in district planning schemes for all approved measures to ensure that design standards and planning goals were safeguarded. Also, factors taken into account in setting the rate of grant for any works included:

- the effectiveness of local planning in the past as a contributory cause of present problems and as evidenced by the measures now being adopted to ensure the situation is not repeated in future;
- the extent to which structural controls (eg, stopbanks) are complemented by non-structural measures (eg, flood plain zoning, building controls);
- economic, environmental and social considerations.

In 1982, some 15 years after it had stated that the numerous floods could only be controlled by more river control works, the NWASCA in its natural hazards policy stated:

> "The National Authority's long experience with ... natural hazards has led it to the belief that avoidance is frequently the best strategy. In many cases the cost of works to protect developed land for urban use is prohibitive. Development in hazardous situations poses a risk to life and creates the potential for inconvenience and loss of investment." (our emphasis)

To avoid or mitigate the damage to assets caused by natural hazards and to encourage land uses compatible with the identified hazards, NWASCA gave nationwide promotion to the following policies.

- General identification of lands subject to hazards (such as flooding) and the promotion of the inclusion of such information in regional planning schemes.
- The promotion of the inclusion in district planning schemes of maps and information describing the location, type and extent of each hazard.
- The promotion of the inclusion in regional and district planning schemes of provisions for land uses compatible with the type and extent of the hazard identified.
- The continued development of standardised techniques for identifying and quantifying flood and other hazards.
- To acquaint local authorities with NWASCA policies during planning scheme reviews and to remind them of their responsibilities under the planning and local government legislation.

The extent to which local authorities have taken up their responsibilities is indicated in the following figures by Copinga (1982). In a survey of 103 floodable urban places, less than ½ had mapped flood hazard zones. Of these, 22 places reported that they had done so in response to planning and local government legislation. Evidence of floor levels having been raised is found in most floodable towns, but mostly as individual decisions, rather than by regulation.

Increasingly, minimum floor levels are being imposed in floodable towns. In many cases, low lying areas have been infilled to above flood levels. However, like most design levels, they may be overtopped or subjected to failure in some way and hence a false sense of security develops in a similar way as for flood control.

One of the reasons why some local authorities have been reluctant to use flood information in planning schemes relates to the manner in which the information is presented. Return periods mean very little to lay people and are often misunderstood. Hence, councils are reluctant to "cry wolf" over flooding, especially if experience of flooding has become hazy with time.

Another reason for concern over the use of flood hazard information relates to a misunderstanding by some local authorities of their liability. Some believe that if flood information is provided they become liable for compensation for loss of property values if flooding does not occur where indicated. Others believe they are liable for damages if areas outside those indicated are flooded subsequently. This is not the case, as the duty of care in New Zealand law must be exercised on the best available information at any point in time. However this does mean that flood information must be continuously updated.

3.4 Flood Warning

An integral part of flood control systems in New Zealand is a comprehensive flood forecasting and warning system which is currently being put into place. The system involves data sensors, a telemetry system, a database archiving management system and a suite of data handling and application programmes. Government through NWASCA gives grants up to 60% for the purchase and installation of flood forecasting and warning systems to the regional catchment authorities who issue flood warnings to other organisations and the public.

Catchment authorities are now preparing flood manuals which set out the procedures to be followed during a flood event. For an example, refer to Koutsos, 1986.

Despite the fact that these flood control and planning links have been made, flood damages still occur. It has become increasingly apparent that there is a need to coordinate the activities of the various agencies involved in any way with flood plain management or which increase the flood damages through its decisions.

4. OTHER AGENCY INVOLVEMENT

Many agencies in addition to NWASCA and regional catchment authorities are involved in aspects of the flood problem, often unwillingly increasing the damage potential in the long term.

Often during a flood, a civil emergency is declared and central and local government funding is made available through the Civil Defence Emergency Fund.

At the same time the public at large frequently contribute to relief funds for locally affected communities. In the last four years $2.2 million has been given through such funds which are commonly matched dollar for dollar by government in addition to the civil defence monies. Mostly, these funds have gone to uninsured or underinsured people for both rural and urban losses and there has been a failure to consider long term protection when administering immediate relief and recovery.

These funds have enabled reinvestment in the flood plain to take place, thus increasing potential damages. In addition, local erroneous perceptions of flooding have been reinforced by such relief.

Up until 1984 an Earthquake and War Damages Fund provided cover for flood damages to property if such properties were covered by normal fire insurance.

The philosophy behind this system was twofold:

- Damage would be random and unpredictable so the levy was uniform throughout the country.
- Costs would be so large that the insurance industry could not be expected to provide affordable cover.

However, flood frequency and magnitude is statistically predictable and damages can in many cases be avoided, or mitigated. As at July 1985 such cover ceased and the private industry now covers it. The Earthquake and War Damage Fund had become a relief fund, and eroded significantly as a response to the increasing number of floods and the size of the damages. Uniform application of the fund and under insurance led to this erosion. In one flood alone, in Wellington in December 1976, the Commission's share of the total amount of assessed flood loss was almost 50%, which represented $19 million ($1984 value).

Insurance Council of New Zealand estimates in 1977 and 1978 noted that lack of insurance could conceivably exceed 30% of floodable properties. This imposed an increasingly heavy burden on voluntary public and government relief funds.

The implications of the change to private industry can only be speculative but two effects could result. Premiums will rise considerably, based on actual risk, and many properties will become uninsurable because the industry will not accept the obvious risks.

If properties became uninsurable or are not insured, added pressure will also be placed on relief funds to compensate. However, it might conceivably put pressures on local bodies to use flood information in their planning so that uneconomic or unwise reinvestment in flood plains is not prepetuated.

Other agencies with an involvement in flood plains that have over the years, increased damage potential include those government and non-government lending institutions and those agencies disposing of crown property.

Each of the agencies have for many years pursued their objectives out of context with the overall flood problem with the result that often actions conflict and the flood problem is exacerbated.

In New Zealand NWASCA has an overview function which has historically been directed at providing structural works as defences against flood waters, ie modifying the flood event with some attention to modifying flood loss effects through relief and insurance funds.

What is now being addressed in addition to the traditional approach is adjusting people to floods by modifying damage susceptibility. To do this effectively NWASCA and the regional catchment authorities must assume a wider coordinating role and provide the information base on floods that can be interpreted by flood plain dwellers and managers so that a wider range of adjustment choices may be made.

5. A UNIFIED FLOOD-LOSS REDUCTION PROGRAMME

To effect a coordinating role NWASCA adopted a unified approach to the reduction of flood losses in November 1984.

The elements of this policy include:

- preparation of flood plain management plans which identify flood hazards and set out a broad range of remedial measures including using mapping, and non-structural options.
- the communication of the flood risk in a form understandable by

the public.
- an assessment of flood losses according to which sector pays.
- an economic evaluation of alternative adjustments to floods.
- research into flood proofing measures.
- preparation of technical and publicity information for public education on flooding.
- coordination of all agencies involvement in flood plain management.

The key to the effective implementation of this programme is the promotion of a "neutral" perspective, ie, a policy that does not favour one adjustment option over another, as the current river control policy does, by favouring structural works, alone. To do this, neutral legislation, neutral financial incentives and neutral techniques for evaluation of the various adjustments, are required.

If funds are only available from central government for one type of adjustment option then that one will be favoured every time despite the costs to the nation in reinforcing damage potential.

The proposed policy change is to extend catchment works grant assistance to the area of modifying damage susceptibility, eg, to land acquisition, permanent flood proofing of buildings and the relocation of buildings. Availability of grant assistance would be dependent on a full consideration in neutral terms of all possible flood loss adjustments. Consideration should be in terms of specified social aims of economic efficiency, social equity, avoidance of social disruption, human health, regional well being and environmental protection and enhancement.

6. WHAT WE HAVE LEARNT FROM OTHERS

New Zealand is very aware of the body of experience worldwide with a unified approach to flood plain management and is eager not to repeat any mistakes made by others or indeed to reinvent the wheel. As a result, considerable effort has been made to review the literature and discuss common problems with other nations.

One overriding aspect which seems to us as the most important, is the need for a coordinating agency to provide a context within which agencies with very specific objectives for aspects of the "flood problem" can operate to mutual advantage. New Zealand is well placed to provide this, through NWASCA and through being a small nation with well developed central/local agency institutional linkages. Such coordination provides the framework within which financial uncertainties can be tied to all aspects of a flood loss reduction programme.

The universal problem of the isolation of technical solutions from their implementation can also be overcome in New Zealand through local input into flood plain management - our local government structures provide the means to do this, the will, can be motivated through a neutral perspective to flood-loss reduction policies.

On the detail, we have learnt from others the need for flood hazard maps to be an integral part of a flood plain management approach rather than an end in themselves so that their effect and a

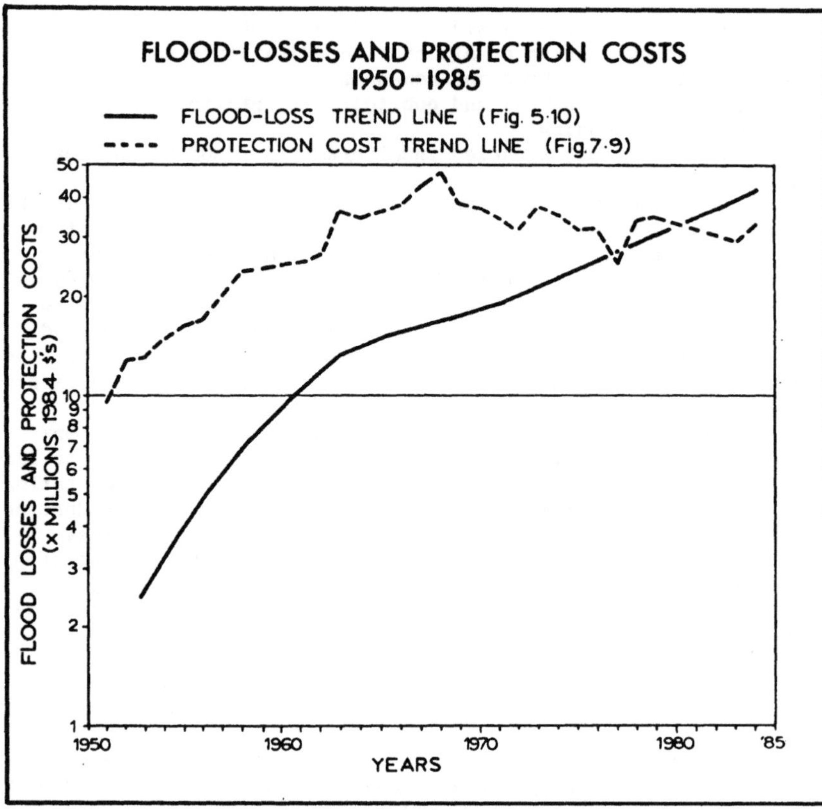

Figure 1. Comparison of flood losses and flood protection costs, 1950-1985 (from Ericksen, N. J., 1985, in press).

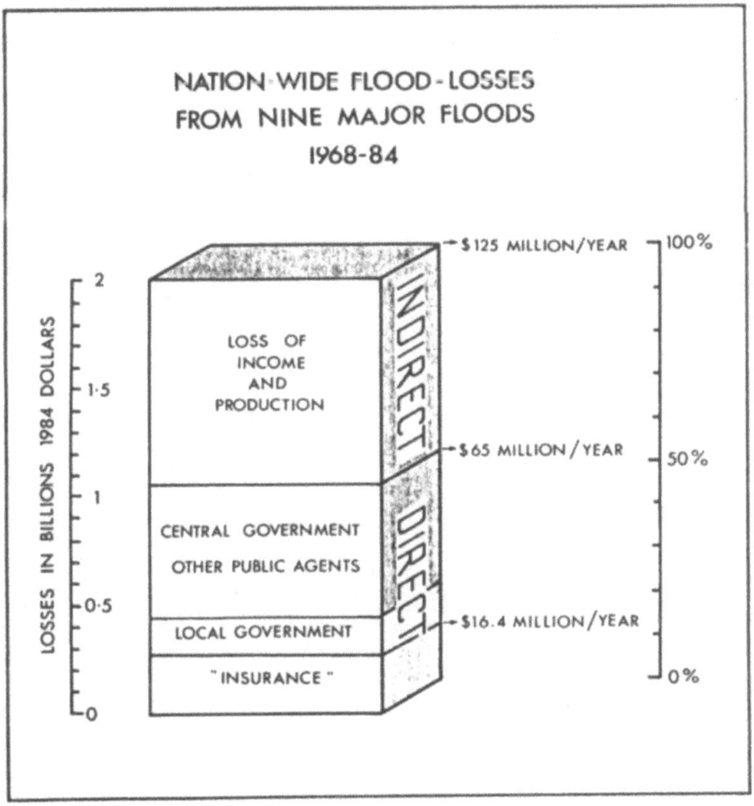

Figure 2. Apportionment of costs resulting from nine major floods, 1968-1984 (from Ericksen, N. J., 1985, in press).

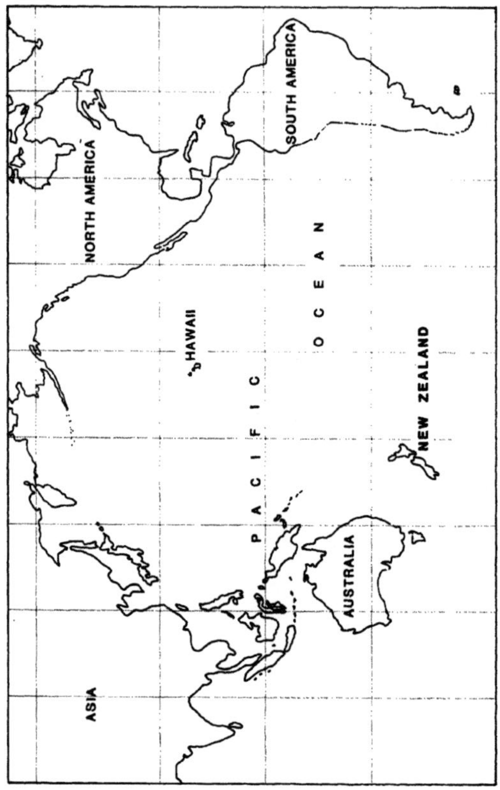

NEW ZEALAND IN THE PACIFIC

Figure 3. Locality Map

Figure 4. Damage by debris movement resulting from the 1985 Thames/Coromandel floods. (NZ Herald).

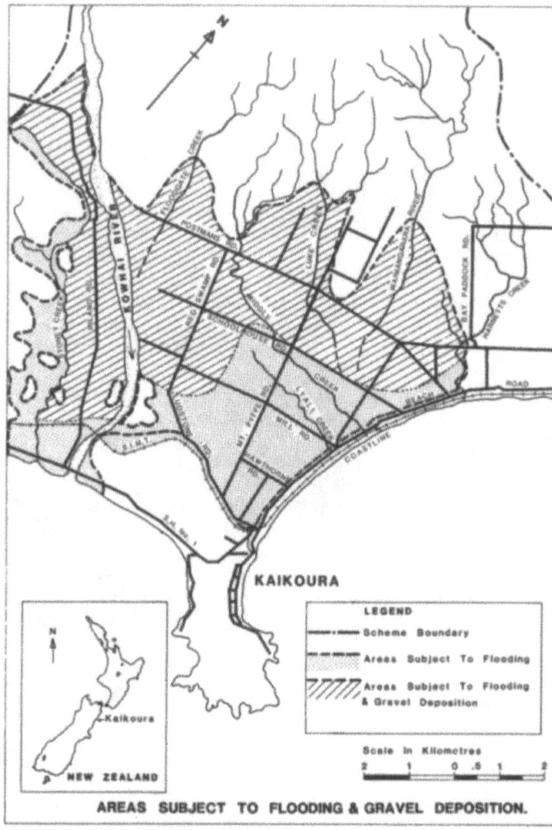

Figure 5. Intensely farmed areas subject to damage by flooding and gravel deposition associated with alluvial fan topography.

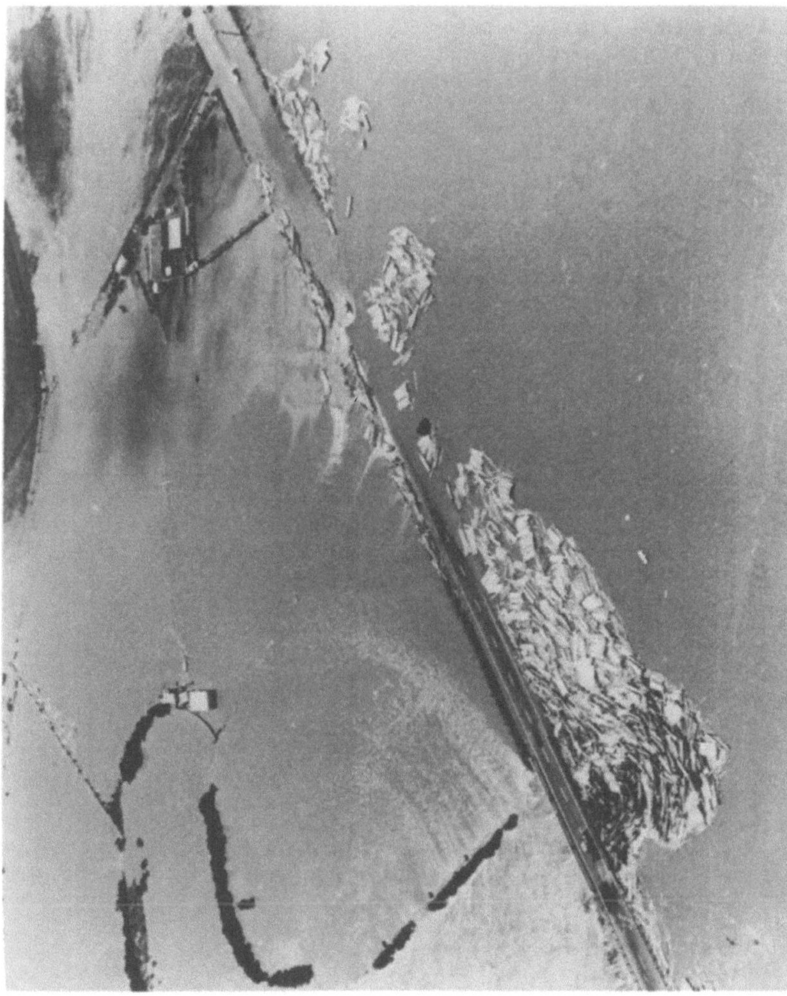

Figure 6. Debris blockage of bridge waterway caused by the 1984 Southland floods. (David Wethey, Otago Daily Times).

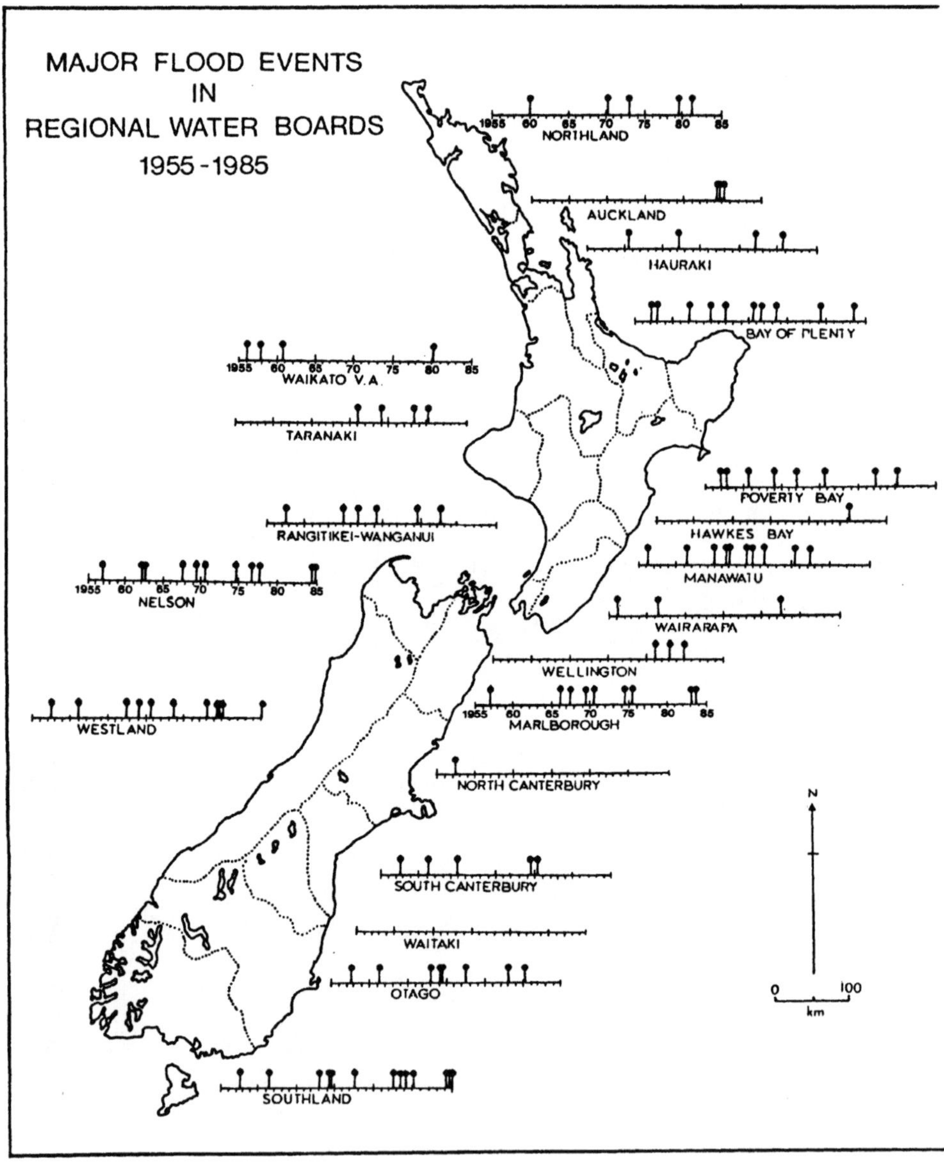

Figure 7. Incidence of major flood events in New Zealand since 1955 (from Ericksen, N. J., 1985, in press).

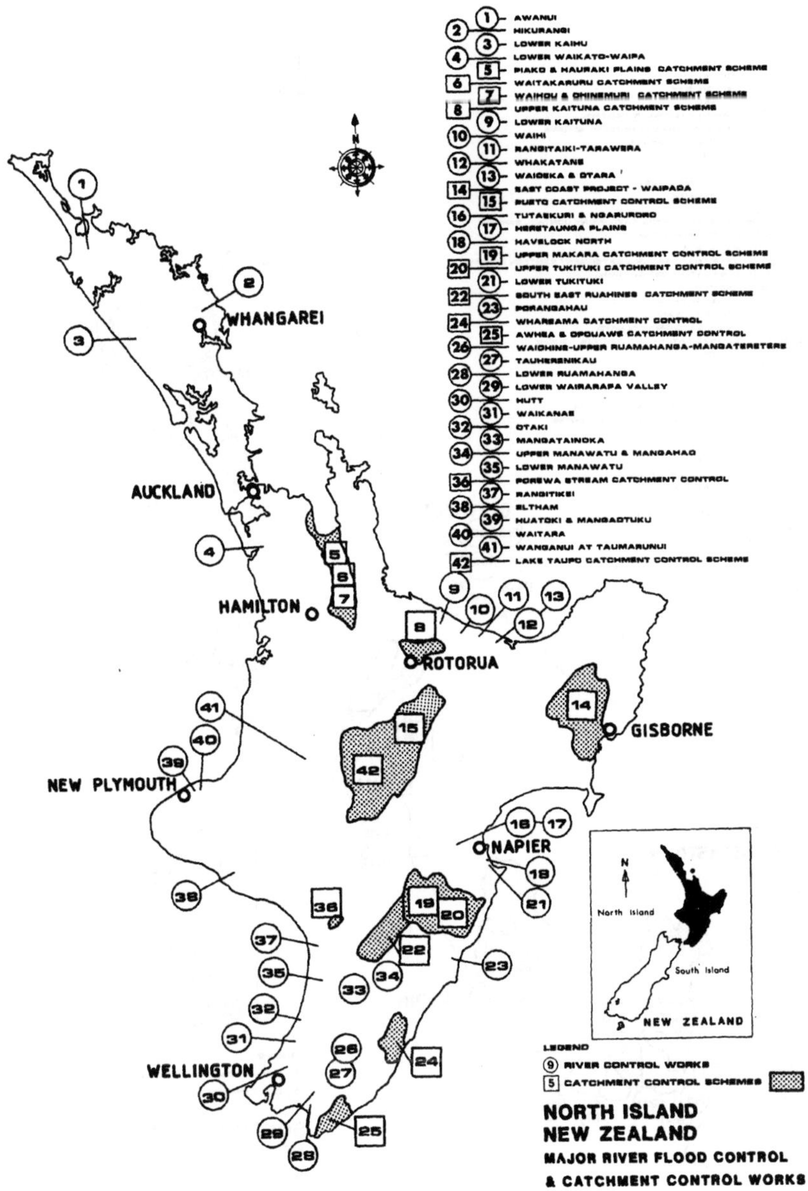

Figure 8. New Zealand, North Island – Major river control, flood control and Catchment Control Schemes since 1941.

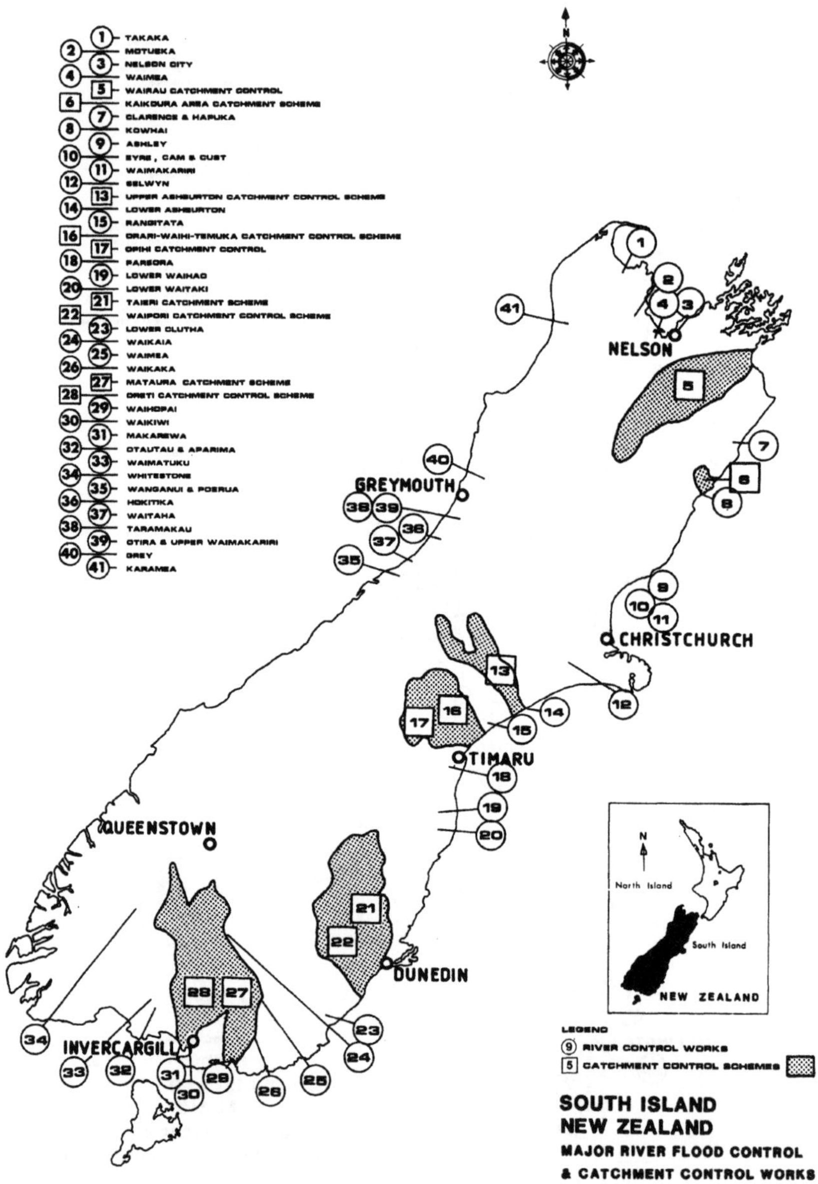

Figure 9. New Zealand, South Island - Major river control, flood control and Catchment Control Schemes since 1941.

range of solutions can be adequately interpreted by the affected public. Also, the need to consider a procedure for future hazard identification and mitigation, simultaneously with immediate recovery actions.

These elements will be integrated with the unified flood loss reduction programme an essential part of which will be its regular review and update as new "problems" arise and new approaches are found.

It is interesting to note that the seeds of our new policy were first sown over 10 years ago, based on Gilbert White's work in USA. Change is slow when there is an historical investment in a particular approach by all sections of the community. It is social inequity which very often triggers a change, and social equity which consolidates it.

ACKNOWLEDGEMENTS

The authors wish to acknowledge the contribution made by Dr Neil Ericksen of the Geography Department of Waikato University, Hamilton, New Zealand to the history of the New Zealand Flood Problem, and the recommendations he has made for the future, and the advice and assistance of Edy Goldberg, Technical Services Manager, Water and Soil Directorate, Ministry of Works and Development, Wellington, New Zealand, a member of the team involved with the new policy directions.

REFERENCES

Copinga, O. 1982. Flood Hazards and Landuse Management in New Zealand. University of Waikato, Hamilton, Department of Geography, Unpublished B. Soc. Sc. Research Paper.

Ericksen, N.J. 1976. "The Role of the Planner in Floodplain Management - An Overview". Town Planning Quarterly (New Zealand), Vol. 43, pp. 13-24.

Ericksen, N.J. 1985 (in press). Urban Flood Problems - New Zealand Flood-loss Reduction Policies - Past Practice and Future Options, Water and Soil Miscellaneous Publication for National Water and Soil Conservation Authority, Wellington.

Henderson, J. and Ongley, M.J. 1920 in McCaskill, L. W. Hold This Land - A History of Soil Conservation in New Zealand. A H and A W Reed, Wellington, New Zealand, pp. 11.

Koutsos, P. 1986. Flood Manual. Hawke's Bay Catchment Board, Napier, New Zealand. (Draft issue)

Wright, A.F. and Koutsos, P. 1969. River Drainage and Erosion Control for Kaikoura, J.A. Hayward (Editor). Proceedings of a

Symposium on Watershed Management in Water Resources Development. Vol. 2, pp. 109-152. Lincoln Papers in Water Resources, Lincoln College, Canterbury, New Zealand.

A STUDY FOR REAL TIME OPERATION OF FOUR FLOOD RESERVOIRS

André **VAN DER BEKEN**, Professor in Hydrology
Laboratory of Hydrology
Georges Louis **VANDEWIELE**, Professor in Statistics
Centre for Statistics and Operations Research
Jan **MARIEN**, Lecturer in Statistics
Centre for Statistics and Operations Research
Ivo **TERRENS**, Staff Member
Laboratory of Hydrology
Guy **DE SCHRIJVER**, Staff Member
Laboratory of Hydrology
Vrije Universiteit Brussel, Brussels, Belgium

ABSTRACT. This paper introduces a second project on real time flood control in Belgium. The studies developed for the automatic operation of the Dijle flood reservoir have been extended for a system of four flood reservoirs in the Demer basin, adjacent to the Dijle basin. The goal of the study is to design a fully automatic operation scheme of all weirs of the four reservoirs by using real time forecasts of the different parts of the flow system. But the on-line calculations and tasks of the process-computer in the future real time system can only be formulated in detail when an acceptable strategy for operating these reservoirs has been developed. Evaluation of this control strategy has been made by an extensive computer simulation study of the complete flood protection scheme.

1. INTRODUCTION

The studies developed for the automatic operation of the Dijle flood reservoir, discussed at the Oxford Symposium on Hydrological Forecasting by Van der Beken et al (1980), Mariën (1980) and Marivoet and Vandewiele (1980), and at the Hamburg Symposium by Bauwens et al (1983), have been extended in the framework of a new project concerning four flood reservoirs in the river Demer basin, situated adjacent to the river Dijle basin.

In this paper some new facets of hydrological modelling developed during the study are reported. The methodology introduced for evaluating appropriate reservoir control policies, by using extensive computer simulations, applies a point rainfall generator (Vandewiele and Mariën, 1986) and a flood reservoir design theory (Mariën, 1984).

V. P. Singh (ed.), Flood Hydrology, 369–378.
© 1987 by D. Reidel Publishing Company.

The river Demer with a total catchment of 2 200 km^2 is a tributary of the river Dijle (3 420 km^2). Both are part of the Scheldt river basin (23 260 km^2), one of the two main river systems in Belgium. Fairly large cities are drained by the two rivers, such as Diest and Aarschot by the Demer and Wavre and Leuven by the Dijle river. Downstream of the confluence, the Dijle crosses the city of Mechelen where tidal effects are observed. The Demer river has a discharge capacity of 95 m^3/s downstream the city of Aarschot. Between Aarschot and Diest the present maximum capacity is 55 m^3/s but an increase of capacity up to 95 m^3/s has been planned. Yet this will not prevent flooding in the upstream part of the Demer basin. Both Dijle and Demer have always been subjected to flooding. This is partially due to the natural topographic situation and flood control therefore exists since the Middle Ages. Nowadays human action and urbanization, especially in the flood plains, made things worse.

2. FYSIOGRAPHY OF THE DEMER BASIN UPSTREAM OF DIEST

In the basin three regions with different soil type can be recognized (Fig. 1). In the south (920 km^2) the top layer consists of loam of several meters of thickness. Wet soils only occur in the valleys and few problems arise. The northern part (500 km^2) is sandy with 25 % of the soils classified as wet or very wet and another 20 % frequently suffers from excess water. The central region is characterised by a sandy loam layer of small thickness, often on impermeable substrate. Most part is low land and the rivers are diked, causing poor drainage of the secondary drainage system. More than 70 % is said to be wet or very wet. Along the Demer river, the altitude is 19 to 20 m above mean sea level, the upper parts in the north have altitudes from 70 to 100 m and in the very south from 130 to 150 m. A computer model calculating groundwater flow from rainfall, hydrogeological and topographical conditions, illustrates the situation quite well : soils suffer from water at places where the piezometric level rises above topographical level (so-called discharge or source areas) (De Smedt and Bronders, 1985).

3. THE FLOOD PROBLEMS

The 100-year return flood for the Demer basin upstream of the city of Diest (1 920 km^2) has been estimated at 200 m^3/s. The permissible flow at Diest, however, is presently only 55 m^3/s and will be maximum 95 m^3/s in the future. The limited discharge capacity at Diest causes back water effects in the Demer and its tributaries, resulting in inundations and increased lack of drainage.

A flood relief programme for the improvement of the water management in the basin has been planned. Four main flood reservoirs were designed (Fig. 1). Table 1 lists some of their characteristics. The design of two of them is still tentative : the final design will be based on an intensive ecological study. The objective of these reservoirs is the limitation of the total flow downstream at Diest to the critical level of 55 m^3/s at present and 95 m^3/s in the future.

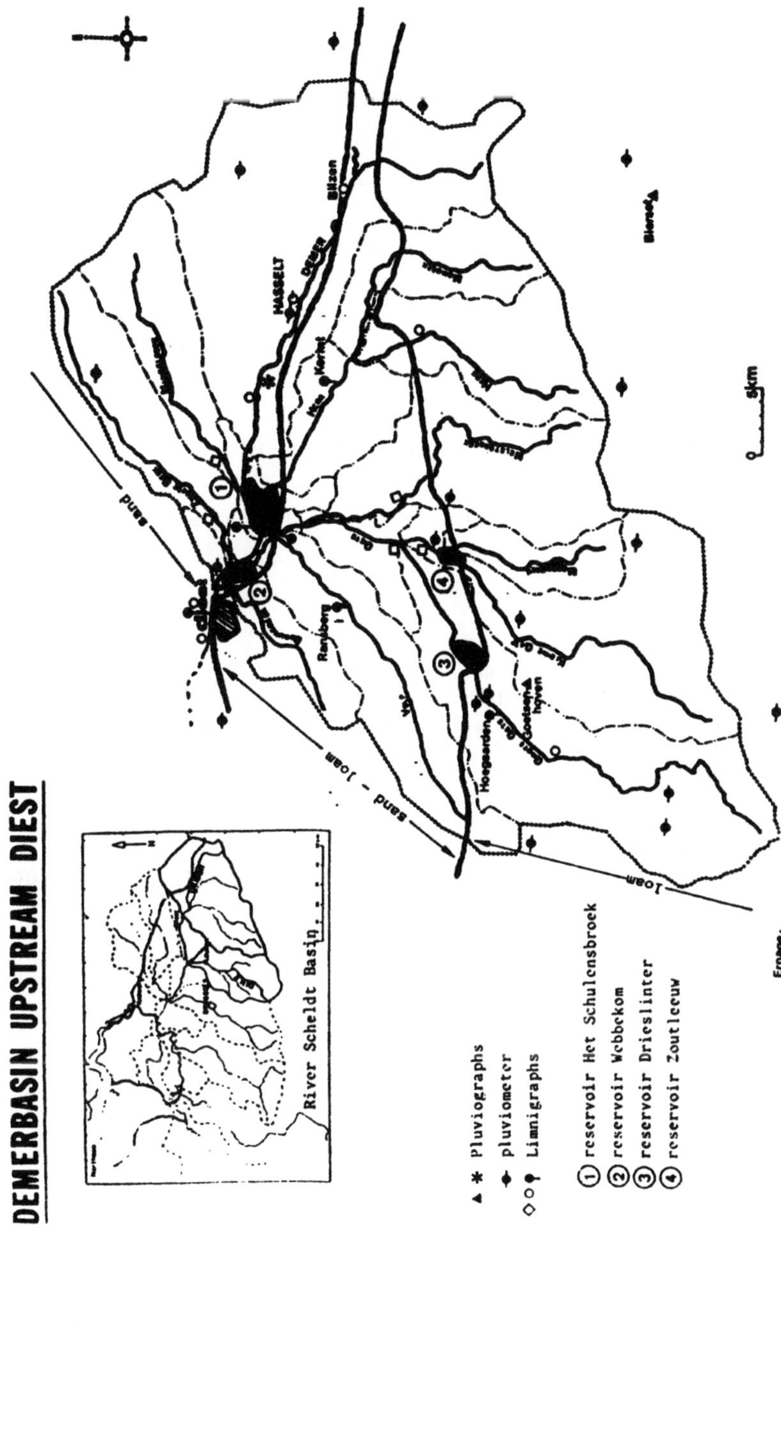

Fig. 1 : The projected flood reservoirs with measurement stations

TABLE 1. Main characteristics of the reservoirs

Flood reservoir	Storage capacity ($10^6 m^3$)	Area of reservoir (km^2)	Controlled river basin (km^2)
1. Het Schulensbroek			
- Inner Part	6.3	1.50	Demer and Mangelbeek (497)
- Outer part	4.6	1.90	Herk (309)
2. Webbekom	3.5	1.44	Velp and Begijnebeek (198)
3. Drieslinter*	4.4		Grote Gete (227)
- Inner part		1.12	
- Outer part		0.72	
4. Zoutleeuw*	2.5	0.90	Kleine Gete(207)
Total	21.3	7.58	1438 km^2

* Tentative design

4. THE STUDY PROGRAMME

The goal of the study, sponsored by the Ministry of the Flemish Region, Department of Agricultural Hydraulics, is to design a fully automatic operation scheme of all weirs of the four reservoirs, which uses real time forecasts of the different parts of the flow system. This should increase the safety of the system and/or reduce the need for filling certain compartments of the reservoirs, which are to be maintained for agriculture or kept in natural conditions. On the other hand, the proposed automatic system should be as simple as possible in order to improve reliability and to balance benefits and costs.

The full study programme contains the following topics :

(1) processing of existing data ;
(2) collection of additional field data ;
(3) formulation and off-line calibration of several models : (a) rainfall-runoff models, (b) flood routing models, (c) operational strategy for the system of reservoirs ;
(4) design and implementation of intelligent measurement stations for future use in the real time system ;
(5) design of the on-line transmission system and process-computer configuration ;
(6) extensive simulation and reliability study and formulation of follow-up requirements, including rules for manual operation.

5. PROCESSING OF DATA

The rainfall-runoff model used in the study was calibrated off-line with rainfall and runoff data from the Dijle basin and different sub-basins in the Demer area. For the Demer area, short-duration rainfall data (10 min) are only available at four stations of the Royal Meteorological Institute , of which only one station is situated in the catchment itself. Evaluation of the areal rainfall in the different sub-basins was therefore improved by using also daily stations. Hourly runoff data are available at stations with water level recorders and good rating curves. Five stations were useful. However, some records were only available for limited periods.

Gradually stations in the basin will be linked in telemetry, even before the reservoirs themselves will be finished completely. Besides a quick access and handling of the data, it will also be possible in a later stage of the study to run parts of the control strategy directly on a mini-computer with data obtained in telemetry (Terrens and Van der Beken, 1986).

6. FORMULATION OF MODELS

6.1. Global Review

The successive on-line calculations and tasks of the process-computer, to be installed at the main reservoir "Het Schulensbroek" in the future real time system, have been formulated in detail after a thorough study of the behaviour of the system under various conditions and the evaluation of an acceptable strategy for operating the reservoirs. With the experience from the previous real time control project on the Dijle river in mind and due to the complex reservoir configuration with predetermined filling priorities, a rather simple control strategy for filling and emptying the compartmentalized reservoirs was formulated. Evaluation of the control strategy is based on extensive simulation studies of the complete flood protection scheme (Fig. 2). After simulating 1000 years (of 6 winter months each) of 2-hour rainfall with a stochastic point rainfall generator (Vandewiele and Mariën, 1986), a rainfall-runoff model was used to obtain flows on a 2-hour time base (this time base is small enough to study flood phenomena in the Demer basin and was chosen to limit computer execution time). Using these flows as reservoir inflows the effects of the developed control strategy on flooding could be calculated.

6.2. The Off-Line Rainfall-Runoff Model

To study runoff in a quantitative way a data-set was built up containing all important characteristics (such as areal precipitation, homogeneity of the rainfall, rainfall intensity, direct runoff, mean baseflow, runoff coefficient, humidity index, temperature,...) of periods of high rainfall, so-called episodes, in several comparable basins in the Dijle and Demer area. An episode is

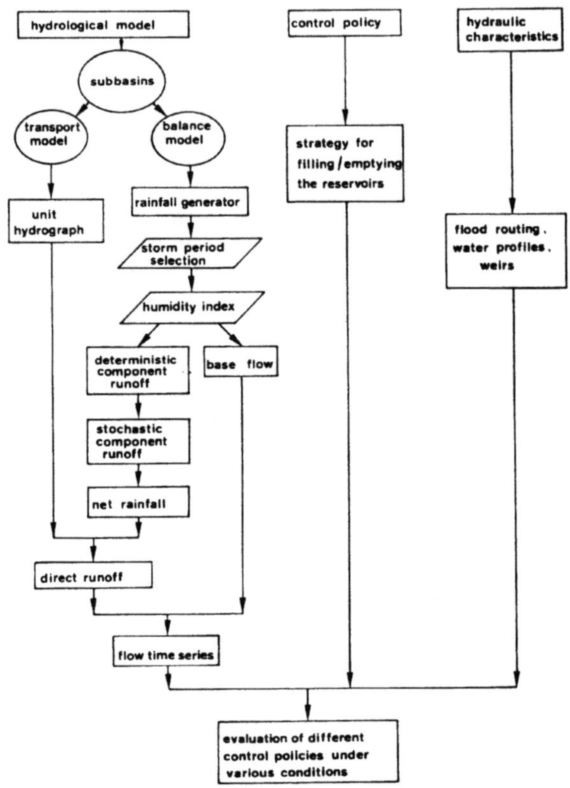

Fig. 2 : Scheme of the complete flood protection simulation study

a rainfall storm which is well isolated from other storms, in such a way that direct runoff can be attributed exclusively to the rainfall during such an episode. More than 200 storm episodes were selected from the available data, divided into two groups according to the region (loam and sand-loam). The selection was mainly based on the amount of rainfall instead of the peaks of flow because the simulation study uses rainfall as input. The analysis was done for winter time only because temperature, not simulated in this study, is not critical in this period. Winter time is also the most important season for flooding.

With two main observed quantities, the runoff coefficient R_o defined as the proportion of total direct runoff in the episode to total areal rainfall, and a mean humidity index V_g over the episode, a probabilistic runoff coefficient model was derived. The humidity index V(t) at time t was defined in a way similar to Marivoet and Vandewiele (1980) :

$$V(t) = a\ V(t-1) + P(t) \qquad \text{(in mm)}$$

where $P(t)$ = areal rainfall from time (t-1) to t
a = time constant (< 1).

Simulated runoff coefficients $\underline{Ro}$ consisted of two parts : a deterministic component, obtained by regression analysis of observed R_o versus V_g, and a stochastic component representing the variability of the observed runoff coefficients, including a correlation between adjacent basins :

$$\underline{Ro} = \exp\left[b_1 - b_2 \exp(-b_3 V_g) + \underline{x} \right]$$

where b_1, b_2, b_3 = regression constants.

The simulated stochastic component $\underline{x}$ is assumed to be normally distributed with zero mean and variance s^2 of the regression, independent of V_g. The runoff coefficient is considered as constant in a storm episode itself. Storm episodes are obtained by splitting up the generated rainfall time series.

The Monte-Carlo simulation of rainfall and, in addition, of the stochastic component of the runoff coefficient results in net precipitation which is to be routed by a transport model. For the transport model a linear model (unit hydrograph) was calibrated wherever possible and adapted to the different rivers. A lognormal transfer function was chosen :

$$h(t) = \frac{1}{t\ c_1 \sqrt{2\pi}} .\exp - \left[\frac{(\ln t - c_2)^2}{2\ c_1^2} \right]$$

with $\sum_{t=1}^{c_3} h(t,c_1,c_2) = 1$

and $h(t)$ = ordinate of the unit hydrograph at time t
c_1, c_2 = constants of the model
c_3 = basetime of the unit hydrograph.

Baseflow was calibrated and simulated as a linear function of the mean humidity index V_g.

For flood-routing a pure translation in time in proportion to a mean travel velocity of 1 m/s was accepted.

With some adaptations this concept could then be used for simulation purposes. The ultimate goal of the flow simulations is the evaluation of different control policies for reservoir action.

6.3. The Reservoir Control Strategy

Evaluation of a reservoir control strategy is done in different steps. In a first step the behaviour of the reservoir system is studied in ideal conditions, when perfect knowledge of all future flows is assumed. The effect of this perfect control can be calculated by a new methodology (Mariën, 1982, 1984) and is expressed as the return period of exceeding the critical discharge at Diest.

The real control policy formulated for operating the reservoirs in the future has following features. The control policy keeps the total volumes of water to be stored as small as possible, of course without exceeding the critical discharge downstream at Diest when possible. At the same time the selected priorities for filling and emptying the different reservoirs are fulfilled as much as possible. The fundamental problem in the control action is due to the position of the reservoir "Webbekom" close to Diest : it is not situated on the Demer River and therefore, it is not possible to store water nor from the important uncontrolled tributary Zwarte Beek close to the city, nor from the previous releases from the main reservoir system "Het Schulensbroek" and the adjacent Gete area, both situated ca. 2-3 hours from Diest. So, a rational distribution of water in the different reservoirs is important. The control strategy is based on real time measurements of rainfall and discharges, but when calculating the forecasts of future flows by using an on-line rainfall-runoff model all future rainfall is assumed to be zero. Hence, this simple control strategy uses no rainfall forecasts. To allow for the inaccuracy of the forecasted flows two adaptations were made. In the reservoir nearest to Diest a buffer volume of 20 000 m^3 is kept available with highest emptying priority. The other reservoirs are controlled as much as possible for a critical discharge in Diest of 55 or 95 m^3/s minus α , α being 10 m^3/s. This parameter α represents in fact the risk factor of exceeding the critical discharge in Diest by releasing water from the reservoir "Het Schulensbroek". The results of this strategy, also expressed as a return period of exceeding the critical discharge downstream, compared with those of the ideal control policy, reveals the good quality of the proposed strategy (Table 2).

TABLE 2. Some return periods from the simulation studies

Situation		Return period (years)	
Critical discharge in Diest (m^3/s)	Critical discharge at conjunction Gete-Demer (1) (m^3/s)	Perfect control strategy	Proposed control strategy
95	25	500	500
95(2)	45(2)	500	333
55(3)	25(3)	100	63
55	45	40	34

(1) design of the Gete reservoirs is still tentative
(2) original proposal
(3) actual situation

6.4. Hydraulic Constraints of the Reservoirs

Up to this point the control strategy was evaluated without taking into account the hydraulic constraints of the reservoirs (i.e. the characteristics of inlets and outlets). Ideally, these constraints should normally not occur in a well-developed technical design. In fact, the implementation of these boundary conditions had no influence on the return period, but the duration of storage as well as the amounts stored in the reservoirs increased considerably. This is for the most part due to restrictive outlet discharges with decreasing reservoir levels.

7. CONCLUSION

The case study, briefly introduced in this paper, is the second project on real time flood control in Belgium. The project has been finalized provisionally with a report proposing a control strategy for two of the four planned reservoirs. The construction of the reservoirs on the river Gete are not yet decided. Furthermore, the critical discharge at Diest is 55 m^3/s at the present time and will become 95 m^3/s in the future. The configuration of the flood reservoirs seems to be sufficiently well-balanced, hence complex real time forecasts can be omitted in the model formulations and control policy. The control policy is a result of flow measuring and flow forecastings schemes and aims to optimize storages and releases in both reservoirs with a minimal risk of exceedance of the critical discharge at Diest. Moreover, requested filling and emptying priorities of the compartmentalized reservoirs have been taken into consideration. The study gives an example of a real time operation system designed by means of computer simulation, using hydrology, hydraulics, statistics and operation research methods.

ACKNOWLEDGEMENTS

This study is supported by the Department of Agricultural Hydraulics, Ministry of the Flemish Region. The authors would like to thank R. Awouters, R. Denys, J. François and W. Goderis for their initiative and support. They also wish to thank G. De Graeve and M. Voet from N.V. Belgroma, Consultants, who planned the reservoirs for their cooperation. Many institutions were helpful in providing data, particularly the Royal Meteorological Institute and the Ministry of Public Works.

REFERENCES

Bauwens, W., Mariën, J., Marivoet, J., Vandewiele, G.L. and Van der Beken, A. 1983. 'A mathematical model for Real Time Control of the Reservoir on the River Dijle'. In : Scientific Procedures Applied to the Planning, Design and Management of Water Resources Systems (Proceedings of the Hamburg Symposium), IAHS at IUGG, Publ. **No. 147**, pp. 253-264.

De Smedt, F. and Bronders, J. 1985. 'A Regional Groundwater Flow Model based upon the Variable Source Area Concept'. IWRA 5th World Congress on Water Resources, Brussels, Paper **No. 127**, pp. 511-521.

Mariën, J. 1980. 'Regulation of a Flood Reservoir with the Use of On-Line Forecasts'. In : Hydrological Forecasting (Proceedings of the Oxford Symposium), IAHS, Publ. **No. 129**, pp. 477-483.

Mariën, J. 1980. 'Over de Dimensionering en de Sturing van Hoogwaterreservoirs' (On the design and regulation of flood reservoirs), Ph.D. Thesis, 359 pp., Vrije Universiteit Brussel, Brussels, Belgium.

Mariën, J. 1984. 'Controllability Conditions for Reservoir Flood Control Systems with Application'. Water Resources Research, Vol. **20**, No. 11, pp. 1477-1488.

Marivoet, J.L. and Vandewiele, G.L. 1980. 'A Real Time Rainfall-Runoff Model'. In : Hydrological Forecasting (Proceedings of the Oxford Symposium), IAHS, Publ. **No. 129**, pp. 409-418.

Terrens, I. and Van der Beken, A. 1986. 'Evolution in the Concept and Realization of Hydrological Measurement Networks as Applied to the Real Time Dijle Flood Control Project'. In : Integrated Design of Hydrological Networks. IAHS General Assembly, Budapest, July 1986, Publ. **No. 158**, pp. 361-370.

Van der Beken, A., Vandewiele, G.L., Tiberghien, J., Mariën, J., Marivoet, J. and Meersseman, M. 1980. 'On-Line Flow Forecasting for Automatic Operation of a Flood Reservoir'. In : Hydrological Forecasting (Proceedings of the Oxford Symposium), IAHS, Publ. **No. 129**, pp. 557-564.

Vandewiele, G.L. and Mariën, J. 1986. 'A Point-Rainfall Generator with Internal Storm Structure'. Water Resources Research, Vol. **22**, No. 4, pp. 475-482.

A MULTI-PURPOSE FLOOD CONTROL AND IRRIGATION RESERVOIR ON A TRIBUTARY OF THE ADIGE RIVER (ITALY)

PAOLO MAZZALAI
Assoc. Prof. University of Padova
Via Gradenigo, 6
35100 Padova

LORENZO ECCHER
Senior Hydr. Eng. ELC Electroconsult
Via Chiabrera, 8
20151 Milano

ABSTRACT. The paper describes briefly the works of the multi-purpose Valda reservoir on the Avisio river one of the major tributaries of the Adige river whose watershed is in Italy the third in extension.

At least for one third of its length the Adige river channel is perched and the possibility of overtopping the dike system creates high hazard of flooding to very densely populated areas.

Flood studies, flood control design criteria and flood routing in the reservoir itself and along the Adige river are among the major topics of the paper.

1. GENERALITIES

The watershed of the Adige river (11.954 Km2) is the third in extension in Italy after Po and Tevere river ones, and is mainly formed by four great tributaries which flow symmetrically inward the main stream (Fig. 1).

The total length of the Adige main collector is 410 Km; in its lower part, 140 Km in length, the channel is perched and the river flow is confined by very large diking systems.

The possibility of overtopping and breaching creates a high hazard of flooding to the adjacent areas where factories, homes, railroad, highways and rich rural areas take place.

The flood events experienced in the Adige basin (43 floods are documented from 1896 to 1939) and the consequent damages and losses of life led to the necessity of efficient flood control works.

First of all in the 1939-59's a two-shell shaped, gate controlled, tunnel (9873 m in length) was designed and completed for temporary diversion of the Adige flood waters into the Garda lake; the maximum flow discharge of the tunnel is 500 m3/s. The Garda lake is large

V. P. Singh (ed.), Flood Hydrology, 379–388.
© 1987 by D. Reidel Publishing Company.

enough (370 Km2) to act as a flood-routing reservoir without much rise in the water level.

Such work assured tangible benefits in the downstream region, as one can note comparing the peak flow from similar storm occuring before and after the activation of the tunnel.

The flood during November 1966 was the greatest recorded. The levees broke near the town of Trento and the city was partly submerged. The event occured upstream the Garda tunnel diversion and downstream the inflow of the four main tributaries, along the densely populated Adige valley.

During the November 1966 flood event the stream flow of Avisio river was estimated at some 1200 m3/s being such value the highest recorded on this river.

The total actual damages of the Trento flood probably exceeded 10 thousand million liras.

This amount includes damages to highways, roads, houses, agricultural and commercial properties, as well as the costs to face the flood. In addition, substantial indirect economic costs occurred as a result of human lives lost.

Studying the entire Adige watershed and particularly the coincidence between flood waves in the main stream and in the four tributaries it was shown that the greatest benefit, among different control alternatives, should be obtained providing two artificial reservoirs (on the Avisio and the Isarco-Rienza rivers).

This paper deals with one of the two reservoirs, being this a multi-purpose flood control and irrigation storage reservoir created by a dam located at an Avisio valley constriction in a low economic value area.

2. DAM AND APPURTENANT WORKS

The general layout of the Valda dam and its appurtenant works is shown in Fig. 2.

The Valda dam is a cupola arch dam 132 m high with a crest 335 m long and a thickness at the key section variable from 25.8 m at foundation level to 5 m at the top crest; it is a symmetric dam whose symmetry axis planimetric position is very variable with elevation; the overall concrete volume including foundation amounts to 460,000 m3.

The reservoir is provided with crest and lateral spillways, low level outlets and a bottom outlet.

The crest spillway and the lateral spillways are ungated structures planned to operate only in case of exceptional events; the lateral spillways, one at each side of the dam, with a crest 30 m long at elevation 617.2 m are designed for a maximum discharge of 375 m3/s each and provided with an open air chute ending with a ski jump; the crest spillway with two sill elevations at 617.7 and 618.2 m and with sill width of 30 and 50 m respectively, is designed for a maximum discharge of 666 m3/s.

The low level outlets consist in two galleries, 6 m in diameter and approximately 300 m in length, located in the right abutment and gate provided at their outlet; the intake elevation for both the galleries is 550 m and the maximum discharge capacity of each gallery is 440 m3/s with the maximum normal water level in the reservoir; the galleries are foreseen not to be operated simultaneously.

The bottom outlet located in the dam foundation will allow the complete reservoir emptying from elevation 550 m in 62 hours.

3. FLOOD STUDIES

The information available for the estimation of maximum discharges of Avisio river at Valda site (855 km2) consists of a series of 15 years of direct flood observations at Stramentizzo site (730 km2) 12 km upstream, where a dam for energy generation only was built in 1955, and a further 15-year series at Lavis (934 km2) about 16 km downstream.

Therefore analysis was made on a 50 year series of rainfall data of short duration (1, 2 and 3 consecutive days) for the three more representative rainfall stations of the basin: Moena, Predazzo and Cavalese (see Fig. 3).

For this analysis the year was divided into two seasons: spring-summer, characterized by moderate floods (generally May-July) and autumn-winter, comprising the peak flood season (October-November).

The distribution of both the spring-summer and autumn-winter extreme yearly precipitation of 1, 2 and 3 consecutive days duration fits the Gumbel distribution.

The results of the statistical analysis obtained in terms of extreme autumn precipitation on the entire basin area (855 km2), giving different weights to the three over mentioned stations, are listed in Tab. 1.

The corresponding values obtained for the estreme spring precipitation are approximately 30 per cent lower.

With the detailed information relevant to the extreme event of 4-6 November 1966, which affected the downstream area producing severe and widespread damages, it was possible to obtain the unit hydrograph at the existing Stramentizzo dam.

The hyetograph and hydrograph of this event with the relevant components (infiltration losses, base flow, interflow, etc.) are indicated in Fig. 4.

The application to the same extreme precipitation event of the IUH according to the Nash model (1951) produced very satisfactory results, as it can be seen in the same Fig. 4.

By convoluting the unit hydrograph, modified to take into account the intermediate basin between Stramentizzo and Valda, with the extreme precipitations on the whole basin, the design floods of Avisio river at Valda site were obtained for several return periods.

The more significant flood figures calculated using the above procedure are listed in Tab. 2.

Fig. 1 - Adige River Watershed.

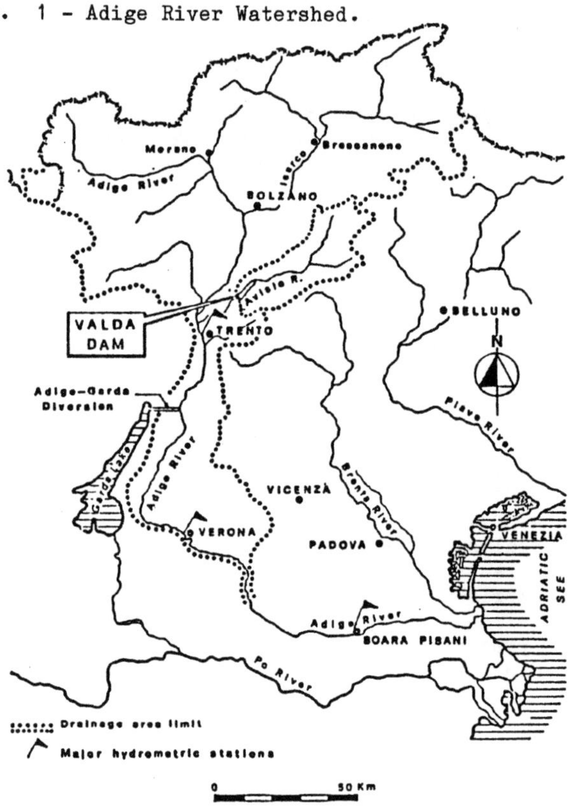

Fig. 2 - Valda Dam and Appurtenant Works Layout.

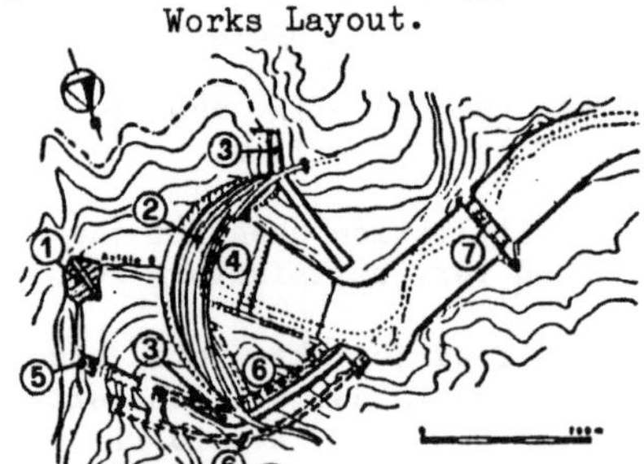

Legend 1-Upstream Cofferdam; 2-Dam; 3-Lateral Spillways; 4-Crest Spillway; 5-Diversion Tunnel; 6-Low Level Outlets; 7-Plunge Pool Dam

Fig. 3 - Avisio River Watershed.

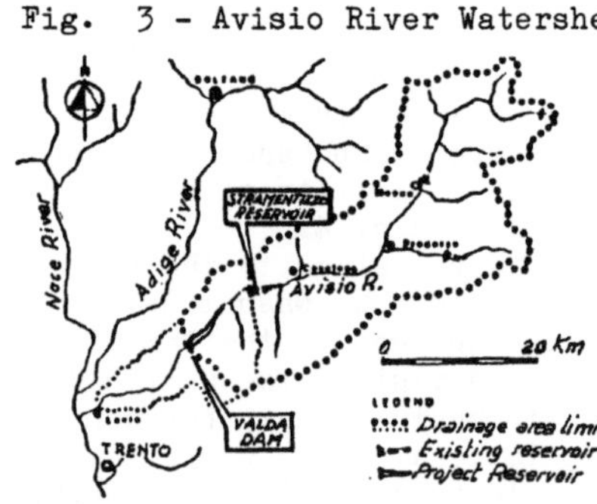

Table 1

EXTREME AUTUMN PRECIPITATION ON THE AVISIO WATERSHED AT VALDA SITE (IN MM)

Return Period	Duration 24 hours	48 hours	72 hours
TR = 100	107.8	159.5	181.4
TR = 1 000	143.6	215.0	240.8
TR = 10 000	179.4	270.5	300.2
PMP (1)	285.4	459.3	502.3

(1) According to Hershfield (1961)

Table 2

DESIGN FLOOD FIGURES OF AVISIO RIVER AT VALDA SITE

Return Period	24 hours precip.		48 hours precip.	
	peak (m3/s)	volume (hm3)	peak (m3/s)	volume (hm3)
TR = 500 s	660	67	578	101
TR = 500 a	1 001	98	876	148
TR = 10 000 a	1 351	132	1 181	198
PMF a	2 200	210	2 030	340

where : s means spring period
a means autumn period

4. FLOOD CONTROL

4.1. Design Criteria

The following leading criteria in the design have been assumed:

- flood control active volume chosen to store completely the 500-year autumn flood by releasing water only through the low level outlets without any surface spilling;
- free surface spillway structure designed to release (without low level outlets contribution) the 10000-year autumn flood with a fixed free board of 1.4 m;
- in the case of the PMF (without low level outlets contribution), free board is reduced to 0.3 m;
- at the beginning of the irrigation period (May-September) maximum reservoir level should enable to storage, without surface spilling, the 500-year spring flood.

The application of such criteria associated to the site morphology and to a low level outlets maximum discharge capacity of 440 m3/s led to the following figures:

- normal autumn flood control active storage 52 hm3;
- minimum normal spring flood control active storage 10 hm3;
- storage for spring-summer irrigation 42 hm3;
- additional exceptional flood control active storage 6 hm3.

For the reasons previously mentioned, a further 6 hm3 devoted to debris and sediment retention have to be added to the above storage figures.

4.2. Flood routing at Valda reservoir

Among the cases analysed, flood routing effects of Valda reservoir on the largest recorded natural floods are shown in Fig. 5.

The return period of the peak value of these floods can be estimated in between 300 and 500 years.

The peak flood reduction of the 500-year autumn flood (generated assuming 2 days uniform continuous rain) is of the order of 50 per cent.

4.3. Flood routing along the Adige river

Valda reservoir effect on flood modification along the Adige river downstream has been analysed for design and natural floods; reference was made to three of the major existing hydrometric stations located near Trento, Verona and Boará Pisani (see Fig. 1).

Fig. 4 - Hyetograph and Hydrograph of the Extreme Event of 4-6 Nov. 1966.

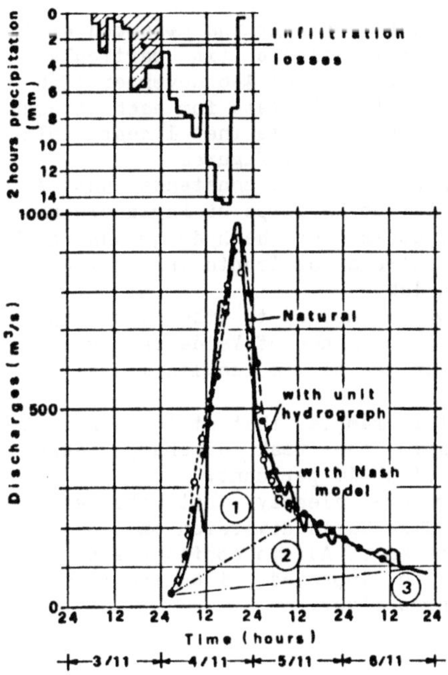

Legend 1-Surface Flow; 2-Interflow; 3-Base Flow

Fig. 5 - Flood Routing at Valda Reservoir.

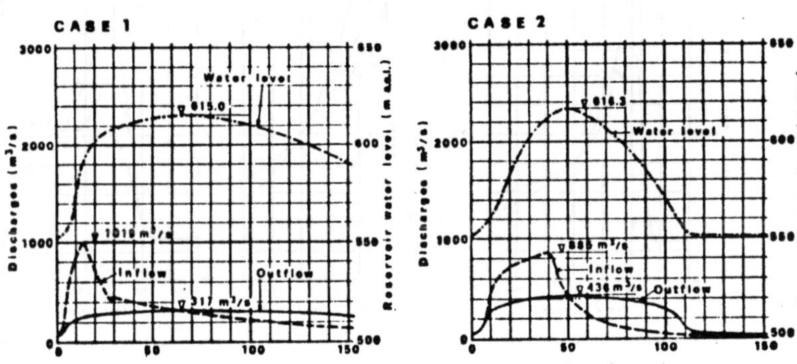

Legend : - Case 1 : Natural Flood of 4-6 November 1966 with a gate opening equal to 0.7;

- Case 2 : 500 yrs Design Flood with gate fully opened.

Between Trento and Verona the Adige-Garda diversion system, whose maximum discharge capacity is approximately 500 m3/s, is operating since 1959.

Utilizing the available hydrographs measured at the three stations during the major observed floods (in particular the two already mentioned of 1-4 September 1965 and 4-6 November 1966) it was possible with the Muskingum method to calculate for each stretch of the river the constants X and K which characterize the linear relation between the storage and the inflow and outflow figures.

The determination of the above constants was satisfactorily made with the HEC1 program of the US Corps of Engineers with the introduction of some simplifying hypothesis on the hydrographs of the tributaries of the Adige river downstream of Trento and on the hydrograph shape the diverted flow in Garda lake.

The application of the method to the two major observed floods modified or not by the presence of Valda reservoir and by the presence of Adige-Garda diversion led to the results listed in Tab. 3 and shown in Fig. 6.

It can be seen, for example, that at Boara Pisani, where the hazard of flooding is higher, the peak discharge reduction of the natural floods reaches values of 35-40 per cent.

It appears therefore clearly that the effects of the Valda reservoir and the Adige-Garda diversion on flood peak reduction along the Adige river downstream of Avisio confluence are quite satisfactory and could furtherly be improved with the presence of another flood control dam on the Isarco river.

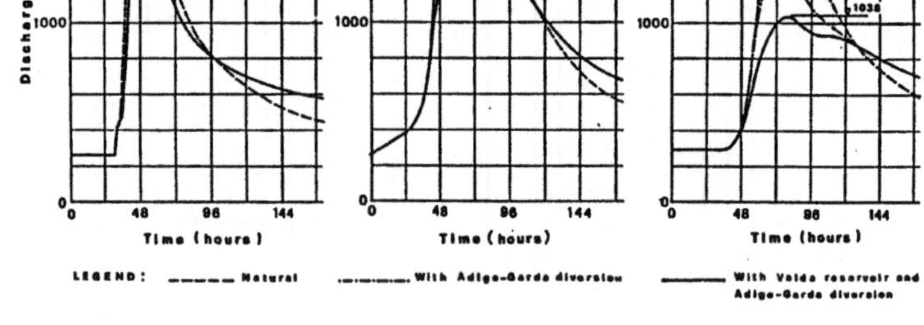

Fig. 6 Flood Routing along Adige River - Natural Flood of 4-6 Nov.1966

Table 3

ADIGE RIVER PEAK FLOOD REDUCTION

DESCRIPTION	HYDROMETRIC STATION					
	Trento		Verona		Boara Pisani	
	Disch. (m3/s)	Level (m)	Disch. (m3/s)	Level (m)	Disch. (m3/s)	Level (m)
1-4 September 1965 Flood						
With Valda reservoir only (gate opening = 0.3)	420	0.85	280	0.49	260	0.70
With Adige-Garda diversion only	-	-	360	0.63	380	1.03
With Valda reservoir and Adige-Garda diversion	-	-	650	1.18	640	1.74
4-6 November 1966 Flood						
With Valda resevoir only (gate opening = 0.7)	740	1.46	350	0.55	310	0.79
With Adige-Garda diversion only	-	-	400	0.64	360	0.90
With Valda reservoir and Adige-garda diversion	-	-	750	1.24	690	1.77

5. ACKNOWLEDGEMENTS

The authors wish to thank Profs. R. Cola and F. Veronese of the Padova University for their contribution to the hydrologic studies, Engs. U. Picozzi, G. Zanella and A. Spada for their contribution to the hydraulic and structural studies, Mr R. Del Signore for the graphical support.

A special acknowledgment is made to Eng. G. Castelli Chief of Public Water Service of the "Provincia Autonoma di Trento" for the support and facilitation of the project studies.

6. REFERENCES

1. Cola, R., Veronese, F. 1979. 'Studio per l'Utilizzazione dell'Invaso del Serbatoio di Valda.' Universita' di Padova.

2. Grigg, Neil S. 1985. Water Resources Planning, McGraw-Hill Book Co., New York.

3. Gumbel, E.I. 1966. 'Extreme Value Analysis of Hydrologic Data.' Proceedings Hydrology Symposium N.5, N.R.C., Ottawa.

4. Hershfield, D.N. 1961. 'Estimating the Probable Maximum Precipitation.' Journal of the Hydraulic Division, Proceedings of ASCE, September.

5. Nash, S.E. 1957. The Form of the Instantaneous Unit Hydrograph. International Association of Scientific Hydrology, Vol. 3.

6. U.S. Army Corps of Engineers 1979. Generalized Computer Program HEC-1, Flood Hydrograph Package. Hydrologic Engineering Center, Davis. California.

A CATASTROPHIC FLOOD AND ITS CONTROL IN AUGUST, 1963 ON HAIHE RIVER BASIN OF CHINA

Hong-Mou Bao
Senior Engineer
Department of Water Management
Ministry of Water Resources and Electric Power
Beijing, P.R.C.

Li-Dan Ke
Deputy Director
Senior Engineer
Water Resource Office
Ministry of Water Resources and Electric Power
Beijing, P.R.C.

ABSTRACT. The Haihe River is one of the seven main rivers in China. The Capital-Beijing and the large industrial city-Tianjin all locate within this river basin. It is a political, economical and cultural center of China. It have frequently occurred flood disasters in the past. In this paper, the salient features of the catastrophic flood in August, 1963 of this region and several key measures adoped are presented. The experiences and lessons learnt in the combine action of structural measures and non-structural measures of flood control are also summarized. For the future, it will have the profound significance for the reconsider of the guide lines and policies of river control as well as the defensive disposition taken to over fulfillment of the Targets Setting by the design flood etc..

The Haihe River is one of the seven major rivers in China. There are five large rivers are tributary to the Haihe and Luanhe Rivers all in the basin. It flows through five provinces of Hebei etc. and Inner Mongolia Autonomous Region respectively. Both the Capital-Beijing and the large industrial city-Tianjin that locate in this river basin too. It is acturally the political, economical and cultural center of China. The drainage area is 319,000 square kilometers, with its population about 100 million and farmland 11.3 million hectars.

There are two mountains in the north and the west of area, which are Yanshan and Taihangshan Mts. Its elevation is between 500 m to 2000 m with a highest peak over 3000 m. The terrian in the basin slops down to the Bohei Sea from its north, west and southwest sides. Owing to the alluvial direct influence by the Haihe River System and the Yellow River, as well it has caused a vast alluvial and coastal plain in the east and southeast of this basin. The plain's elevation is under 50 m and even there are some low-lying lands in it. The mountain and plain areas are half and half.

Haihe River Basin belongs to a region of typical seasonal wind climate. The vast Warmflow and humid air are blowed from the ocean with southeast seasonal wind in summer. It's polar front is steady at nearly forty degree north lalitude, leads to the precipitation mainly

V. P. Singh (ed.), Flood Hydrology, 389–398,
© 1987 by D. Reidel Publishing Company.

concentrated in summer. During the period from June to September in general years, the rainfall amount is about 80 percent or more of yearly total. Owing to the affection of the vertical rise of terrain in northern and western, it often forms a rain storm and flood.

In the Haihe River Basin, the flood and waterlogging disasters have frequently happened in human history. Especially in this century in 1907, 1924, 1939, 1956 and 1963 years there were heavy floods in which from hundreds of counties and 1.3-2.0 million hectars to 3.3-4.0 million hectars of the cultivated were inundated. Ever since seventeenth century. Beijing was water bounded in 1607, 1626, 1737, 1801 and 1890 years and the present Tianjin Urban District was submerged in 1653, 1654, 1668, 1801, 1871, 1890, 1917 and 1939 etc. eight years. All the streets of Teanjin then were water logged for two whole months in 1939 and coused enormous losses. The broad masses of people in the region have already had a long history of the struggle against the flood calamities. But a vast amount of water-conservancy projects were built after the liberation of our country. Since the New China was founded in 1949 until the catastrophic flood happened in 1963. Covering the period of 14 years, 1500 km dilapidated embankments left over by the old society were rebuilt and 34 large and middle scale reservoirs were constructed. The Xingaifang Flood By-Pass, Zhaowangxinhe River, Duliujianhe River and Chaobaixinhe River were excavated as well as the Sinusijianhe River (Zhangweixinhe River) and other waterways into the sea were expanded, thus leading to the flood-relief capacity increased to 4620 c.m.s. and this is twice as large as that before. It gave these flood protection work's condition better. Haihe River Basin occurred "63.8" flood which exceeded the maximum records in the history.

During the first ten days, August 1963, continuous rain gush occurred in the southern areas of Haihe River Basin. Not only the rain's intensity but also the area covered and its duration all exceeded the maximum records in the history. It covered areas account for 56 percent of the total in Hebei Province. The 10 day's rain height exceeded 1000 mm with shround area 5390 square kilometers as well as 400 mm 66 thousand square kilometers. The center of 7 day's rain gush, amounted 2050mm, located at Zhangme of Neiqiu County. The total rainfall volume amounted 57.7 billion cube meters and yielded runoff 30.2 million cube meters in south Haihe River, it tripped the the maximum record of 1939. After the occurance of this continuous rain gushed, the flood increased suddenly in every river. The flood of the west mountainous regions overflowed the railway line from Beijing to Guangzhow, with a flood peak amount to 78,000 c.m.s., and it exceeded 22 times of the discharge in the south river system draining into the sea. Althrough, there were Yuecheng, Gangnan, Wangkuai and other ten large reservoirs traping amount of 46 percent of mountainous flood upper reservoirs, due to the extraordinary flood peak and its overbounded volume, they over-flowed in succession. these flood along with waterlogging had already been exceeding the discharge of middle and lower reaches, so same dykes were overflowed and bursted, it flowed over level ground. It not only inundated the vast plain of the east and middle of Hebei Province but also pressed on towards the outside around Tianjin, the security of Tianjin was serious threatened.

The Tianjin City is located at the lower reaches of Haihe River System. In the past, the Nanyunhe, Ziyahe, Daqinghe, Yongdinghe and Beiyunhe Rivers all flowed collection in Tianjin, then passing the Haihe trunk stream into the sea. The Duliujianhe River was excavated in 1951, but its flood discharge was still limited. In face of the flood nearly 10 billion cube meters pressed forward to Tianjin, if we had only relied on guard keeping a few natural low-lying lands around Tianjin for flood detention, the flood stage of low-lying lands would be higher than the Tianjin District 4 m-5 m, it would undoubtely have created a very serious situation for the fighting against flood around Tianjin. We had ever drawn a bitter leason, that a whole area of Tianjin became submerged immediately due to dyke burst in 1917 and 1939. With the result of these analysis a decisive factor for this flood struggle was become a rational dispatch for this flood.

Owing to adoption a series of immediate and accurate measuares under the effective organizative of our goverment the victory of the fighting against flood finally achieved. Its several key measures will emphatically be introduced in this paper.

1. TO ACCURATELY SELECT THE PLACE FOR OPENING A BREACH IN DYKE, TO MAKE INITIATIVE FLOOD DIVERSION OF THE BAIYANGDIAN LAKE

Baiyangdian Lake is a water storing and flood retaining area. It is situated in the middle reach of Daqinghe River. In the south part of the river there are 10-odd tributaries. In general years, its flood flows into the lake to be retained, and then, steadily flows to the lower reach. In the past, however, once it met with an abundant water year, the flood of the tributaries would exceed the capacity of the retention in Baiyangdian Lake, the "500 km-dyke" around the lake would repeatedly be broken at first, and next, the protection embankment around Tianjin would be broken, too, and finally the flood would flow down into the Tianjin Urban Districts. In the remarkable "63.8" Flood, its general volume flowing in the upper districts to the Baiyangdian Lake reached 6.5 billion cube meters. Except that a part of the flood was retained in Wangkuai, Xidayang, Hengshanling etc. large scale reservoirs in the tributaries of mountain areas, it rushed down into the Baiyangdian Lake, on Aug. 9, p.m. the flood stage in the lake suddenly began to rise, and round the clock on Aug. 10, it rose up 2.7 m, then on Aug. 11, it exceeded the highest level in the historical records. In the meantime, the dyke crest, about 7.2 km in length was even blow the flood level. It relied only on rush-repairing a low bank with earth on the top embankment to keep off the flood water. But, now its inflow was more than its outflow still. According to the estimate, the whole dykes line would have been overflowed or burst if a measure of an increase in the outflow had not been adopted. In order to ensure the success of fighting against the flood around Tianjin a breach must be made in the Baiyangdian Dyke for initiative flood diversion. Its object was that the breach could both release the superfluans retardative flood water and control its outflow volume. It was, therefore, necessary to accurately select a place for breaching in the earth dyke which was without any controlling cnostruct. According to the historical data of flood burst of the dykes in the past, we analysed a possible site for occurrence of dangerous section in the dyke. Lastly, a hard clay section of the dyke was selected. It was Xiaoguan Hamlet, located in the north of Rengiu County. Its terrain outside dykes was higher and it had not any dangerous section, so it had a finer controlling condition. The break, about 300 m long, was made at 3 o'clock p.m. on Aug. 11, and mean while the flood flowed into the Wenanwa Detention Basin. Afterwards, another break was made in Ronghuashu Dyke on Aug. 13 with its flood diverting to the Yiluwa Detention Basin. Yet, the flood level sustained steady rising in the Baiyangdian Lake still. The width of the first breach at Xiaoguan had to be expanded again to 430 m with its maximum outflow increasing to 1070 c.m.s. until 3 p.m. on Aug. 14, the flood level of Baiyangdian Lake reached its maximum height with its relevant volume of flood storage 4.172 billion cube meters (Fig. 1). In brief, during the whole period, the breach at Xiaoguan outflowed total 1.166 billion cube meters and the other breach at Ronghuashu outflowed total 1.593 billion cube meters, and consequently, the crisis of bursting the dyke around the lake was relieved.

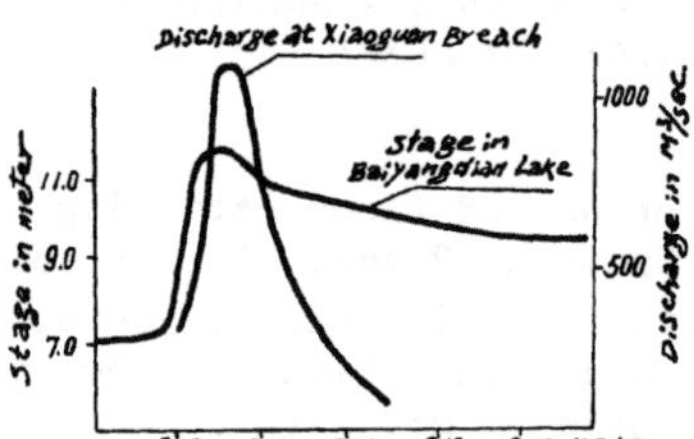

Fig. 1. Hydrographs of Stage in Baiyangdian and Discharge at Xiaoguan Breach.

2. TO APPLY THE TIANJIN-PUKOU RAILROAD VIADUCT TO MAKE THE FLOOD DIVERT TOWARDS TUANBOWA

In the Baiyangdian Lake, an initiative flood diversion had already made. However, for the sake of safeguarding the Tianjin City. Tainjin-Pukou railroad and making the flood havoc reduced as far as passible, another flood diversion must be made. According to the flood prediction, we estimated that the total flood volume converged on around Tianjin would be 10 billion cube meters. But the capacity of the Jiakouwa Detention Basin located in the west to Tianjin, was only 1.2 billion cube meters. If the flood arrived, it might be resulted on the following two passible aspects; firstly, the partitive dyke between Jiakouwa and Dongdian Detention Basins would be burst and then, the flood joining togather would scatter, it was bound to exert a largest pressure upon the right dyke of the Ziyahe River and the left dyke of the Duliujianhe River, then Tianjin would be heavily threatened.

(Fig. 3). Next, the flood possibly burst in dykes of the Nanyunhe River and this would make a pounce upon the Tianjin-Pukou Railway with a break off the main line. In face of such an emergent condition and after a reconnaissance of scene, we decided to draw up a plan for another flood diversion towards Tuanbowa Detention Basin passing through the viaduct of the Tianjin-Pukou Railway, which was build in 1939, the abundant flood year. Then we set about some comparative designs for making breachs in both dykes on each side of the Nanyunhe River to straighten out the flowing direction in order to make it safety pass the railway. At last we chose the position near Wangjiaying Village which was at the west dyke of the Nanyunhe River and just opposite to the viaduct. The two breaches faced each other across the river with each width of 250 m and its base was in elevation of 5.5 m. In addition, on both sides from the breaches to the viaduct, it was necessary to construct diversion dykes with the total length of 700 m. These flood-relief work started on Aug. 13. There were 20 thousand soldiers and people mobilized to fight bravely for eight days, and in the end, a volume of earthwork sumed up 211 thousand cube meters had been achieved. The flood began to enter the Jiakouwa Detention Basin on Aug. 17, and within two days (Aug. 19-20) it entered amounting 1.78 billion cube meters. Just before the crisis of a dyke burst, which faced the Nanyunhe River, occurred, the flood-relief project started to make the flood divert in time.

From the effect in the implementings the temporary flood diversion measure, we can get such important experiences: a rational selection of topography; and adroitly guide action according to circumstances. Best using the flood erosion topography taken place in 1939, a rational of a place for breach in dykes; and keeping the flood within the bounds of narrow and deep type, all had played a main role in controlling the flood direction of flowing to straight (Fig. 2). At the same time, adopting the loose demolition blast meassure in construction, was another victory for wining over the time.

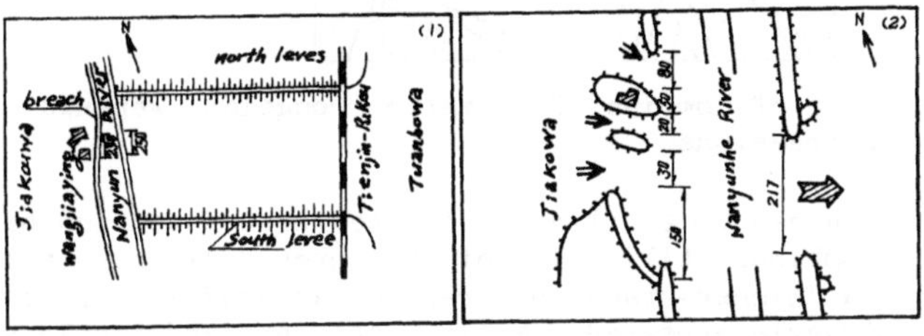

Fig. 2. (1) Desing scheme of the breach made in Nanyunhe River Dyke
(2) Actual situation of the breach after diversion

3. A COMBINED UTILIZATION OF THREE DETENTION BASINS FOR A LARGE-SCALE AND MORE EFFECTIVE WATER DETENTION

There are three detention basins in the western area around Tianjin: Dongdian, Wenanwa and Jiakouwa Basins. These are historically the main flood detention basins of the Daqinghe and Ziyahe Rivers. The capacity of Wenanwa Basin is the largest of the three and its volume is even more than the sum of the two others. But, at this time, from our analysis of the flood discharge we found that the inflow of Jiakouwa Basin was also the greatest of the three, it made up nearly 50-60 percent of the total volume of the upper stream flood. In order to bring flood detention effect into full play and lighten the burden of Jiakouwa Basin, we made a thorough research and then determined to adopt a combining utilization of the three detention basins. On Aug. 20, at 12 o'clock a.m. as soon as the fortified stage of Jiakouwa Basin was to be exceeded, a breach was immediately made in dykes of the Ziyahe River nearby Wangkouzhen Town to be the flood diversion go down towards Wenanwa Detention Basin (At 4 o'clock p.m. another breach was also made in both dykes on each side of Nanyunhe River to let the flood flow passing the viaduct into Tuanbowa Basin). Hereafter, a breach was again made in a partitive dyke between two detention basins near Qibao village, and the flood flowed northward into Dongdian Detention Basin. In order to reduce the pressure of the Ziyahe River flowed against the outflow of Jiakouwa Basin, a breach was made in the safe-dyke of Ziyahe River at Yaomatu Village, Dacheng County, then the flood of Ziyahe River might flow into Wenanwa Detention Basin at first. (Fig. 3). Then, the flood of the upper-stream was still 7000-8000 c.m.s., until Aug. 25

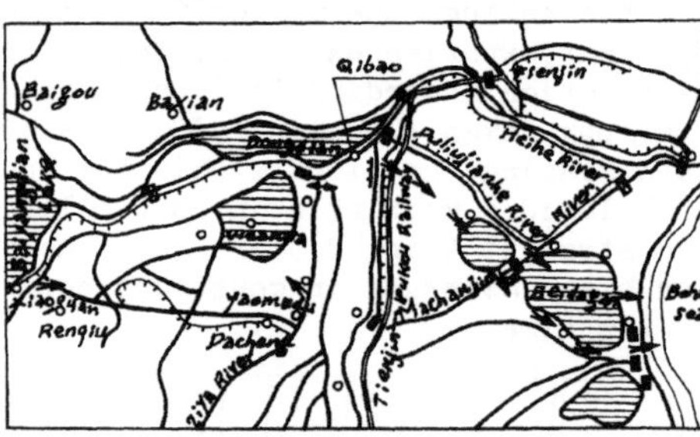

Fig. 3. Sketch Map of Combined Utilization Dongdian, Wenanwa and Jiakouwa Depressions

the flood stage tended to equilibrium in the three detention basins. Only after the combined utilization of the three detention basins was carried out did a "higher-stage, protracted warfare" in the fighting against flood struggle appeared. Million soldiers and people standing on more than 250 km long dykes fought against flood, dykes got highten and strengthen, the back prop of embankment was made, a ditch with drainge percolation was dug etc.. Thus the safety of the embankment around Tianjin got ensured. The three detention basins were being reached the highest stage on Sept. 1, and there it started to be steady and falling. (Fig. 4). The detention volume in the three detention basins sumed to 7.28 billion cube meters.

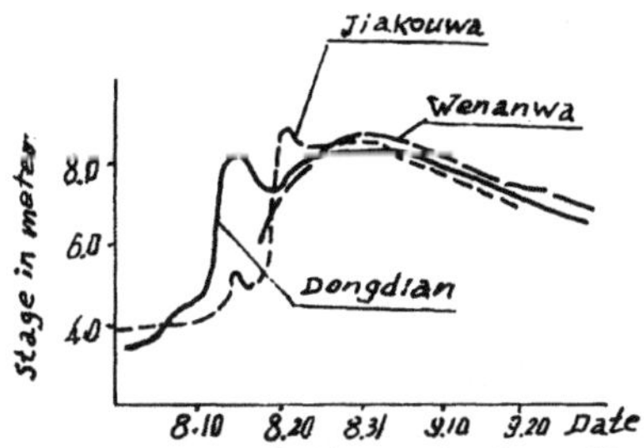

Fig. 4. Hydrographs of Stage in Dongdian, Wenanwa and Jiakouwa Depressions

4. TO RESCUE THE TIANJIN-PUKOU RAILWAY FROM DANGER

Since the flood diversion passed through the Tianjin-Pukou Railway Viaduct and poured into Tuanbowa Detention Basin, the water level in the Jiakouwa Basin, located in the eastern suburban counties of Tianjin, was quiekly going up. This led generally severe threats to a section, about 42 km in length, of the roadbed of Tianjin-Pukou Railway. It ever overflowed some section of the roadbed and even entered the Duliu Railway station. If the roadbed was destroied, not only the main line of the communication from north to south in our country would have been cut off but also the flood of the three detention basins and flood in Tuanbowa Basin would have jointly given a pincer attack to both dykes on each side of the Nanyunhe River and then the security of Tianjin City would severely have been threatened. Thus, we deternined to adopt the following measures: (1) to rush-build sublevee by straw bags which were filled with earth; (2) to dig a cut off groove on the bank side of the sublevee with its deep to the soil of the roadbed, to thoroughly clear away the pervious materials as ballast etc. and backfill it with earth until it was 0.5-1.0 m higher above the surface of the roadbed; (3) to riprap the caves in the bank of the roadbed with a large amount of stones and to make a prop here as well as a filter at the piping place; (4) to deep guard the front of the levee against storm waves' lash with adopting such a measure that a row of piles of logs with 10 cm in diameter, 3-4 m in length, was driven into the front toe of levee, and reed boundles in diameter of 20-30 cm were put into the row of piles (Fig. 5). As more than 20 thousand soldiers and people had been engaging in the indomitable struggle, this salvage work was at long last achieved on schedule. After ten days' break, the traffic on the Tianjin-Pukou Railway resumed on Sept. 7.

5. DIVERT THE TUANBOWA FLOOD AWAY INTO THE SEA

Tuanbowa is a low-lying land surrounded by the Nanyunhe, Duliujianhe and Machangjianhe Rivers. While we decided to make the flood diversion of the "three detention basins" flow into Tuanbowa, we also determined to open up an outlet to the sea for divert the flood of Tuanbowa away (into the sea).

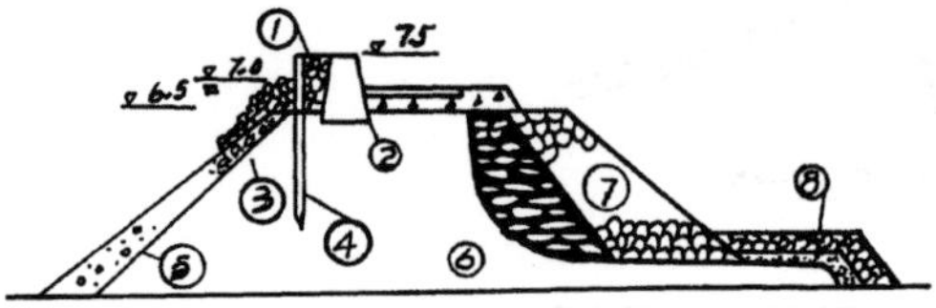

Fig. 5 Sketch of Rush to Deal with an Emergency in Tianjin-Pukou Railbed

1. retaining dike with straw bag filled cinder.
2. Laomy cut-off wall.
3. throw stone-block for protection from the wind wave
4. piles with reed bundles for protection from the wind wave.
5. throw stone dust or cinder.
6. embankment prop bystraw bag filled earth material or stone ballast.
7. throw stone-block or stone ballast.

The project of divert for Tuanbowa flood was set about by 5 aspects in process as: (1) to breach both dykes on each side of Machangjianhe River and the southwest levee of Beidagang plain Reservoir; (2) to remove the obstructive dykes and clear reeds from the flood by-pass area for dredging up the waterway; (3) to build a sublevee on the northern dyke top of the Beidagang plain Reservoir; (4) to clear away some low banks of earth in salina and to raze on third of saltern; (5) to break open the eastern dyke of Beidagang Reservoir (i.e. the Haidadao tidal barrage) by blasting technique for divert the flood discharge into the sea (Fig. 6). For all these measures we worked intensely but orderly for a whole month and on Sept. 21 it was achieved after all. According to the discharge measured, the flood outflowing from Tuanbowa through the breach of Machangjianhe River amounted to 5.631 billion cube meters.

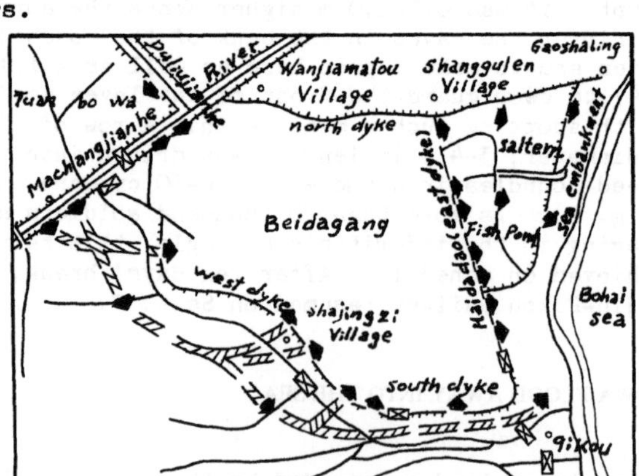

Fig. 6 Sketch of Divert the Tuanbowa Flood away into the Sea

CONCLUDING REMARKS: The concrete practice in this struggle of fighting against the flood may prove that such catastrophic flood as the "63.8" flood can hardly be resisted by normal flood-control measures. It can only be won by the following: adroitly guide action according to circumstances as well as the natural law, a combination of flood control-structural measures and flood control-nonstructural measures, resolutely adopting an emergent measure if necessary, and daring to throw off parts of an area so as to preserve the whole situation. Timely and actively adopting the above-mentioned key measures was proved to be completely effective and correct. It led to the unified dispatch and management of flood storage, flood retardation and flood discharge measures which had yielded notable results.

These results can be explained in detail: (1) Causing 9 billion cube meters of flood volume retained in reservoirs in the mountainous districts and rivers (amounting about 30 percent of the total) and crippling the momentum of the flood peak and reducing its pressure to low reach areas: (2) The waterway entering to the sea got expanded by demolition blast and 10 billion cube meters of the flood was diverted into the sea (amounting to 33 percent of the total), thus the threat of inundating Tianjin Urban was lightened and the flooded area with its inundated fields was reduced, and a favourable condition for an earlier water removal in the upstream region was also created; (3) Especially, the eight detention basins and the depressions (open up flood plain) with total area about 2000 square kilometers were used, and at the same time we strengthened the joint work of flood-relief dispatch and management, so retarding the flood totaled up to 11.2 billion cube meters (amounting to 37 percent of total).

In order to achieve the flood-relief measures, million soldiers and people fought bravely for a month and more under the unified leadership and centradized direction of Government, successively completed the earth work of digging and filling totaled more than 5 million cube meters and the stone work totaled 150 thousand cube meters as well. Lives and properties of the several hundred thousand inhabitants living in the detention basins and depressions were safely guaranteed and their production and daily life got well arranged under the effective organization of the governments at all levels. Owing to correctly adopting flood dispatch and relying on the strength of masses, the greatest victory over their resisting against flood struggle achieve in the end.

The success consists of (1) these measures had ensured safety in Tianjin and avoided the possible heavy losses in an inundated Tianjin which was flooded for two whole months in 1939; (2) The safety in our rational trunk line, the Tianjin-Pukou Railway, was guaranteed; (3) The flood damage losses had been decreased to a minimum--such decreasing losses estimated about RMB billions Yuan.

Therefore, the Haihe "63.8" flood and its control as a tremendous event, has been written in the history of fighting against flood struggle in China. Simultaneously, the practice and the experience of this struggle will even more open a gorgeous new chapter for permanently harnessing the Haihe River and eliminating the scourge of floods.

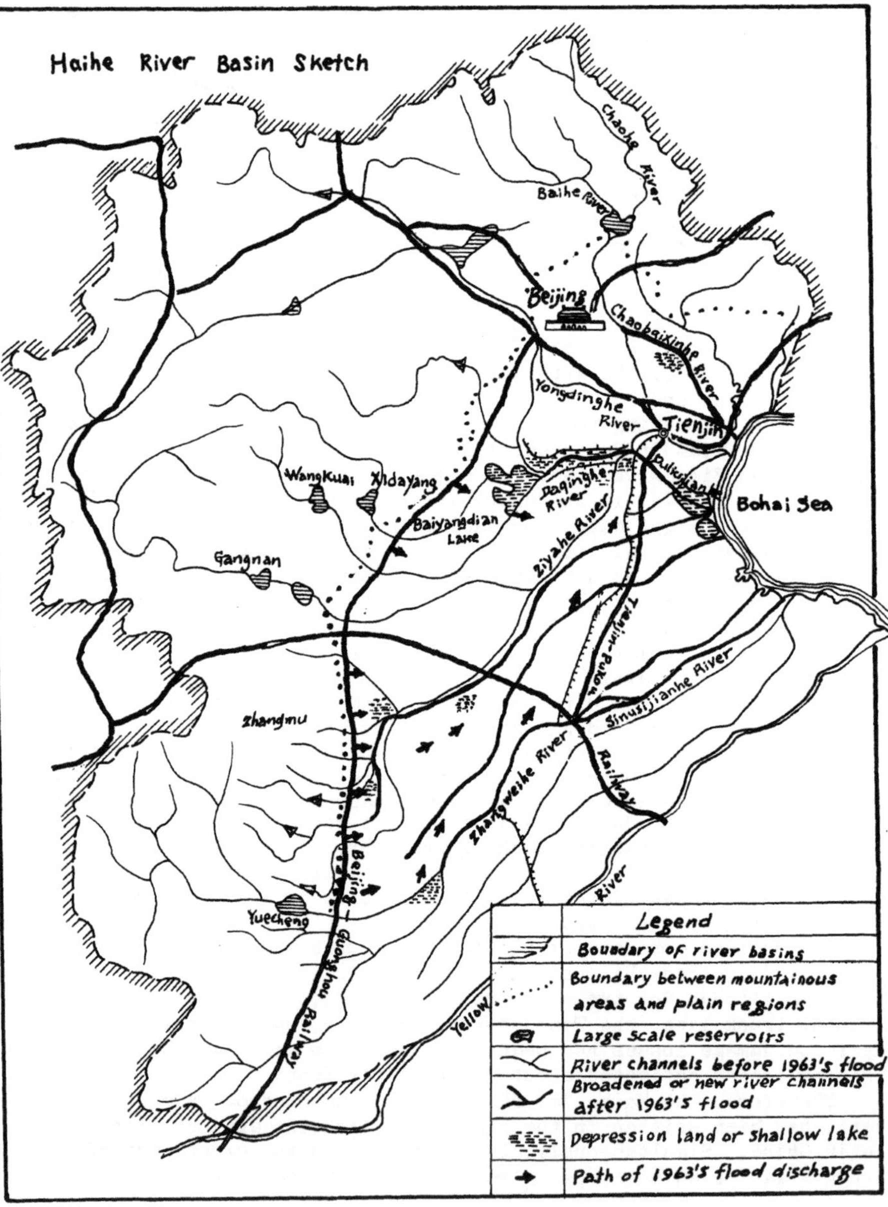

CENTRIFUGE MODELING OF RIVER BANK FAILURES DUE TO SEEPAGE FLOW

Osamu Kusakabe, Associate Professor, Ph.D
Department of Civil Engineering
University of Utsunomiya
Ishii, Utsunomiya, 321, Japan

Yasuhiro Okumura, Former Research Student
Department of Civil Engineering
Tokyo Institute of Technology
Ookayama, Meguro, Tokyo, 152, Japan

Akio Nakase, Professor, Dr.Eng.
ditto

ABSTRACT. A series of centrifuge model tests was carried out in an attempt to study river bank failures due to seepage flow. The necessity of centrifuge modeling and the similarity law were discussed. The results of the model tests showed that capillary tension above phreatic surface plays an important role in stability, and that there seems to exist a critical value of velocity of fluid flow to trigger an initial failure. An analytical model to simulate the development of failure was also proposed.

1. INTRODUCTION

The total length of major river banks, managed by governmental organizations, amounts to over 100 thousand kilometers, which form one of the most important soil structures in Japan. These river banks must be stable against floods which occur due to frequent attacks of typhoon in this country. It has been considered that three major causes to river bank failures are erosion, overtopping and seepage flow in the banks. Erosion is closely related to shear stresses on the surface of the bank slope, which is considered to be the function of mainly depth and velocity of flood flow. Phenomena of overtopping are obviously resulted from excessive discharge of the flood flow. The type of river bank failure due to seepage flow, however, may occur, even when the water level remains below the planned high water level for a long time. Therefore, the failure due to seepage flow may have some relation with the duration of the flood, thus the permeability of the river bank materials.

One of the recent examples of the river bank failure due to seepage flow took place at the River of Nagara located in the middle of Honshu,

V. P. Singh (ed.), Flood Hydrology, 399–408.
© 1987 by D. Reidel Publishing Company.

the biggest island among the Japanese archipelago, in September, 1976. A typhoon, designated by No.17 typhoon, brought a massive total rainfall of 1000 mm over the watershed of the River of Nagara, and the water level at the failure location remained about 1.0 meter below the planned high water level - three-quaters of the total height of the river bank - for as long as 70 hours. Three hours before the total failure, the first crack appeared near the toe of the river bank on the inland side. The failure length along the river bank was initially about 30 meters, and gradually spread up to 80 meters, resulting 3500 houses being under water. The soil of the river bank was classified into silty sand (Tsuchiya et al., 1976).

The complexity of such phenomena, involving saturated and unsaturated soil zones, in which water is seeping through, requires accumulated data of close observations on the phenomena in the field as well as reproduced in the laboratory, to understand such physical events and to deduce physical models and analytical methods to predict the risk of the failure. To achieve this objective is a difficult task, in a sense that prototype tests would be too expensive and even dangerous to investigators, and that small scaled model tests would not be meaningful without correct similarity laws.

However, centrifuge modeling technique is now available to overcome such difficulties, to some extend. This paper introduces the method of centrifuge modeling technique to study the failure of soil structures due to seepage flow, with particular interest in phenomena of river bank failure due to seepage flow.

2. CENTRIFUGE MODELING

Recently, it has been shown in geotechnical enginnering field that fruitful information can be obtained from geotechnical centrifuge technique, which , as explained by Schofield (1980), requires the creation within the model of stress and strain field everywhere identical to those of the prototype, and from the study of the models behavior under a scaled perturbation, useful information about the behavior of an equivalent prototype can be obtained.

Looking back the histrical development of the technique, it is found that the first group of problems tackled was stability problems, such as stability of slopes, tunnels and excavations, where loading is predominantly due to the ground self weight. Application has been expanded more widely in recent years, ranging from silo engineering to earthquake engineering. Goodings (1979) studied stability of mine slopes under seepage flow. Padfield and Schofield (1983) modeled a flood bank failure due to uplife presssures. Sutherland and Rechard (1984) examined the phreatic surfaces developed in stable embankments. Ko (1984) investigated phenomena of overtopping.

Similality laws should be examined in any type of model tests. As pointed out by Croce et al. (1984), a single scale factor cannot be achieved simultaneously for all kinds of forces, such as viscous force and seepage force, in centrifuge modeling of phenomena involving soil and fluid flow. We must , therefore, be statisfied with the concept of partially similarity. Physical quantity that has a primary influence on the problem of the river bank failure due to seepage flow, is considered to be soil strength, thus the distribution of effective stresses, mainly due to selfweight of soil and partially due to capillary tension.

Figure 1 illustrates the need for centrifuge modeling technique for this particular problem. Since a soil behaves essentially a frictional material, the strength of soil is directly proportional to normal stress acting on a failure plane. Therefore, the soil strength mobilised in the scaled model under gravitational acceleration (1 g model) is 1/N times that of prototype. In the prototype, below the phreatic surface, there is a Darcian flow, and above the phreatic surface, there exists a zone which retains water by capillary tension. The tension creats negative pore water pressure, which increases in effective stress , thus the shear strength. In the scaled 1g model, the height of capillary rise will be the same of that in the corresponding prototype, provided that the same material is used. However, the height will be out of proportion to the relative dimension of the model compared to the prototype. If the same model is subjected to centrifuge acceleration of N times the gravitational acceleration, it will have the same normal stress, thus the soil strength and also a capillary rise in height 1/N times that of the prototype. By doing so, correct geometrical and mechanical modeling is achieved.

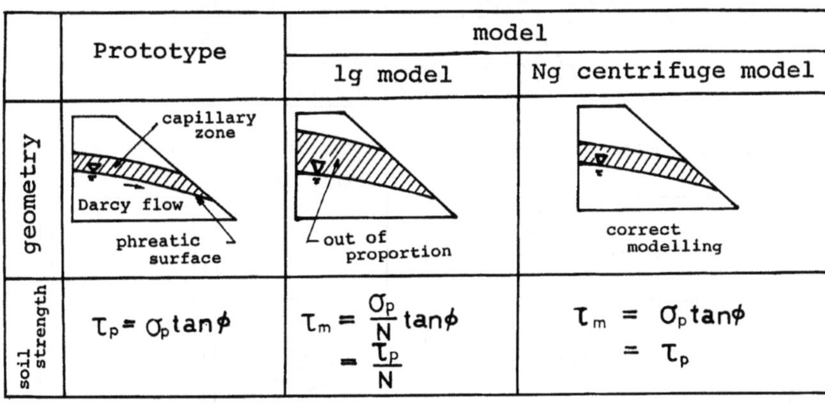

Figure 1 Explanation of the need for centrifuge modeling

Common practice of selecting the similarity law in model tests of hydraulics adopts the similarity of Reynolds for a laminar, Darcian flow. The Reynolds number to hold the Darcian flow is said to be less than 10 (Scott, 1965). Care must be, therefore, taken to adopt an appropriate centrifugal acceleration in order to ensure the Darcian flow within the model. Under the flow regime of this kind, it has been concluded (for example, Schofield, 1980) that the velocity of fluid flow in the model is N times faster than that of prototype and the time in the prototype is reduced by a factor of N square, compared to that in the model. In this paper, tests results will be presented with both model scale and corresponding prototype scale in brakets.

3. TEST SYSTEM AND PROCEDURES

The centrifuge used in this study is the Mark II centrifuge of the Tokyo Institute of Technology, which is 3 meter in diameter and it can spin a pair of packages of 250 kg in mass under the centrifugal acceleration of 150 g (Figure 2). The facilities of the Mark II centrifuge are well documented by Kimura et al. (1985).

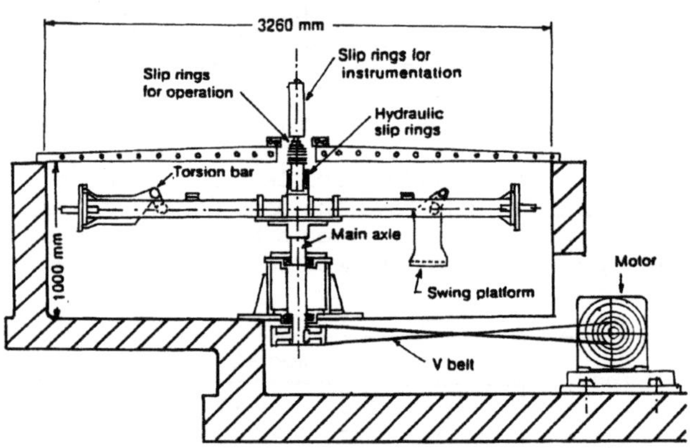

Figure 2 Mark II centrifuge of Tokyo Institute of Technology

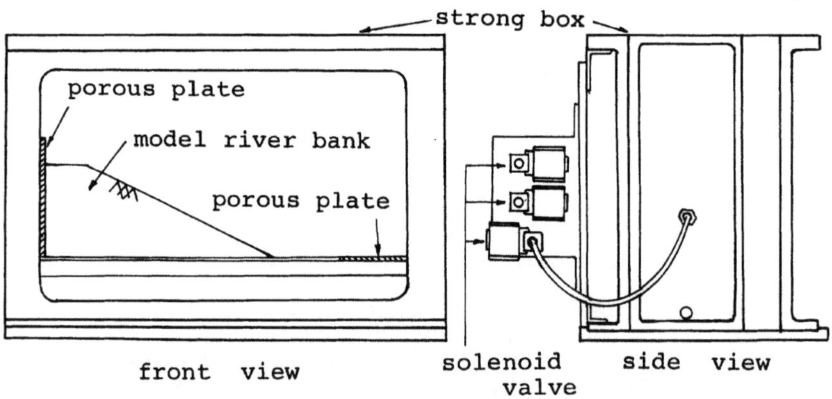

Figure 3 Test system

The test system used is illustrated in Figure 3. The model container is a rectangular strong box, of which main parts are made of steel and the front face is made of a highly shock resistance plastic plate, so that visual observation can be made during the rotation of centrifuge. The internal dimensions of the strong box are 0.4 meter long, 0.15 meter wide and 0.4 meters in height. The model river bank consists of a mixture of dry Toyoura sand and crushed sand. Three different mixture ratios of Toyoura sand to the crushed sand were used, of which material properties are given in Table I.

TABLE I Material properties of mixtured soils

	Mixtured soils		
	n = 1 : 1	n = 1 : 2	n = 1 : 4
Specific gravity	2.66	2.66	2.66
Effective radius (mm)	0.10	0.09	0.06
Coefficient of uniformity	2.00	2.22	3.37
Sand content (%)	94.0	93.1	88.9
Silt content (%)	6.0	6.9	11.1
Coefficient of permeability(cm/sec)	3.1×10^{-3}	1.9×10^{-3}	1.7×10^{-3}
Frictional angle (°)	38.3°	39.9°	42.8°

A steel box is attached to the strong box, which serves a water reservoir. The water is continuously supplied to the reservoir through a hydraulic slip ring. The water feeding to the model bank is controled by a series of solenoid valves. Porous plates are placed on left side wall of the strong box and on the bottom plate of the strong box, except the part on which the model river bank is placed (see Figure 3).

The model preparation was carried out after having mounted the strong box on the swing platform of the centrifuge arm horizontally. The sand was carefully rained down into the box by manual with a maximum height of pore of about 50 mm. This ensured consistency productions of the model bank with a density of 15.5 kN/m^3. The model slope was then formed by removing the excess sand by suction with a device specially made for this purpose. The centrifuge was gradually accelerated up to predetermined acceleration. The process of seepage was then carried out by operating the solenoid valve. A series of photographs was taken during the flight. Among various factors influencing the phenomena, two factors were selected to exmine in this series of tests : bank slope(β) and soil type, that is to say, the mixture ratio (n). Therefore, other parameters, such as the centrifuge acceleration (40 g) and the height (120 mm) and the top width (60 mm) of the river bank were kept the same throughout the tests. Thus the corresponding prototype behavior is the behavior of silty and river bank of 4.8 meter high, through which water is seeping. The whole test programme is listed in Table II.

TABLE II Test programme

Test code	Soil type (n)	Angle of slope (β)
HH-10-5	1 : 1	35°
HH-10-3	1 : 1	25°
HH-18-1	1 : 1	15°
HH- 9-4	1 : 2	35°
HH-17-4	1 : 2	25°
HH-16-2	1 : 4	35°
HH-19-1	1 : 4	25°
HH-17-2	1 : 4	15°

4. TEST RESULTS AND DISCUSSIONS

Figure 4 (a) and (b) presents the data showing the development of phreatic surface and the failure during the course of the test. The phenomena observed in this series of centrifuge tests were, in most cases, such that water seeps into the model bank and reaches at the toe of the bank and the phreatic surface gradually rises until it reaches at certain height without any visible deformation. And then the initial failure takes place near the seeping surface, and as the phreatic surface rises further, the failure zone progressively creeps up along the slope, together with the occurrence of a secondary failure in the unsaturated zone above the phreatic surface. Phenomena reproduced in the centrifuge tests are considered to be quite similar to those often observed in the field.

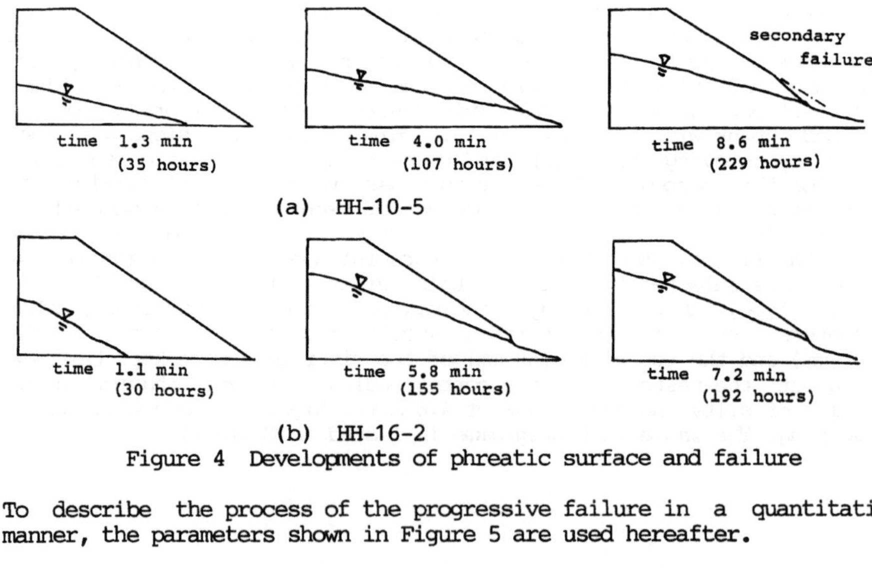

(b) HH-16-2

Figure 4 Developments of phreatic surface and failure

To describe the process of the progressive failure in a quantitative manner, the parameters shown in Figure 5 are used hereafter.

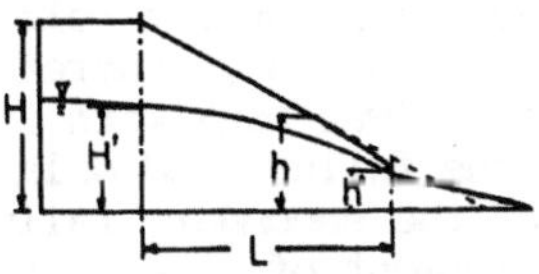

$$I = \frac{H' - h'}{L}$$

H : total height of river bank
H': depth of phreatic surface
h : height of failure
h': height of seeping point
I : average hydraulic gradient

Figure 5 Key sketch for parameters used

Changes in the values of the parameters H', h', and h are plotted against the elapse test time for the cases of n = 1 : 1, and 1 : 4 with the angle of slope $\beta = 35^\circ$ in Figure 6 (a) and (b). It is apparent by comparing the two figures that difference in silty content by 5 % causes a significant effect on the stability and failure process of the river bank. For the case of n = 1 : 1, the initial failure occurred at the elapse time of 2 minutes (53 hours in the prototype scale), when H' is 33 mm (1.2 meter). It should be remembered that the initial failure at the River Nagara occurred at the elapse time of 70 hours when the water level was three- quaters of the total height of the river bank. The secondary failure in the unsaturated zone took place at the elapse time of 7 minutes (190 hours). The value of h increases up to 8 cm (3.2 meter), two-thirds of the total height of the model bank, although H' remains in the order of 50 - 60 mm (2.0- 2.4 meter). In the case of n = 1 : 4, however, h' and h are levelled off under the relatively large value of H' = 8 cm (3.2 meter), two-thirds of the total height of the bank at around 7 minutes (190 hours), which indicates that the condition became a steady state.

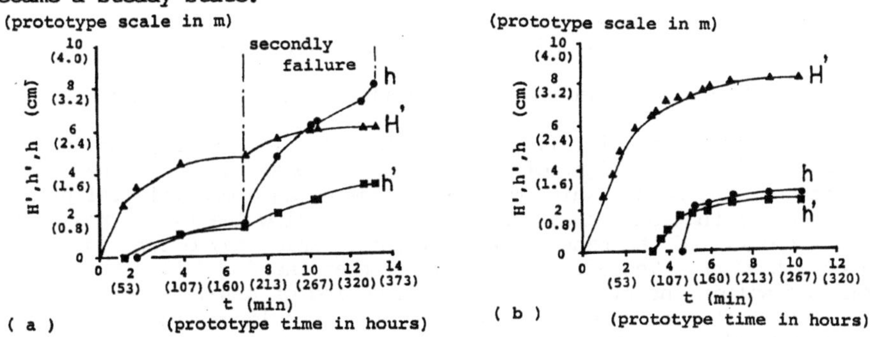

Figure 6 Changes in values of H', h' and h with the elapse time

Another feature of Figure 6 (b) is that no secondary failure takes place in this case. The reason why the size of secondary failure is small for the soil having more finer contents, would be that the height of capillary rise is indirectly proportional to the soil pore radius and thus the capillary tension zone of finer soils is larger than that of coarse soils. A separate series of centrifuge tests, where the river bank was subjected to rainfall without seepage flow from the river bank, showed that the secondary failure was hardly observed, since the rainfall created a zone of capillary tension, as the wetting front went down. The fact again supports that capillary tension plays an important role in stability of unsaturated zone. Engineering implication is , therefore, that mixing the soil near the upper surface of the river bank with finer particles is effective in reducing the size of the failure and delaying the failure and even preventing the occurrence of the secondary failure.

In an attempt to relate the onset and development of the failure with the depth of phreatic surface measured at the edge of the bank slope shoulder, the values of H' are plotted against the values of h for various cases in Figure 7. It is seen that the values of H' at the initial failure becomes larger with the increase in finer content and with the decrease of the angle of the bank slope, supporting an empirical judgment that larger contents of finer particles and small angle of slope the river bank has, more stable the bank is, aganist the seepage flow. In fact, no failure took place for the case of $\beta = 15^\circ$.

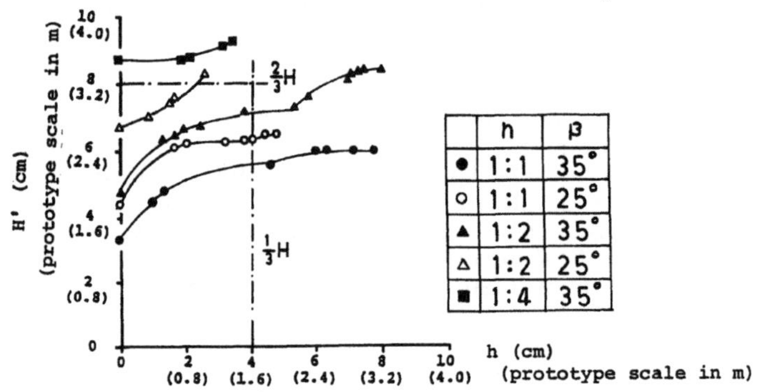

Figure 7 Relationship between the values of H' and h

For the cases of ($n = 1 : 1$, $\beta = 35^\circ$), ($n = 1 : 1$, $\beta = 25^\circ$) and ($n = 1 : 2$, $\beta = 35^\circ$), the height of failure exceeds one- third of the total height of the bank, which may be considered as a crucial failure in practical situation, under the value of H' below the two-thirds of the total height of the bank for a long time, which may occur in actual flood situation. Therefore, the policy that the flood flow should be retained within the river for a certain period of time, rather than flow it down to the sea as quickly as possible, should be reconsidered, as far as the risk of river bank failure due to seepage flow is concerned.

If experimental curves of this type are obtained for a particular river bank site, by conducting a series of centrifuge model tests, then it may become possible to predict the approach of the risk of failure by monitoring the water level of flood flow. In this case, it may be a safe side judgement that H' is assumed to be equal to the height of flood flow.

Close observation of the model tests have led to develop an analytical model to simulate the process of the progressive failure, as illustrated in Figure 8. We start the condition, where the initial failure had already taken place. From the observations, it is assumed that at the edge of sliding body of soil, the failure plane coincides with a part of phreatic surface at which angle measured from the horizontal plane was observed to be approximately a half of the frictional angle of the material. Above the phreatic surface the zone has some amount of apparent cohesion due to capillary tension, whereas the zone below the phreatic surface is fully saturated, thus purely frictional with no apparent cohesion.

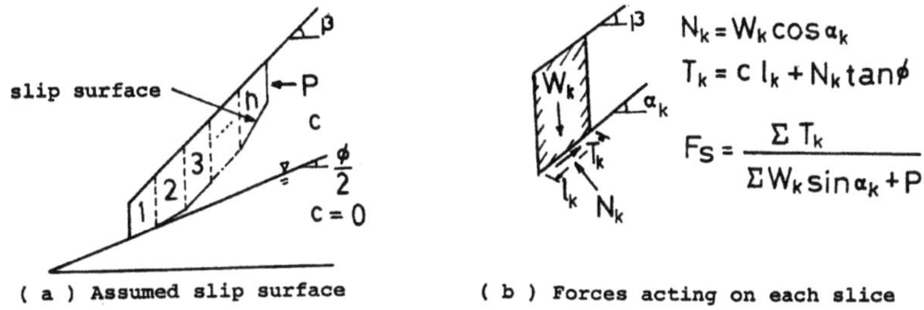

Figure 8 An analytical model for progressive failure

By adopting these assumptions, an iterative calculation on equilibrium of forces acting on each slide along an assumed failure plane is performed, until the factor of safety against slide becomes unity, which implies that the failure occurs at that particular failure plane. As the phreatic surface rises further, the next failure may occur, which can be also calculated in the same manner. Here, it should be said that it was observed in the experiments that before the next failure took place, the soil particles at the toe of the slope newly made as a result of the failure, were necessary to be eroded and transported, making the slope steeper. Therefore, the same process was assumed in the calculation. An example of the calculation results is given in Figure 9, which may be compared with Figure 4 (b).

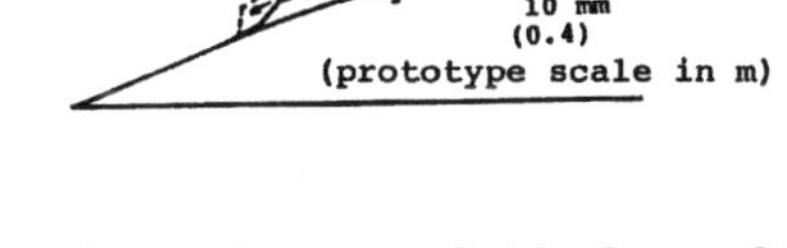

Figure 9 An analytical result

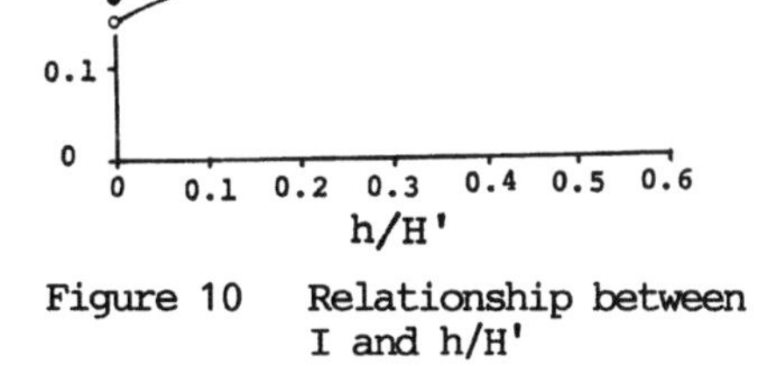

Figure 10 Relationship between I and h/H'

The proposed analytical model requires a cause triggering the first failure. Figure 10 is a replot the data from Figure 7, in terms of I , an index of average hydraulic gradient as defined in Figure 5, against the ratio of h to H'. It is seen that the data of different slope angles are grouped into a single line for a given particular soil type. The velocities of seeping water at the time of the first failure were calculated from the values of permeability (given in Table I) and the values of I read from Figure 10. The velocities are found to be in the range of $6\sim7 \times 10^{-4}$ cm/sec for the case of $\beta = 35^\circ$. It implies that there is a critical value of velocity for a given soil to trigger an initial failure. A combination of the idea shown in Figure 10 with the proposed analytical model to simulate the progressive failure may give a good chance to predict the risk of failure due to seepage flow.

5. CONCLUSIONS

A series of centrifuge tests was carried out to collect the data on river bank failure due to seepage flow. The following points are concluded;

(1) Finer contents of river bank material have a significant effect on the stability and failure process of the river subjected to seepage flow. This is a strong indication that the capillary tension plays an important role in the stability of unsaturated slopes.

(2) The secondary failure took place for a bank made of materials having lesser finer contents. The size of the secondary failure became smaller with the increase in finer contents.

(3) Based on the close observations of the model tests, an analytical model to simulate the development of the failure due to seepage flow was proposed.

REFERENCES

Croce,P., Pane,V., Znidarcic, D., Ko, H.Y., Olsen,H.W. and Schiffman,R.L. 1984 'Evaluation of consolidation theories by centrifuge modelling' Proceedings of Synposium on Application of centrifuge modelling to geotechnical design. pp. 380-401.

Goodings,D.J. 1979. 'Centrifugal modelling of slope failures' Ph.D Thesis, University of Cambridge.

Kimura,T., Kusakabe,O. and Saitoh,K. 1985. 'Geotechnical model tests of bearing capacity problems in a centrifuge' Geotechnique Vol. 35, No.1, pp.33-45.

Ko.H.Y. 1984. 'University of Colorado centrifuge operation' Proceedings of International symposium on Geotechnical centrifuge model testing, pp 35-42.

Padfield,C.J. and Schofield, A.N. 1983. 'The development of centrifugal models to study the influence of uplift pressures on the stability of a flood bank' Geotechnique Vol.33, No.1, pp 57-66.

Schofield A.N. 1980. 'Cambridge geotechnical centrifuge operations' Geotechnique, Vol. 30, No.3., pp 227-268.

Scott, R.F. 1963. 'Principles of Soil Mechanics' Addison-Wesley Publishing Co. U.S.A.

Sutherland, H.J. and Rechard, R.P. 1984. 'Centrifuge simulations of stable tailing dams' Journal of Geotechnical Engineering, A.S.C.E., Vol. 110, No.3, pp 390- 402.

Tsuchida, A., Suga,K.,Yabe,M. and Ishizaki,K. 1976. 'A report on disaster of the River Nagara during No.17 typhoon' Technical report of Public Works Research Institute, Vol. 18, No. 12, pp 641-648.

THE BRAZILIAN ELECTRIC SECTOR EXPERIENCE IN FLOOD CONTROL

A.M. Vieira, P.R.H. Sales and L.A.L. Barretto
Centrais Elétricas Brasileiras S.A. - ELETROBRÁS (DEOP)
P.O. Box 1639
Rio de Janeiro, RJ
Brazil

ABSTRACT. The objective of this work is to report the Brazilian experience in the Flood Control Problem Analyses by the determination and allocation of flood control space in its reservoirs. The Volume x Duration Curve Method (Beard, 1963) of flood control space determination is still the current solution for the problem. The research of new methodologies was tried in order to avoid difficulties, such as allocation, in the use of this method. As a matter of fact this method just gives the maximum flood control storage for the wet season and a constant allocation of it could reach a higher energetic deficit risk. The Surplus Volumes Method, also presented here, is an alternative for the flood control storage evaluation without overcoming, however, the allocation question. Another alternative is the Critical Trajectories Method (Kelman, 1983), which is an extension of the Surplus Volumes Method. This one solves the problem of allocation by indicating the flood control storage variation through the rainy period. However, this method demands synthetic daily mean streamflow series. The results presented by the three methods for the Sobradinho reservoir (34.116 km3) on the São Francisco River on the Northeastern Region of the country, are compared ones.

INTRODUCTION

The Brazilian Energy Generation System is understood to be a large hydrothermal system with several owners. In terms of its installed capacity, the hydro power plants represent 85% of the total capacity of 41,661 MW. In 1984, the hydro generation was responsible for more than 94% of total energy production of 176,464 GWH.

The Brazilian reservoirs were first built for power generation objective. However, one of the most usual effects is the occupation of downstream areas which became safer from floods. In Brazil, this effect has been increased so much that many reservoirs started to keep releases restriction to avoid any damages downstream during the rain-flood season.

In the system operation planning studies the allocation of flood

V. P. Singh (ed.), Flood Hydrology, 409–418.
© 1987 by D. Reidel Publishing Company.

mitigation space is admitted. However, the contribution given by the Electric Sector in order to avoid and/or reduce the damage caused by floods has a conjunctural character. The protection level influenced by the reservoirs varies with system energetic space, that is, the protection decision is taken based on the results of the energetic simulation of the system.

Actually, the flood mitigation storage studies in the country started in the last decade when the Volume x Duration Curve Method (GCOI, 1977) was firstly proposed. The solution presented by this method is still used, even though some difficulties have been found in its implementation. The main one can be characterized as follows: it just gives the maximum flood control space for the whole rainy period and does not consider the implicit potential variation of flood as this period passes. Thus, because no more information is available, the flood control storage remains constant towards the end of the rain-flood season.

What can be said is that this solution is not efficient since it has different goals. As a matter of fact the flood control space risk only occurs at the beginning of the rainy period. In this way, the constant allocation of the flood mitigation storage, overcomes the wanted protection. On the other hand there is a possibility of not occurring the flood event. In this case the energy generation aim will not be reached on the following dry season.

In order to consider the risk evolution, the adopted strategy evaluates the flood mitigation storage after blocking successively the mean daily flow sequences. In other words, the flood control space on the j day is determinated only in the period between the j day and the end of the wet season. In this case the risk keeps steady whereas the flood control storage becomes changeable.

The indicated procedure above manages a fast refilling of the reservoir. Nevertheless, this solution faces some estimation problems. That is, it is expected the continous fall of the flood mitigation storage in the sample. But, sometimes, this does not occur at all due to the sample variation.

To overcome these points the Electric Sector through the Coordinating Group for the Interconnected System Operation (GCOI) started to invest in the research of new methodologies. The Surplus Volumes Method came out as an alternative. This method has its basis on the Rippl's Method of active storage capacity determination, applied in an inverse scheme. Even though having the same limitations of the previous method, it enabled the Critical Trajectories Method appearance (Kelman, 1983). This last one represents a significant evolution and solves jointly the determination and allocation problems.

The three methods mentioned above are presented below and applied in Sobradinho reservoir (34.116 km3), which is located on the São Francisco River on the Northeastern Region of the country.

VOLUME x DURATION CURVE METHOD

The maximum runoff volume for each time interval with d consecutive days of duration is defined as:

$$va = \max_{0<t<h-d+1} \left[1/d \, . \sum_{j=0}^{d-1} (q(t+j) \, . \, \Delta t) \right] \tag{1}$$

Where: va - maximum runoff volume;
d - duration in days;
q(t+j) - mean daily flow on the t+j day;
Δt - discrete time interval;
h - the very last day of the wet season;
t - day.

Admitting a release restriction not to cause any damage downstream and taking into account the mean daily flows of the historical data, it can be defined for the rainy season of each hydrological year, the necessary empty volume in order to contain peak flood of any duration. It can be represented by the following expression:

$$ve = \text{Max} \left[(va(d) - d.qr.\Delta t), \; d = 1,2,3,\ldots h \right] \tag{2}$$

Where: ve - flood control storage;
qr - release restriction.

The duration associated to this volume is called critical duration, cd.

One possible criterion, in order to obtain the flood control space to be allocated, is to take into account the maximum value of the historical sequence. Nevertheless, there is no guarantee that this series will be repeated; not even the feasible allocation of this volume. Thus, one suitable solution leads to a frequency study as well as to a probability distribution.

By fitting a theoretical distribution such as log-Pearson type III, for instance, for each sequence with duration of d days, it is created the volume x duration curve which is associated to a p' fixed probability of the flood control storage, ve, not being enough.

In this case, va(d) is:

$$P\left[VA(d) \; > \; va(d)\right] = p' \tag{3}$$

This curve is equivalent to the spillway design floods whose flows have the same return period for each d duration.

This method has shown not to be consistent for practical purposes. Due to sample variations the va(d) curve is not always convex as it is expected. To avoid this kind of inconsistence, the adopted procedure is to fit, by the least square method, an analytic function to the pairs (volume, duration). If this solution still gives inconsistent results, the graphical one can be tried. On the other hand, it was shown by Kelman et al. (1980) that this method is not precise at all. In reality, the given solution is only approximated. That is, the flood control space does not satisfy the equality: $P\left[VE > ve\right] = p'$. As a matter of fact, $P[VE > ve] > p'$. This happens, because an emergency can occur due to an unbalance between the inflow and outflow volumes to any other duration, not only for the critical duration, cd.

THE SURPLUS VOLUMES METHOD

This method can be abridged in the two following steps. Firstly, the sequence of annual maximum cumulative exceeding volumes (AMCEV) should be analysed for a selected restriction discharge. In this case, the Rippl's Method (1883) is applied in an inverse way, so that instead of getting the maximum cumulative deficit, one obtains the maximum cumulative exceeding volume.

Secondly, having the AMCEV sequence, a probability distribution can be tried in order to estimate the AMCEV values to different probabilities. These volumes are called flood control storage.

Exceeding volumes are supposed to happen only in rainy years, so the AMCEV sequence should contain several zero volumes. Thus, a mixed distribution, where there is a fixed α probability of occurrence this event should be considered.

Based on this idea, a Mixed Exponential Distribution was chosen. Its expression is:

$$P[X > x] = (1 - \alpha) . \exp(- \lambda . x) \tag{4}$$

The α and λ estimators are given by the following expressions:

$$\hat{\alpha} = n1/N \tag{5}$$

$$\hat{\lambda} = n2/\Sigma X(t) \tag{6}$$

Where: n1 - number of AMCEV equal to zero;
N - total number of observations;
$[X(t), t = 1,2,.......,n2]$ - series of AMCEV higher than zero;
n2 - number of AMCEV higler than zero.

The flood control storage, ve, for a fixed p' probability of emergency can be estimated as:

$$ve = \frac{\ln[p' / (1 - \hat{\alpha})]}{\hat{\lambda}} , \quad \hat{\lambda} > 0 \tag{7}$$

Again, this method presents sample variation problems. That is, if the restriction discharge is not that severe, the AMCEV series will contain only a few positive volumes. When it occurs, the theoretical distribution fitting becomes unfeasible.

In order to reduce this problem a synthetic daily mean streamflow series could be used. Thus, the empirical series distribution, which should contain several positive volumes, replaces the theoretical one.

THE CRITICAL TRAJECTORIES METHOD

This method was presented by Kelman (1983) and enables, on each day of the wet season, the determination of the flood control storage for a previously p' fixed probability.

The method uses the Rippl's modified algorithm in a recursive way on mean daily streamflow series and can be summarized as follows:

Letting an annual rainy period of h days. At the beginning of the very last day, the flood control storage should be equal to:

$$ve(h,i) = \text{Max}\left[\,0\,,\ (q(h,i)-qr).\Delta t\,\right] \qquad (8)$$

Where: ve - flood control storage;
h - very last day of rainy period;
i - year;
q(h,i) - mean daily streamflow on h day and i year;
qr - restriction discharge;
Δt - number of seconds in a day.

At the beginning of the penultimate day, the flood control storage should be equal to:

$$ve(h-1,i) = \text{Max}\left[\,0\,,\ (q(h-1,i)-qr).\Delta t+ve(h,i)\,\right] \qquad (9)$$

In a general way:

$$ve(t-1,i) = \text{Max}\left[\,0\,,\ (q(t-1,i)-qr).\Delta t+ve(t,i)\,\right],\quad t = 1,2,\ldots\ldots,h \qquad (10)$$

As time passes, the evolution of the flood control storage is called critical trajectory. It defines a limit to the reservoir operation. See figure 1.

It can be observed at this point that the flood control storage, ve(j), which is the maximum of the critical trajectory shown in figure 1, is equal to the flood control space indicated by the Surplus Volumes Method for the i year.

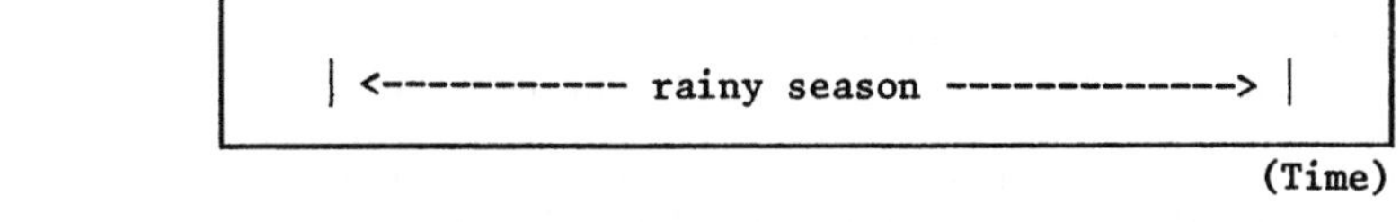

Figure 1. Day by day Evolution of the Flood Control Storage for the i year.

Suppose now 3 different trajectories as shown in figure 2. Their cover, which is represented by the broken line will be called Limit Curve. This curve guarantees, no matter the selected year, there will not be an outflow higher than the restriction discharge, if this outflow is equal to qr when the stored volume lies on it.

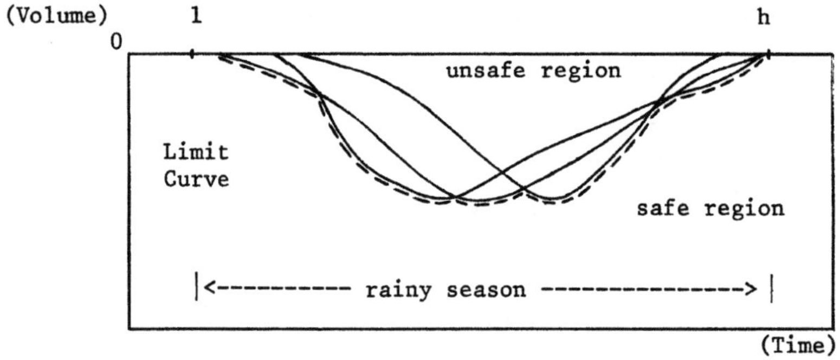

Figure 2. Definition of Limit Curve as Envelope Curve of all Critical Trajectories.

In a historical sequence of n years, and so n trajectories, the Limit Curve will guarantee the non occurance of the discharge restriction violation, in the case of any "hydrological situation" observed in the past.

The Limit Curve is defined as:

$$v'(t) = \text{Max}\left[(ve(t,i)),\ i = 1, \ldots n \right];\ t = 1, \ldots, h \qquad (11)$$

As the Limit Curve allocation would be unfeasible for practical purposes, it is necessary to accept some p' emergency probability. It is defined, then, a Cp' as the k critical trajectories "non safe" set by the flood control storage to be allocated. Such set should minimize the effect of the flood control space in relation to the "energy generation" aim. The following function should be minimized:

$$B\left[V_{p'}(1), V_{p'}(2), \ldots\ldots, V_{p'}(h) \right] \qquad (12)$$

Where: $V_{p'}(j)$ - flood control storage on j day for a p' probability of emergency.

In practical terms, two criteria can be chosen to the minimization of the B [.] function. The first one could consider the Cp' set which implied in a minimization of the average flood storage. In the second one, the Cp' set would be formed by the critical trajectories with the maximum flood control storage in the considered interval.

By any of the criteria above, the obtained solution determines, day by day, the flood control storage evolution, for a previously especified p' emergency probability.

Nevertheless, the above algorithm needs a synthetic daily streamflows generator. As a matter of fact, suppose historical data of 30 years and a p' probability equal to 0.02 (return period of 50 years). The number of "non safe" critical trajectories, in this case, would be k = np' = 0.6, which is not enough for the Limit Curve determination.

CASE STUDY

The algorithms for the three methods were developed and used to test their merits. These were coded in FORTRAN IV, and run on an IBM - 4341.

The application case considers the reservoir system of the lower São Francisco River. This system is composed by the reservoirs of Sobradinho, Moxotó and Paulo Afonso. Sobradinho has the highest regularization capacity and is supposed to be one of the largest reservoirs in Brazil.

The flood mitigation exercised by the set of reservoirs considers only the flood control space allocation in Sobradinho. The hydraulic operation of this plant shows, during the wet season, one maximum outflow of 8000 m3/s in order to protect not only the cities of Juazeiro and Petrolina but also other downstream cities. Table I displays the principal characteristics of the Sobradinho reservoir.

TABLE I - THE SOBRADINHO RESERVOIR CHARACTERISTICS

Regulated flow	2060 m3/s
Long-term average flow	2800 m3/s
Surface area	4241 km2
Drainage area	498425 km2
Maximum pool level	393.50 m
Minimim pool level	380.50 m
Normal pool level	392.50 m
Maximum storage	38.541 km3
Dead storage	5.477 km3
Active storage	34.116 km3
Maximum outflow capacity	22835 m3/s

The mean daily unregulated streamflow series of Sobradinho varies from 1929 to 1985. The worst critical rainy period goes from December to May.

By using the DIANA Model (Kelman et al., 1985), which was developed to flood studies, 2000 rainy periods were generated (December - May) in order to apply the Critical Trajectories Method.

Table II shows the flood control storage in two-week intervals, evaluated from the three methodologies. The solutions from the Volume x Duration Curve Method and from the Surplus Volumes Method took into account the variable allocation procedure described in the introduction.

The Critical Trajectories Method solution from the synthetic sequences considered the two criteria previously mentioned.

TABLE II - FLOOD CONTROL SPACE FOR THE SOBRADINHO RESERVOIR (RETURN PERIOD = 25 years and RESTRICTION RELEASE = 8000 m3/s)

DAY	METHODOLOGY			
	VOLUME x DURATION CURVE	SURPLUS VOLUMES	CRITICAL TRAJECTORIES	
	Historical Streamflow Sequences	Historical Streamflow Sequences	Synthetic Streamflow Sequences	
			Criterion	
			1	2
01/Dec	10.168	11.240	2.543	7.035
16/Dec	9.537	11.240	5.859	9.799
01/Jan	9.161	11.240	6.774	9.924
16/Jan	9.859	11.240	9.164	10.855
01/Feb	10.064	11.056	12.211	10.070
15/Feb	9.443	10.081	13.834	10.322
01/Mar	7.515	8.231	12.349	10.890
16/Mar	4.097	4.053	7.397	9.178
01/Apr	1.362	0.817	3.060	6.565
16/Apr	0.305	0.0	1.258	5.720
01/May	0.0	0.0	0.263	1.012
16/May	0.0	0.0	0.116	0.533

obs.: Volumes in km3

The first criterion considers the minimization of the average flood control space. On the other hand, the second criterion is based on the elimination of the trajectories which have the maximum flood control space in the interval.

In order to get the appraisal of the solutions redemption, it was used a simplified simulation model which observes the flood control operation rules. However, this model does not consider induced surcharge. The results of the simulations allowed the emergency risks verification throughout the rainy period as well as the reservoir refilling probabilities at the end the wet season.

Taking into account that the solutions should start from the same point, it was considered two sets of simulations: one, utilizing the first criterion and the other, the second one. In the solutions with the Methods of Volume x Duration Curve and Surplus Volumes, the variable allocation procedure was considered from January to May. The allocation in December was done considering the initial space equal to that one given by the Critical Trajectories Method. The remaining days of December were obtained by linear interpolation.

The abridgement of the simulations with the historical series, is shown on table III. From the results it can be noticed that the variable allocation from the Volume x Duration Curve Method as well as from the Surplus Volumes Method does not reach the objective of protection for floods with recurrence intervals up to 25 years. As a matter of fact, the expected value of the number of emergencies from one historical sample of 57 years and one risk of 0.04 is 57 x 0.04 = 2.28. Using the Binomial Distribution, this solution has only 0.07 probability of being true.

The Critical Trajectories Method performing was better with the first criterion. With this solution, only three emergencies happened. There is 0.4 probability of this solution being true. In reality, the first criterion seems to be the best one considering the aim of the flood control space minimization.

It is important to point out that the best performance of the Critical Trajectories Method has the onus of having less probability of refilling the reservoir at the end of the wet season. However, one should keep in mind the objective of keeping the risk steady. Sometimes this level of protection could imply in a high energetic deficit risk. In this case, the level of protection in the flood control should be diminished.

TABLE III - SIMULATION RESULTS FOR THE SOBRADINHO RESERVOIR - HISTORICAL STREAMFLOW SEQUENCES (57 years)-

METHODOLOGY	EMERGENCY (year)	MAXIMUM RELEASE (m3/s)	REFILLING PROBABILITY (May,31 - % A.S.)		
			100%	90%	80%
CRITICAL TRAJECTORIES (Criterion 1)	1945 1949 1979	9869 14983 18039	0.47	0.58	0.74
VOLUME x DURATION CURVE	1945/1949 1957/1979 1980	9869/14983 8122/18088 11330	0.54	0.82	0.93
SURPLUS VOLUMES	1945/1949 1957/1979 1980	9869/14983 8193/18088 10520	0.53	0.79	0.91
CRITICAL TRAJECTORIES (Criterion 2)	1945/1949 1979 1980	9869/14983 18039 10520	0.26	0.49	0.74
VOLUME x DURATION CURVE	1945/1949 1957/1979 1980	9869/14983 8122/18088 11330	0.54	0.82	0.93
SURPLUS VOLUMES	1945/1949 1957/1979 1980	9869/14983 8193/18088 10520	0.53	0.79	0.91

obs.: A.S. - Active Storage

CONCLUSIONS

- The Critical Trajectories Method presents a relatively simple and convenient solution for the flood control storage determination and allocation problem. Also, in this method, the seasonality of the events is better described throughout the rainy period. Yet, it should be noticed that some well known allocation problems are surpassed too. As a matter of fact, at the beginning of the wet season, the allocation of the maximum flood control space can be avoided, and so, the spillage risk at the end of the dry season is diminished. Another point to jut out is related to the gradual reduction of the flood control storage allocated as time passes;
- In some regions of the country, where the seasonality is not well defined, the Volume x Duration Curve Method performs well. On the other hand, the Surplus Volumes Method seems to be extremely vulnerable to sample variations and its redemption is not that good for not severe restriction discharges;
- The simulation results show that the variable allocation of the flood control storage procedure can not guarantee the aimed protection level.

REFERENCES

BEARD, L.R. 1963. 'Flood Control Operation of Reservoirs', *Journal of the Hydraulics Division*, ASCE, vol. 89, no. HY1, pp. 1-23

GCOI, 1977. *Flood Control Storage Determination for the Grande River Reservoirs*, Technical Report SCEN-CECCA 02-77, Rio de Janeiro. (in Portuguese).

KELMAN, J.; DAMAZIO, J.M.; COSTA, J.P.; PEREIRA, M.V.F.; OLIVEIRA, G.C.; PINHEIRO, M.C. 1980. *Flood Mitigation Storage*, CEPEL Technical Report no. 193/80, Rio de Janeiro. (in Portuguese).

KELMAN, J. 1983. *Floods and Hydroplants*, thesis submitted in the competition for the Full Professorship in the Hydraulics Department of the Federal University of Rio de Janeiro. (in Portuguese).

KELMAN, J.; DAMAZIO, J.M.; COSTA, J.P. 1985 'A Multivariate Synthetic Daily Streamflow Generator', *Fourth International Hydrology Symposium*, Fort Collins, Colorado, USA.

RIPPL, W. 1883. The Capacity of Storage Reservoirs for Water Suppy, *Proc. Inst. Civil Engrs.*, vol. 71, pp. 270-278.

SUBJECT INDEX

AUTHOR INDEX

MIX
Papier aus verantwortungsvollen Quellen
Paper from responsible sources
FSC® C105338

If you have any concerns about our products,
you can contact us on
ProductSafety@springernature.com

In case Publisher is established outside the EU,
the EU authorized representative is:
Springer Nature Customer Service Center GmbH
Europaplatz 3, 69115 Heidelberg, Germany

Printed by Libri Plureos GmbH
in Hamburg, Germany